Advanced Dynamics Modeling, Duality and Control of Robotic Systems

Advanced Dynamics Modeling, Duality and Control of Robotic Systems

Edward Y.L. Gu

CRC Press
Taylor & Francis Group
Boca Raton London New York

CRC Press is an imprint of the
Taylor & Francis Group, an **informa** business

MATLAB® is a trademark of The MathWorks, Inc. and is used with permission. The MathWorks does not warrant the accuracy of the text or exercises in this book. This book's use or discussion of MATLAB® software or related products does not constitute endorsement or sponsorship by The MathWorks of a particular pedagogical approach or particular use of the MATLAB® software.

First edition published 2022
by CRC Press
6000 Broken Sound Parkway NW, Suite 300, Boca Raton, FL 33487-2742

and by CRC Press
2 Park Square, Milton Park, Abingdon, Oxon, OX14 4RN

© 2022 Edward Y.L. Gu

CRC Press is an imprint of Taylor & Francis Group, LLC

ISBN: 978-0-367-65371-2 (hbk)
ISBN: 978-0-367-65373-6 (pbk)
ISBN: 978-1-003-12916-5 (ebk)

Typeset in Nimbus font
by KnowledgeWorks Global Ltd.

*To my family Sabrina,
Heather and Jacob*

Contents

Preface

This book presents novel research and development incorporated with computer simulation and 3D animation studies. It focuses on the modeling of nonlinear complex dynamics for a variety of robotic systems from common open serial-chain manipulators to closed serial/parallel hybrid-chain robots. The book also covers a cascaded dynamic model development with backstepping control design procedures for robot-environment interactive systems. Dynamics is a study of motion response to force and torque. From this perspective, the kinematics is an infrastructure of the dynamics for every motion-related system. Thus, a set of general theories and advanced representations on robot kinematics will be introduced and reinforced in the fundamental preliminary chapter of the book in addition to the mathematical preparations.

Recent robotics research and technological advances have bolstered many robotic systems design innovations. As our capabilities increase, so do the demands for more efficient designs. Following the rapid development of industrial robots from innovative design, mass production to wide-range applications over the past half a century, many new dexterous and mobile robotic systems have been created. These include: dual-arm and triple-arm industrial robots, wheel-mobilized human-like robots, legged crawling robots, humanoid walking robots, and many more bio-inspired and bio-mimic robotic systems.

The application requirements in development of these robotic systems demand better and more accurate solutions. The pressing challenges center around how to better model their dynamics and further develop effective nonlinear control strategies. Therefore, this research book is to provide the readers with new powerful mathematical tools. The advanced theories on modeling the nonlinear complex dynamics are presented and the compact formulas and algorithms that can be incorporated into the further development of new control systems are provided.

In addition, this book never tries to dodge the complexity and difficulty by imposing unreasonable approximation or deduction on the lengthy derivations. It is the nature of dynamics modeling for such large-scale and nonlinear complex robotic systems. By taking advantage of both the symbolical and numerical recursions, all the derivation and computation difficulties are tackled. Using the mathematical tools presented, new compact procedures, formulas and algorithms can be achieved. Once a compact dynamic model for each type of robotic systems becomes available, the design of its specific control scheme makes the objectives become more achievable. Because of the innovative dynamic model compaction, the principle of duality between the open serial-chain and closed parallel-chain robotic systems are unveiled and can be further developed to cover a wider range of robot dynamics modeling and control applications.

Therefore, this book is intended to be an indispensable research reference for scientists, researchers, engineers, and college senior undergraduate and graduate students who are interested in the research and study areas of kinematics, statics, dynamics and control of robotic systems.

I hereby acknowledge my indebtedness to the people who helped me with different aspects of collecting knowledge, experience, data and programming skills toward the book completion. The author is also grateful of the university colleagues for their encouragements and idea exchanges. A special appreciation is giving to my long-term friend John Porter from Traverse City, Michigan for helping me to polish my English writing. In addition, the author is under obligation to Fanuc America Corporation for their courtesies and permissions to include their photographs into the book.

Edward Y.L. Gu, *Troy, Michigan, U.S.A.*
guy@oakland.edu
March 2021

MATLAB® is a registered trademark of The MathWorks, Inc. For product information, please contact:

The MathWorks, Inc.
3 Apple Hill Drive
Tel: 508-647-7000
Fax: 508-647-7001
E-mail: info@mathworks.com
Web: www.mathworks.com

Author

Edward Y.L. Gu received his M.S. degree in Physics and Electronics from the Graduate School of Chinese National Institute of Science in 1981, and Ph.D. degree in Electrical and Computer Engineering from Purdue University in 1985. He is currently a Professor with the Department of Electrical and Computer Engineering (ECE), Oakland University, Michigan. He taught multiple ECE courses in the past and recent three decades, and supervised many Ph.D. and M.S. students towards their graduations with degrees.

His research interests are in the areas of robotics, nonlinear control theories and applications, digital human modeling and realistic motion generation, human-machine dynamic interactions, learning and intelligent control systems and electric machines with their control and drive electronics.

Dr. Gu has also won and collaborated numerous funded joint R&D projects with the automotive industry, such as GM and Chrysler, and worked closely with Chrysler as a summer professor intern and consultant for longer than a decade. He co-founded the OU-Chrysler Robotics and Controls Lab at OU ten years ago. He has written and published a book and a large volume of research papers in the leading and prestigious transactions, journals, and conference proceedings, many of which have made remarkable impacts on the research communities.

1 Introduction

1.1 KINEMATICS, STATICS AND DYNAMICS

In general, kinematics is a study of motions, which concerns relationships between motion displacements or positions and their time-derivatives: the velocities, accelerations, and jerks without considering causes by force and/or torque for a particle, a body or a multi-body system. In contrast, dynamics is to study motion response to force and/or torque. Both are the key branches in classical mechanics. In history, classical mechanics was divided into statics and dynamics. Statics is to treat the equilibrium of force and/or torque structures, where no motion and no change in position over time are involved. Dynamics, on the other hand, is further subdivided into kinematics and kinetics. Kinematics, as described above, is to study motions without referring to any underlining force/torque structures, while kinetics is to describe the force/torque structures [1–8].

The studies of kinematics, statics and dynamics, as parts of classical mechanics in modern times, are all represented by their variables, parameters and equations. In kinematics study, the common variables are the linear displacement or position, orientation and arclength, and the linear velocity and angular velocity, linear acceleration and angular acceleration, etc. The common kinematic parameters include the body dimensions, structure offsets and twist angles, all of which are constants. Whereas in dynamics study, the common variables are the forces and torques in addition to all the kinematic variables. The common dynamic parameters are the mass, moment of inertia and mass center location, all of which are also constants if the object is a rigid body. The study of statics is similar to the dynamics, but every time-derivative should be zero at the equilibrium status except for the principle of virtual work, where, as a theoretical notion, the work as well as its related positions are virtually varying.

It is observable that the variables between kinematics and dynamics are often overlapped due to their common interest in motion analysis. The kinematic parameters are always measurable, and thus they are certain in most cases. In contrast, the dynamic parameters are often imprecisely measurable or immeasurable and thus they are uncertain in many cases, such as for a multi-link robotic system, where each link mass, moments of inertia and the mass center coordinates are difficult to be determined accurately.

In the studies of statics and dynamics, one thing is forever in common, that is **the principle of energy conservation**. In fact, the energy conservation law not only holds in a cumulative sense, but also instantaneously. Therefore, the power at each time-instant must be conserved as well.

The above descriptions on kinematics and dynamics are given at a standpoint of physics. At a mathematical perspective, however, the concepts of both kinematics and dynamics can be further interpreted in a more abstract fashion. For each single rigid body or multi-body system, any kind of motion can be interpreted to be constrained on an invisible super-surface, called a **configuration manifold (C-manifold)** of the body or system, at each point of which the total kinetic energy K of the body or system is endowed as a Riemannian metric. Then, *the topology of this C-manifold is structured by the kinematics, while the geometry of the C-manifold is determined in every detail by the dynamics.* In other words, two robotic systems with different structural configurations have different kinematic models so that their C-manifolds, as their individual motion constraints, possess different topological structures. In order to determine the dynamic model for each of the two different robots, we need to combine each kinematic structure, such as the positions and orientations, by the corresponding dynamic parameters in a certain individual way. The detailed exploration, development and discussion on this new concept and its realizations will be given in Chapter 2 and Chapter 4 of this book [17–19].

Based on this mathematical interpretation, *the architecture of dynamics is virtually a linear combination of the kinematic structure by dynamic parameters.* Therefore, in order to model the dynamics for either a single-body or a multi-body system, the first step must develop its detailed kinematic model. One of the most typical examples is a robot manipulator that is a multi-body system. It would be hard or almost impossible to find its dynamic formulation without a complete kinematic model.

1.2   DYNAMICS MODELING AND MODEL COMPACTION

After the kinematic model is well-determined for a given system, the dynamics modeling can be performed in many ways, depending on how complicated the system is. To model the dynamics for a single-body, the Newton's second law is sufficient enough in either a translation or a rotation case. However, for a multi-body system, it is more efficient to start finding the total kinetic energy K, and also finding the total potential energy P at beginning, and then to substitute $K - P = L$ as a Lagrangian into the Euler-Lagrange equation. Because the Euler-Lagrange equation is a 2nd-order differential equation, we must have the Lagrangian $L = K - P$ available as an explicit function $L(q, \dot{q})$ of the generalized coordinates q and $\dot{q}$. This is a traditional approach to finding the dynamic formulation for a multi-body system in classical mechanics [9–16].

If we alternatively follow the advanced mathematical interpretation by adopting the C-manifold concept, we need first to define a set of local coordinates q in a small neighborhood of each point on the smooth C-manifold M^n. For a robotic system, the joint positions of the robot can be directly used as the most convenient set of local coordinates q.

Because the C-manifold for a multi-link robotic system is globally non-Euclidean in general, the Riemannian metric that is endowed on the C-manifold by using the total kinetic energy K in terms of the local coordinates is often a complicated quadratic form of the tangent space. In order to simplify the representation, the compact but non-Euclidean C-manifold is intended to be isometrically embedded into an ambient Euclidean space. Namely, in easier words, the compact C-manifold is one-to-one sent into a Euclidean space with the metric preserved. Once the compact and non-

Euclidean C-manifold can be seen in the ambient Euclidean space, its Riemannian metric becomes a constant. Under such an innovative notion of the **isometric embedding**, the Euler-Lagrange equation-based dynamic model for any multi-link robotic system can be reduced apparently to a compact form that looks like the Newton's second law or like the robot statics equation [19].

Therefore, the ultimate step to realize the compact dynamic formulation is to find an isometric embedding. Chapter 2 provides a mathematical foundation behind the dynamic model compaction. Chapter 4 describes a detailed procedure to find the isometric embedding for any types of multi-link mechanisms, including open serial-chain and closed parallel or hybrid-chain robotic systems.

1.3 THE PRINCIPLE OF DUALITY FOR ROBOT KINEMATICS, STATICS AND DYNAMICS

The principle of duality has been frequently referred to, extensively studied and further developed for almost every system in many different fields of mathematics and physics as well as a numerous areas of applied science and engineering since hundreds years ago. Duality is not a theorem, but is a principle. It, though, does not require to prove, is commonly recognizable that almost every system can find a dual system as its dual counterpart [20, 21]. In many cases, the dual systems and their duality properties can reveal beautiful dual relations to enhance mathematical and physical insights into deeper understanding of the systems characteristics and behaviors, but may not always be substantially useful in many application circumstances. However, in some special cases, the principle of duality can not only demonstrate an intrinsic connection between the dual phenomena in nature but also become an indispensable tool to solve the problems.

Over the past decades, the exploration of duality principle between serial and parallel robots has drawn a great deal of attentions in robotics research. Most of the research investigations focus on the dualism in robot kinematics and statics [22–24]. One of the most representative duality relations was discovered in formulating the differential-motion kinematic equations between serial and parallel-chain mechanisms. By defining a similar Jacobian matrix that is, however, a function of the Cartesian positions and orientations, the differential-motion Jacobian equation for a parallel-chain robotic system can be resemble in appearance to the robot statics equation. Comparing it to the Jacobian equation for a serial-chain robotic system, the Jacobian matrix of which is a function of the joint variables, they clearly form a beautiful duality pair. Since solving the forward kinematics (FK) for a parallel-chain robot is recognizable to be extremely difficult, by taking advantage of the duality relations, such an almost hopeless FK problem can be resolved by the Jacobian equation in differential-motion for the parallel-chain robotic systems [19].

In this book, the duality research between open serial-chain and closed parallel-chain robotic systems has been further extended to manifest as a theoretically beautiful and practically useful principle to meliorate the analysis of robot kinematics, statics and dynamics, see Chapter 5 of the book. The principle of duality is discovered not only in robot kinematics and statics, but also in robot dynamics. It is also evidently convinced that the explicit form of duality in robot dynamics between the open serial-chain and closed parallel-chain mechanisms, in general, cannot be emerged until the dynamic formulation is reduced into a compact form by the isometric embedding approach developed in Chapter 4. Under such a compact formulation, the clear dualism has been unveiled in all the areas of robot kinematics, statics and dynamics. Every explicit duality form developed and discussed in Chapter 5 can also ease and facilitate the dynamics modeling and control schemes design for both types of robotic systems.

1.4   ADAPTIVE AND INTERACTIVE CONTROL OF ROBOTIC SYSTEMS

Control of a robotic system with inaccurate or uncertain dynamic parameters may still keep converging if the controller is designed in a more robust manner, even though the parameter inaccuracy or uncertainty is ignored. However, the degree of robustness is limited in a certain range and we wish to have a parameter adaptation law added to associate with the control law in order to allow a wider range of tolerance against the parameter inaccuracy or uncertainty. Adaptive control strategies have been extensively investigated and developed in the past decades, especially for robot manipulators such nonlinear, high-coupled and high-dimensional systems. There are direct and indirect adaptive control schemes. In the direct method, parameters are directly adjusted at each time-instant by using an adaptation law. In contrast, the indirect method is to estimate the parameters based on the control error to update the required system parameters. It is in common that to successfully adapt every dynamic parameter while the robotic system is controlled to perform a task, all the dynamic parameters have to be linear in the dynamic equation. The direct adaptation law is just developed by taking advantage of the parametric linearity [25–27].

There is a large volume of the literature available on a variety of adaptive control methods, schemes and algorithms. In Chapter 7 of the book, however, we will only focus on the direct adaptive control scheme by using the Lyapunov direct method to develop both the control law and adaptation law for the applications to robotic systems in both the open serial-chain and closed parallel-chain mechanisms.

The interactive control is to effectively control the interactions between a robot and its environments. The realistic interactions can be modeled as a k-stage cascaded system. Then, the interactive control of such a cascaded model is designed by using the backstepping recursion [28].

1.5   THE ORGANIZATION OF THE BOOK

Since the objective of this book is aiming to share with readers the three innovative ideas: *an advanced dynamics modeling approach, the principle of duality, and the interactive control of robotic systems*, we must first give a mathematical foundation to prepare for the advanced modeling and representations. Chapter 2 offers a mathematical preliminary that includes an introduction to Lie group and Lie algebra with their topological structures, the manifolds and their topology and geometry attributes with Riemannian metric and embedding, fundamental differential geometry with geodesic equations, and the dual number algebra and calculus. In addition to the mathematical preparation, Chapter 2 also covers detailed fundamentals of robot kinematics and statics.

Starting Chapter 3, the basic theories and procedures of dynamics modeling and formulation will be scanned and reviewed along with examples of a variety of robotic systems. The innovative dynamics modeling methods based on the concepts of C-manifold and its isometric embedding for a robotic system are introduced and discussed in a fairly detail in Chapter 4. The procedures of finding an isometric embedding are also provided with a number of illustrative examples and formulas. They are followed by the principle of duality with applications to kinematics and dynamics modeling for a large variety of robotic systems in open serial-chain, closed parallel-chain or hybrid-chain mechanisms in Chapter 5.

After the theories, representations and procedures of dynamics modeling are completed, the fundamental nonlinear control theories and algorithms are introduced and further developed in Chapter 6. The control design and application schemes and strategies are investigated, explored and discussed based on both the basic dynamic models and the advanced dynamics modeling of robotic systems. A number of typical examples with their MATLAB^{TM} simulations and 3D graphical animation studies are included. The last two chapters: Chapter 7 and Chapter 8 will give more specific focuses, explorations and case studies on direct adaptive control and interactive control strategies and methodologies.

Many examples in the book are 3D full-size robotic systems, not just a simple planar robot with only single or two links. In addition to the full-size modeling, the control processes of both position and orientation for a full-size robotic system will also be illustrated to validate and verify the effectiveness of the newly developed control algorithms based on the basic and advanced dynamic models.

REFERENCES

1. Becker, R.A., (1954) Introduction to Theoretical Mechanics. McGraw-Hill, New York.
2. Prentis, J.M., (1980) Dynamics of Mechanical Systems, 2nd Edition. John Wiley, New York.
3. Bingham, G.P., (1987) Dynamic Systems and Event Perception: A Working Paper. Parts I-III. *Perception/Action Workshop Review*, 2, 4-14.
4. Marmo, G., Saletan, E.J., Simoni, A. and Vitale, B., (1985) Dynamical Systems: A Differential Geometric Approach to Symmetry and Reduction. John Wiley, New York.
5. Featherstone, R., (2007) Rigid Body Dynamics Algorithms. Springer, Boston.
6. Arnold V., (1978) Mathematical Methods of Classical Mechanics. Springer-Verlag, New York.
7. Abraham, R. and Marsden, J., (1978) Foundations of Mechanics. The Benjamin/Cummings Publishing Company.
8. Marsden, J. and Ratiu, T., (1994) Introduction to Mechanics and Symmetry. Springer-Verlag, New York.

9. Brady, M., Hollerbach, J.M., Johnson, T.L., Lozano-Perez, T. and Mason, M.T., (1982) Robot Motion. MIT Press, Cambridge, MA.

10. Spong, M., Vidyasagar, M., (1989) Robot Dynamics and Control. John Wiley & Sons, New York.

11. Murray, R., Li, Z., Sastry, S., (1994) A Mathematical Introduction to Robotic Manipulation. CRC Press, Boca Raton, London, New York.

12. Craig, J., (2005) Introduction to Robotics: Mechanics and Control. 3rd Edition, Pearson Prentice Hall, New Jersey.

13. Siciliano, B., Khatib, O. (ed) (2008) Springer Handbook of Robotics. Springer.

14. Sciavicco, L, and Siciliano, B., (1996) Modeling and Control of Robot Manipulators. McGraw-Hill.

15. Lenarčič, J., Husty, M., (ed) (1998) Advances in Robot Kinematics: Analysis and Control. Kluwer Academic Publishers, the Netherlands.

16. Siciliano, B., Sciavicco, L., Villani, L., Oriolo, G., (2009) Robotics, Modeling, Planning and Control. Springer.

17. Gu, Edward Y.L., (2000) Configuration Manifolds and Their Applications to Robot Dynamic Modeling and Control. *IEEE Transactions on Robotics and Automation*, Vol.16, No.5, October, pp. 517-527.

18. Gu, Edward Y.L., (2000) A Configuration Manifold Embedding Model for Dynamic Control of Redundant Robots. *International Journal of Robotics Research*, Vol.19, No.3, March, pp. 289-304.

19. Gu, Edward Y.L., (2013) A Journey from Robot to Digital Human. Springer, Heidelberg, New York.

20. Kostrikin, A.I., (2001) Duality, *Encyclopedia of Mathematics*, EMS Press.

21. Gowers, Timothy, (2008) III.19 Duality, *The Princeton Companion to Mathematics*, Princeton University Press, pp. 187-190.

22. Shai, O. and Pennock, G.R., (2006) A Study of the Duality Between Planar Kinematics and Statics. *ASME Journal of Mechanical Design*, May 2006, Vol. 128, pp. 587-598.

23. Pandilov, Z. and Dukovski, V., (2014) Comparison of the Characteristics Between Serial and Parallel Robots. *Acta Technica Corvininesis-Bull Eng* 7(1), pp. 144-160.

24. Iqbal1, H., Umair, M., Khan1, A. and Yi, Byung-Ju, (2020) Analysis of Duality-Based Interconnected Kinematics of Planar Serial and Parallel Manipulators Using Screw Theory, *Intelligent Service Robotics* 13, pp. 47-62.

25. Slotine, J. and Li, W., (1988) Adaptive Manipulator Control: A Case Study. *IEEE Transactions on Automatic Control*, Vol. 33-11, pp. 995-1003.

26. Slotine, J. and Li, W., (1989) Composite Adaptive Control of Robot Manipulators. *Automatica*, Vol. 25-4, pp. 509-519.

27. Sastry, S. and Bodson, M. (1989) Adaptive Control: Stability, Convergence, and Robustness. Prentice Hall.

28. Kristic, M., Kanellakopoulos, I. and Kokotovic, P., (1995) Nonlinear and Adaptive Control Design. John Wiley & Sons, New York.

2 Fundamental Preliminaries

2.1 MATHEMATICAL PREPARATIONS

2.1.1 LIE GROUPS, LIE ALGEBRAS AND THEIR TOPOLOGICAL STRUCTURES

A collection, or a set of elements in mathematics is often studied by associating it with a certain mathematical operation. This was the primary notion for the **group theory** that distinguishes it from the set theory [1,2]. The formal definition of group is given as follows:

Definition 2.1. A **group** is defined as a set G along with a binary operation $\circ$ such that all the following conditions hold:

1. *Closure:* For any a, $b \in G$, $a \circ b = c \in G$;
2. *Associativity:* For every a, b, $c \in G$, $(a \circ b) \circ c = a \circ (b \circ c)$;
3. *Identity:* There is an identity element $\iota \in G$ such that $\iota \circ g = g \circ \iota = g$ for all $g \in G$;
4. *Inverse:* For each $g \in G$, there exists an element $h \in G$ such that $g \circ h = h \circ g = \iota$.

There are a lot of group examples, and the following just show a few of them:

- All the real integers associated with the addition form an additive group if the identity $\iota = 0$, but they cannot be a multiplicative group because the reciprocal of a real integer $k \neq 1$ is no longer an integer, which violates the inverse condition;
- All the real rational numbers with either the addition or the multiplication can constitute an additive group or a multiplicative group, respectively;
- All the real or complex numbers under either the addition or the multiplication can form an additive or a multiplicative group, and they are further qualified to constitute an algebraic field, called the real field $\mathbb{R}$ and complex field $\mathbb{C}$, respectively;
- A finite set of n symbols under the mathematical permutation can form a symmetric group S_n;
- All the n primitive roots of the complex number 1, i.e., $\sqrt[n]{1} = e^{j\frac{2k\pi}{n}}$ for $k = 0, \cdots, n-1$ under the multiplication form a cyclic group $\mathbb{Z}_n$.

If a set of elements along with a certain operation satisfies every condition but the inverse, even though the identity condition can still hold, then, this set just forms a semigroup. For example, all the dual numbers $a + \varepsilon b$ under $\varepsilon^2 = 0$ and all the double numbers $a + jb$ under $j^2 = +1$ can constitute an additive group in each case, but they can only form a multiplicative semigroup due to the inverse lost if $a = 0$ for the dual numbers and $a = b$ for the double numbers. The formal introduction and detailed discussion on both the dual number and double number can be found in the later section of this chapter. In contrast, the set of all 3D real vectors in the space $\mathbb{R}^3$ under the cross product $\times$ is neither a group, nor a semigroup, because it violates the associativity, identity and inverse conditions.

Depending on the number of elements and their variation properties, a group can be classified into finite and infinite groups, and discrete and continuous groups. A Lie group is a typical infinite, continuous and smooth group to make itself a differentiable manifold with a certain topological structure. In contrast to the well-known Galois group theory that was developed by the French mathematician Evariste Galois (1811-1832) to study the discrete symmetry of algebraic equations, the great Norwegian mathematician Marius Sophus Lie (1842-1899) created and introduced his Lie groups to model the continuous symmetry of differential equations in history. Because of the differentiable and topological nature, the Lie group theories are widely used in many areas of modern mathematics and physics [1–3, 7, 8].

Typical and commonly useful Lie groups are given but not limited in the following representative examples:

- All the non-singular n by n real matrices with the multiplication form a general linear Lie group $GL(n, \mathbb{R}) = \{A \in \mathbb{R}^{n \times n} \mid \det(A) \neq 0\}$;
- If the determinant of each member of $GL(n, \mathbb{R})$ is restricted to $+1$, then the Lie group $GL(n, \mathbb{R})$ becomes a special linear group $SL(n, \mathbb{R}) = \{A \in \mathbb{R}^{n \times n} \mid \det(A) = +1\}$;
- All the n by n real orthogonal matrices with the multiplication form an orthogonal Lie group $O(n) = \{A \in \mathbb{R}^{n \times n} \mid A^T A = AA^T = I\}$, where we omit $\mathbb{R}$ for $O(n, \mathbb{R})$ in most cases of application as a default;
- Similarly, by further imposing the determinant of each member of $O(n)$ to be $+1$, it becomes a special orthogonal group $SO(n) = \{A \in \mathbb{R}^{n \times n} \mid A^T A = AA^T = I$ and $\det(A) = +1\}$. Clearly, $SO(n) = SL(n) \cap O(n)$. Every rotation matrix in 3D space is a member of the $SO(3)$ group, which is also a 3-dimensional topological space;
- All the real even-dimensional square matrices with the multiplication under the certain skew-symmetric condition form a simplectic group $Sp(2n, \mathbb{R}) = \{A \in \mathbb{R}^{2n \times 2n} \mid A^T SA = S\}$, where the $2n$ by $2n$ non-singular skew-symmetric matrix S is defined as

$$S = \begin{pmatrix} O & I_n \\ -I_n & O \end{pmatrix}$$

with the n by n zero matrix O and identity I_n. The simplectic groups are useful in classical mechanics and quantum physics.

Among the most popular Lie groups, the special orthogonal group $SO(n)$ plays an essential role in uniquely representing rotations in Euclidean n-space $\mathbb{R}^n$. $SO(n)$ is not only a group, but is also a topological space. Since $SO(n)$ is represented by an n by n matrix, the net rotation degrees of freedom (d.o.f.) m can be determined by the total number n^2 of the elements in the n by n matrix minus the total number of constraints. For an orthogonal matrix with each column being unity, there are n constraints for each column normalization. Also, every pair of the columns must be orthogonal to each other so that the additional number of constraints should be the combination of two out of the n columns, i.e.,

$$\binom{n}{2} = \frac{n!}{2!(n-2)!} = \frac{n(n-1)}{2}.$$

Therefore, the net d.o.f. m for the special orthogonal group $SO(n)$ can be determined as

$$m = n^2 - \binom{n}{2} - n = \frac{n(n-1)}{2}.$$

For example, there is no rotation chance at all in the 1-space $\mathbb{R}^1$ so that the d.o.f. $m = 0$. In the 2D plane $\mathbb{R}^2$, the rotation can only be performed on the plane about the axis normal to the plane, and obviously $m = 1$. In the visualizable 3D space $\mathbb{R}^3$, $m = \dim(SO(3)) = 3$, while in the non-visualizable 4D space $\mathbb{R}^4$, the d.o.f. of rotation are jumped to $m = 6$, instead of just $3 + 1 = 4$. Actually, the single d.o.f. of rotation on the 2D plane for $SO(2)$ can be simply referred to as a spin. If we suspend the spin from the 3 d.o.f. rotations for $SO(3)$, then based on the space quotient operation, we obtain

$$SO(3)/SO(2) \simeq S^2,$$

where S^2 is a 2-dimensional spherical surface in $\mathbb{R}^3$, i.e., $S^2 = \{x \in \mathbb{R}^3 \mid \|x\| = \text{const.}\}$, and $\simeq$ is a homeomorphism (continuous one-to-one and onto mapping) to represent the topological equivalence between the two spaces. The physical meaning is so clear that *an universal (U-type) joint will be reduced to a spherical (S-type) joint if the spin is suspended* in a mechanical system.

In addition, for the Lie group $O(n)$ of any finite n, every member $A \in O(n)$ is an n by n orthogonal matrix so that $\det(A^T A) = (\det(A))^2 = \det(I) = 1$ and thus, $\det(A) = \pm 1$. Because the continuous mapping $\det : A \to \mathbb{R}$ can map A to either $+1$ or -1, the group $O(n)$ is closed. Furthermore, since all the n column (or row) vectors of A can form an orthonormal basis in $\mathbb{R}^n$, each element a^i_j of A holds $|a^i_j|^2 \leq 1$ so that $O(n)$ is also bounded. Therefore, based on the following well-known Heine-Borel Theorem, the topological space $O(n)$ is compact:

Theorem 2.1. (Heine-Borel) – *A subset of the Euclidean n-space, $M \subset \mathbb{R}^n$ is compact if and only if it is closed and bounded.*

Now, because $SL(n) \cap O(n) = SO(n)$, the topological space $SO(n)$ is a closed subset of the compact space $O(n)$. Hence, $SO(n)$ is also compact. The compactness of $SO(3)$ will play an important role in the future discussions on the advanced dynamics modeling and control in Chapter 4.

However, many useful sets under the certain binary operations violate either one or more conditions of the group definition, and they are disqualified to be a group, though they are quite useful. In order to keep them for further study and application, we have to relax the conditions. Lie algebra is one of the most typical and important approaches to rescuing the useful sets that have been ruled out by the group definition [1–3].

Definition 2.2. A **Lie algebra** over the real field $\mathbb{R}$ or the complex field $\mathbb{C}$ is a vector space $\mathcal{G}$, in which a bilinear mapping $(X,Y) \to [X,Y]$ is defined from $\mathcal{G} \times \mathcal{G} \to \mathcal{G}$ such that

1. $[X,Y] = -[Y,X]$, for all $X,Y \in \mathcal{G}$, and
2. $[X,[Y,Z]] + [Y,[Z,X]] + [Z,[X,Y]] = 0$ for all $X,Y,Z \in \mathcal{G}$.

The first condition in the above definition implies that $[X,X] = 0$, and the second one is called the *Jacobi Identity*, which reflects the closed cyclic nature of skew-symmetry. Commonly $[X,Y]$ is called the *Lie bracket* between X and Y as a kind of bilinear mapping.

When Lie created the Lie algebra concept, he also discovered that every finite-dimensional real Lie algebra is the Lie algebra of some simply connected Lie group. This connection is often referred to as the **Lie's Third Theorem** that reveals the intrinsic correspondence between Lie groups and Lie algebras. More specifically, if G is a Lie group, then we can always define a following set:

$$g = \{X \mid e^{aX} \in G, \forall a \in \mathbb{R}\}$$

to be a Lie algebra that is associated with the Lie group G, denoted as $\text{Lie}(G) \cong g$, where $\cong$ is an isomorphism. This means that through the **exponential mapping**, every finite-dimensional real Lie algebra can be simply connected to a certain Lie group, and some of the following Lie algebra examples will also show such connections:

1. Let the Lie bracket be defined as a *commutator*, i.e., for two arbitrary n by n square matrices A and B, $[A,B] = AB - BA$. Obviously, all the n by n square matrices under the commutator satisfy every requirement to constitute a Lie algebra. The commutator of matrices also reveals that the multiplication between two matrices are not commutable in general, and the result of $[A,B]$ is just their commutation difference. This implies that for two no-zero matrices A and B, $[A,B] = 0$ if and only if they are commutable.
2. As we mentioned earlier, all the 3D real vectors under the cross product $\times$ cannot form a group, but they are well-qualified to be a Lie algebra. It is not difficult to show that if a and b are two non-zero vectors in $\mathbb{R}^3$, then $c = a \times b \in \mathbb{R}^3$ as well. In addition, based on the cross-product definition, $a \times b = -b \times a$, while the following cyclic cancellation is also valid:

$$a \times (b \times c) + b \times (c \times a) + c \times (a \times b) = 0.$$

In fact, if we substitute the following well-known triple cross product formula

$$a \times (b \times c) = (a \cdot c)b - (a \cdot b)c \tag{2.1}$$

into the left-hand side of the Jacobi identity, all the terms will be immediately canceled out.

Since the triple cross-product is not associative in general, i.e., $a \times (b \times c) \neq (a \times b) \times c$, we want to know what is the difference between them? Based on the above Jacobi identity, it is now clear that

$$a \times (b \times c) - (a \times b) \times c = a \times (b \times c) + c \times (a \times b) = (c \times a) \times b.$$

For instance, let

$$a = b = \begin{pmatrix} 2 \\ -3 \\ 1 \end{pmatrix}, \quad \text{and} \quad c = \begin{pmatrix} 0 \\ 4 \\ -2 \end{pmatrix}.$$

Then,

$$a \times (b \times c) - (a \times b) \times c = \begin{pmatrix} -28 \\ -14 \\ 14 \end{pmatrix},$$

which exactly agrees with the result of $(c \times a) \times b$.

3. All the 3 by 3 skew-symmetric matrices under the Lie bracket of commutation form a Lie algebra $so(3)$. In fact, if S_1 and S_2 are two 3 by 3 skew-symmetric matrices, i.e., $S_1^T = -S_1$ and $S_2^T = -S_2$, then $S = [S_1, S_2] = S_1 S_2 - S_2 S_1$, and

$$S^T = (S_1 S_2 - S_2 S_1)^T = S_2^T S_1^T - S_1^T S_2^T = S_2 S_1 - S_1 S_2 = [S_2, S_1] = -S.$$

This implies that $S = [S_1, S_2]$ is still a 3 by 3 skew-symmetric matrix, plus both conditions in the Lie algebra definition hold as well. Furthermore, every 3 by 3 skew-symmetric matrix S is actually a 3D vector cross-product operator. Namely,

$$S = \begin{pmatrix} 0 & -z & y \\ z & 0 & -x \\ -y & x & 0 \end{pmatrix} = \begin{pmatrix} x \\ y \\ z \end{pmatrix} \times = s \times.$$

Let $\phi = \|s\| = \sqrt{x^2 + y^2 + z^2}$ be the norm of the 3D vector s. Then, $S = \phi K$, and the new 3 by 3 matrix K becomes a normalized skew-symmetric matrix. It can be shown that K satisfies the following three properties [13]:

a. $K^T = -K$ with $\text{tr}(K) = 0$;
b. K^2 is a 3 by 3 symmetric matrix with $\text{tr}(K^2) = -2$; and
c. $K^3 = -K$.

Let us now substitute $S = \phi K$ into the exponential mapping $e^S = e^{\phi K}$, and then expand it to the Taylor series around the zero with noticing the above properties of K, we obtain

$$\begin{aligned} \exp(\phi K) &= I + \phi K + \frac{\phi^2}{2!} K^2 + \frac{\phi^3}{3!} K^3 + \frac{\phi^4}{4!} K^4 + \frac{\phi^5}{5!} K^5 + \cdots \\ &= I + \left(\phi - \frac{\phi^3}{3!} + \frac{\phi^5}{5!} - \cdots \right) K + \left(\frac{\phi^2}{2!} - \frac{\phi^4}{4!} + \cdots \right) K^2 \\ &= I + \sin\phi K + (1 - \cos\phi) K^2 = R, \end{aligned} \tag{2.2}$$

where the sine and cosine Taylor expansions have been applied:

$$\sin\phi = \phi - \frac{\phi^3}{3!} + \frac{\phi^5}{5!} - \cdots \quad \text{and} \quad \cos\phi = 1 - \frac{\phi^2}{2!} + \frac{\phi^4}{4!} - \cdots.$$

According to the properties of the unity skew-symmetric matrix K, the above new 3 by 3 matrix R has its transpose $R^T = I - \sin\phi K + (1 - \cos\phi)K^2$, which can be used to show $R^T R = RR^T = I$ readily. By further invoking the following well-known identity for any n by n square matrix A in linear algebra:

$$\det(e^A) = e^{\text{tr}(A)}, \tag{2.3}$$

we have

$$\det(R) = \det(e^{\phi K}) = e^{\text{tr}(\phi K)} = e^0 = +1.$$

Therefore, the matrix R from (2.2) is an orthogonal matrix and $R \in SO(3)$. This evidently demonstrates that the Lie algebra $so(3) = \text{Lie}(SO(3))$, a clear connection between the Lie algebra $so(3)$ and Lie group $SO(3)$ [1,2,9,13]. Moreover, let $S(v) \in so(3)$ be the cross-product operator for an arbitrary vector $v \in \mathbb{R}^3$, i.e., $S(v) = v\times$. We want to know if $[S(a), S(b)] = S(c)$ for some non-zero 3D vectors a, b and c, what are the relationship among the three vectors? To find the answer, let the Lie bracket operates on the arbitrary vector v, i.e.,

$$[S(a), S(b)]v = S(a)S(b)v - S(b)S(a)v = a \times (b \times v) - b \times (a \times v),$$

which is the difference between two triple vector products. By applying the triple product formula in (2.1), we obtain

$$[S(a), S(b)]v = (a^T v)b - (a^T b)v - (b^T v)a + (b^T a)v$$
$$= (a^T v)b - (b^T v)a = (a \times b) \times v.$$

Because v is arbitrary, the above equation implies that

$$[S(a), S(b)] = (a \times b)\times = S(a \times b), \tag{2.4}$$

i.e. the cross-product operator of $a \times b$. The above justification can also be applied in the reversed direction. Therefore, we prove the following theorem:

Theorem 2.2. *Let $S(a) = a\times$, $S(b) = b\times$ and $S(c) = c\times$ be the three non-zero 3 by 3 skew-symmetric matrices, all of which are the members of the Lie algebra so(3). Then, $[S(a), S(b)] = S(c)$ if and only if $a \times b = c$.*

This theorem indicates that the 3D vector cross-product algebra should be the *subalgebra* of the Lie algebra $so(3)$.

4. A collection of all the n by n square matrices with zero trace under the Lie bracket of commutation forms a Lie algebra $sl(n)$. We can clearly see that if $A, B \in sl(n)$, then $C = [A, B] = AB - BA$, and its trace $\text{tr}(C) = \text{tr}(AB) - \text{tr}(BA) = 0$, because $\text{tr}(AB) = \text{tr}(BA)$. In addition, based on the identity in (2.3), for an n by n matrix $M = e^A$ with $A \in sl(n)$, $\det(M) = \det(e^A) = e^{\text{tr}(A)} = e^0 = +1$ so that $M \in SL(n)$. Therefore, $\text{Lie}(SL(n)) = sl(n)$, another beautiful connection between the Lie algebra $sl(n)$ and Lie group $SL(n)$.

5. All the n-dimensional real vector fields, each of which is a smooth function of point $x \in \mathbb{R}^n$ under the following Lie bracket of differential commutation:

$$[f,g] = \frac{\partial g}{\partial x}f - \frac{\partial f}{\partial x}g \tag{2.5}$$

can also constitute a Lie algebra. This Lie bracket definition in (2.5) is referred to as a *Lie derivative of vector fields*. For example, if two 3-dimensional vector fields $f(x)$ and $g(x)$ are given by

$$f(x) = \begin{pmatrix} x_1^2 + x_2 \\ -x_3^2 \\ x_2^2 - x_1 \end{pmatrix} \quad \text{and} \quad g(x) = \begin{pmatrix} x_2 + x_3 \\ 0 \\ x_1^2 - x_2 \end{pmatrix},$$

then,

$$[f,g] = \frac{\partial g}{\partial x}f - \frac{\partial f}{\partial x}g = \begin{pmatrix} 0 & 1 & 1 \\ 0 & 0 & 0 \\ 2x_1 & -1 & 0 \end{pmatrix}f - \begin{pmatrix} 2x_1 & 1 & 0 \\ 0 & 0 & -2x_3 \\ -1 & 2x_2 & 0 \end{pmatrix}g$$

$$= \begin{pmatrix} x_2^2 - x_3^2 - x_1 - 2x_1x_2 - 2x_1x_3 \\ 2x_1^2x_3 - 2x_2x_3 \\ 2x_1^3 + 2x_1x_2 + x_3^2 + x_2 + x_3 \end{pmatrix},$$

which is still a 3-dimensional vector field.

6. The *Lie derivative of scalar fields*:

$$L_f h(x) = \frac{\partial h}{\partial x}f(x)$$

is a generalized directional derivative for the scalar field $h(x)$ with respect to $x \in \mathbb{R}^n$ along the direction of the vector field $f(x)$. The directional derivative operator L_f under the Lie bracket of commutation can form a Lie algebra. Namely, $[L_f, L_g] = L_f L_g - L_g L_f$ is still a directional derivative operator. For example, using the previous data for the two vector fields $f(x)$ and $g(x)$, if we define a scalar field $h(x) = x_2 - x_3$, then,

$$L_f h(x) = (0 \; 1 \; -1)f(x) = x_1 - x_2^2 - x_3^2,$$

and

$$L_g h(x) = (0 \; 1 \; -1)g = x_2 - x_1^2.$$

Thus,

$$L_f L_g h(x) = (-2x_1 \; 1 \; 0)f(x) = -2x_1^3 - 2x_1x_2 - x_3^2,$$

and

$$L_g L_f h(x) = (1 \; -2x_2 \; -2x_3)g(x) = x_2 + x_3 - 2x_1^2x_3 + 2x_2x_3.$$

Finally,

$$[L_f, L_g]h(x) = L_f L_g h(x) - L_g L_f h(x)$$
$$= -2x_1^3 - 2x_1 x_2 - x_3^2 - x_2 - x_3 + 2x_1^2 x_3 - 2x_2 x_3.$$

Furthermore, we can also show that the directional derivative operator has a close relation to the differential commutation in (2.5):

$$[L_f, L_g] = L_{[f,g]}. \tag{2.6}$$

In fact, after the left-hand side operates on an arbitrary scalar field $h(x)$, it becomes

$$[L_f, L_g]h(x) = L_f L_g h(x) - L_g L_f h(x) = L_f \left(\frac{\partial h}{\partial x} g \right) - L_g \left(\frac{\partial h}{\partial x} f \right)$$

$$= f^T \frac{\partial^2 h}{\partial x^2} g + \frac{\partial h}{\partial x} \frac{\partial g}{\partial x} f - g^T \frac{\partial^2 h}{\partial x^2} f - \frac{\partial h}{\partial x} \frac{\partial f}{\partial x} g = \frac{\partial h}{\partial x} [f, g] = L_{[f,g]} h(x),$$

where $\frac{\partial^2 h}{\partial x^2} = \left\{ \frac{\partial^2 h}{\partial x_i \partial x_j} \right\}$ is a Hessian matrix that is symmetric so that $f^T \frac{\partial^2 h}{\partial x^2} g = g^T \frac{\partial^2 h}{\partial x^2} f$.

    The above Lie algebra examples, though only a few, are quite typical and useful in modern mathematics, physics and many other fields of science. The profound and insightful concepts of the Lie algebra and Lie group also underline the recent advanced engineering research and development as powerful and indispensable mathematical tools.

2.1.2   MANIFOLDS, RIEMANNIAN METRICS, AND EMBEDDINGS

Manifold is a generalized concept for a high-dimensional non-Euclidean (curved) hyper-surface. It was first formulated in mathematical terms by the great German mathematician Carl Friedrich Gauss (1777-1855) to underline the procedures used in cartography (drawing the maps for the earth surface) at that time. The motivation to introduce the manifold concept was to better study a globally non-Euclidean manifold M^n by making it locally Euclidean. This implies that a manifold M^n is a topological space, each point on which has an open neighborhood U that can be homeomorphic to an open region U' of the Euclidean n-space, i.e., $U' \subset \mathbb{R}^n$. This homeomorphism $\phi : U \to U'$ along with U is called a *chart* of the manifold [3–7]. Since the small open neighborhood U at a point of the manifold M^n can be topologically equivalent to a Euclidean region, we may thus define a coordinate system in U, called the **local coordinates**. In fact, the concept of local coordinates is very common in calculus and differential equations. Typical examples are given as follows with belief explanations:

1. In a 3D cylinder-type region, we define a set of cylindrical coordinates (r, ϕ, z) referred to the 3D orthogonal coordinate system of the basis $\{e_r, e_\phi, e_z\}$ at each

point of the region. These variables (r, ϕ, z) forms a set of local coordinates in the small open neighborhood U of a point under the following relationship with the Cartesian coordinates:

$$\begin{cases} x = r\cos\phi = rc\phi \\ y = r\sin\phi = rs\phi \\ z = z, \end{cases} \qquad (2.7)$$

where r and ϕ are the polar coordinates on the x-y plane under the constraints $r \geq 0$ and $0 \leq \phi < 2\pi$. Here and later, we denote $c\phi = \cos\phi$ and $s\phi = \sin\phi$ to shorten the equations and derivations as a common abbreviation in robotics.

2. In a 3D sphere-type region, we define a set of spherical coordinates (R, θ, ϕ) referred to the 3D orthogonal coordinate system of the basis $\{e_R, e_\theta, e_\phi\}$ at each point of the region. The variables (R, θ, ϕ) form the local coordinates in the small open neighborhood U of a point under the following relationship with the Cartesian coordinates:

$$\begin{cases} x = R\sin\theta\cos\phi = Rs\theta c\phi \\ y = R\sin\theta\sin\phi = Rs\theta s\phi \\ z = R\cos\theta = Rc\theta, \end{cases} \qquad (2.8)$$

where R is the radial variable, θ is the elevation angle, and ϕ is the azimuth angle under $R \geq 0, 0 \leq \theta \leq \pi$ and $0 \leq \phi < 2\pi$. Both the cylindrical and spherical coordinate systems are shown in Figure 2.1, which have been frequently used in calculus to deal with the differentiation and integration over a cylinder-type or sphere-type region.

3. A 2-dimensional surface of torus T^2 is topologically equivalent to $S^1 \times S^1$, the product of two 1-circles, or to the quotient space $\mathbb{R}^2/\mathbb{Z}^2$ as the Euclidean 2-space $\mathbb{R}^2$ modulo an action of the integer lattice $\mathbb{Z}^2$. Thus, a 2-torus surface contains two "holes". The local coordinates (θ_1, θ_2) on the open neighborhood U at each point of T^2 can be defined to associate with the 3D Cartesian coordinates as follows:

$$\begin{cases} x = (R + r\cos\theta_2)\cos\theta_1 = (R + rc_2)c_1 \\ y = (R + r\cos\theta_2)\sin\theta_1 = (R + rc_2)s_1 \\ z = r\sin\theta_2 = rs_2, \end{cases} \qquad (2.9)$$

where $R > r > 0$ are two constant parameters, called the major and minor radii, respectively, and the two local coordinates $q^1 = \theta_1$ and $q^2 = \theta_2$ are depicted in Figure 2.2.

4. On so-called the **configuration manifold (C-manifold)** M^n that is a dynamic constraint hyper-surface for a robot arm with n joints, all the joint angles in the joint position vector $q = (\theta_1 \cdots \theta_n)^T$ can be defined as a set of local coordinates

to study the topology, geometry, and trajectory-tracking on such non-Euclidean C-manifold M^n. The local coordinate basis may not be orthogonal to each other in this case, but they must be linearly independent. Using all the joint angles to be a set of local coordinates has an exclusive advantage that every joint angle is directly measurable without any conversion needed. The concept of C-Manifold and its isometric embedding theory will be introduced and discussed in Chapter 4.

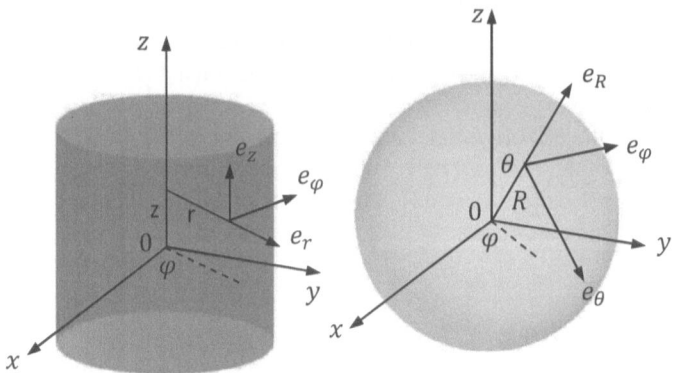

Figure 2.1   The cylindrical and spherical local coordinate systems

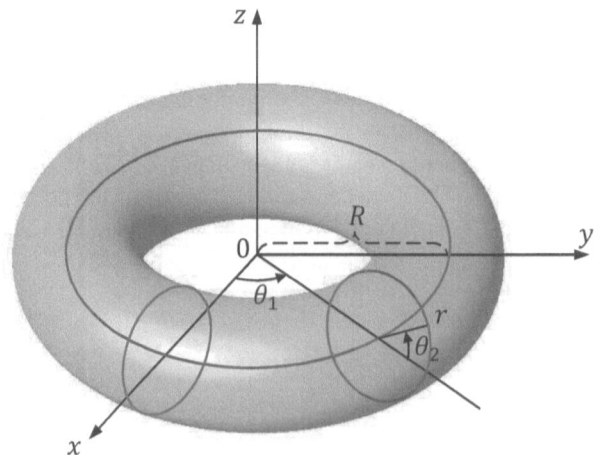

Figure 2.2   A 2D torus T^2 situated in the Euclidean space $\mathbb{R}^3$

If a manifold M^n is differentiable at each point, it is called a smooth manifold. We can then determine a **Riemannian metric** that is endowed on the smooth manifold to study its geometrical details, such as arc length, angle at an intersection, curvature, and area of a surface, etc. The great German mathematician Bernhard Riemann (1826-1866) first created the metric concept. He extended the Gauss's theory of manifold to higher-dimensional spaces in a way that allows distances and angles to be measured and the notion of curvature to be defined. He also discovered that all the measures were intrinsic to the manifold and independent of its embedding into higher-dimensional spaces. In general, a Riemannian metric on M^n is a point-dependent and positive-definite quadratic form on the tangent vectors at each point, which smoothly depends on the local coordinates of the point.

Now, consider an n-dimensional smooth manifold M^n that is situated in an m-dimensional ambient Euclidean space $\mathbb{R}^m$ with $m \geq n$. Let $q = (q^1 \cdots q^n)^T$ be a set of local coordinates defined on M^n along with a homeomorphism ϕ that maps q in an open neighborhood U of each point on M^n to an open region U' of the Euclidean n-space $\mathbb{R}^n$ [3,6]. Let $z = (z^1 \cdots z^m)^T$ be a coordinate system of the ambient Euclidean m-space $\mathbb{R}^m$. Then, this M^n can be represented by

$$z = \zeta(q). \tag{2.10}$$

The Jacobian matrix of (2.10) is given by

$$J = \frac{\partial \zeta}{\partial q} = (g_1 \cdots g_n), \tag{2.11}$$

which is m by n and $g_i = \partial \zeta / \partial q^i$ is the i-th column of J for each $i = 1, \cdots, n$. The manifold M^n is said to be **non-singular** if its Jacobian matrix has full rank, i.e., $\text{rank}(J) = n$ at every point on the manifold. Based on differential geometry [3,6], all the n columns $g_1, \cdots, g_n$ of the Jacobian matrix J for a non-singular manifold M^n span a **tangent space** $T_q(M^n)$ at each point $q \in U \subset M^n$.

In order to determine the arc length l of a curve given by $z(t) = \zeta(q(t))$ for $t \in [a,b]$ on M^n situated in $\mathbb{R}^m$, we have to integrate the following inner product of the tangent vectors:

$$\dot{z}^T \dot{z} = \langle \dot{z}, \dot{z} \rangle_\delta = \sum_{i,j=1}^{m} \delta_{ij} \dot{z}^i \dot{z}^j = \sum_{i,j=1}^{n} w_{ij} \dot{q}^i \dot{q}^j = \langle \dot{q}, \dot{q} \rangle_w, \tag{2.12}$$

where the Kronecker delta in the first summation is defined by

$$\delta_{ij} = \begin{cases} 1 & \text{if } i = j \\ 0 & \text{if } i \neq j. \end{cases}$$

The second summation in (2.12) is represented in terms of all the tangent vectors of the local coordinates on M^n, and thus it produces a set of new weights w_{ij} in the last inner product. Equation (2.12) is obviously a positive-definite quadratic form, and is thus the Riemannian metric for a differentiable manifold [3,5].

With such a coordinate transformation between the local coordinates q on M^n and the coordinates z in $\mathbb{R}^m$, and noticing that $\dot{z} = \frac{\partial \zeta}{\partial q} \dot{q}$ according to (2.10), we obtain

$$w_{ij} = \sum_{k,l=1}^{m} \delta_{kl} \frac{\partial \zeta^k}{\partial q^i} \frac{\partial \zeta^l}{\partial q^j}, \tag{2.13}$$

for each $i, j = 1, \cdots, n$. Equation (2.13) is written in a tensor form and can be rewritten in the following matrix form:

$$W = J^T J. \tag{2.14}$$

This reaches a significant conclusion that *the Riemannian metric endowed on a non-singular smooth manifold M^n situated in an ambient Euclidean space $\mathbb{R}^m$ with $m \geq n$ is given by the symmetric matrix W that can always be factorized by the Jacobian matrix J of the smooth manifold M^n.*

Using equation (2.14), let us revisit the cylindrical, spherical and the torus local coordinates cases. All the three equations (2.7), (2.8) and (2.9) are actually the manifold representations $z = \zeta(q)$ given in equation (2.10) with $m = n = 3$ in the first two cases and $m = 3 > n = 2$ in the third case. In other words, all the cylindrical and spherical region of dimension $n = 3$ as well as the 2-torus surface are situated in the Euclidean 3-space $\mathbb{R}^3$. This means that $z^1 = x$, $z^2 = y$ and $z^3 = z$. The local coordinates for the cylindrical region are $q^1 = r$, $q^2 = \phi$ and $q^3 = z$, for the spherical region, $q^1 = R$, $q^2 = \theta$ and $q^3 = \phi$, while for the 2-torus surface, $q^1 = \theta_1$ and $q^2 = \theta_2$.

Therefore, the 3 by 3 square Jacobian matrix in the cylindrical case can follow (2.11) to be calculated by

$$J = \frac{\partial(x,y,z)}{\partial(r,\phi,z)} = \begin{pmatrix} c\phi & -rs\phi & 0 \\ s\phi & rc\phi & 0 \\ 0 & 0 & 1 \end{pmatrix}.$$

While the 3 by 3 square Jacobian matrix in the spherical case is calculated by

$$J = \frac{\partial(x,y,z)}{\partial(R,\theta,\phi)} = \begin{pmatrix} s\theta c\phi & Rc\theta c\phi & -Rs\theta s\phi \\ s\theta s\phi & Rc\theta s\phi & Rs\theta c\phi \\ c\theta & -Rs\theta & 0 \end{pmatrix}.$$

Similarly, the 3 by 2 tall Jacobian matrix in the 2-torus surface case is determined as

$$J = \frac{\partial \zeta}{\partial q} = \begin{pmatrix} -(R+rc_2)s_1 & -rs_2 c_1 \\ (R+rc_2)c_1 & -rs_2 s_1 \\ 0 & rc_2 \end{pmatrix}. \tag{2.15}$$

Now, based on equation (2.14), the Riemannian metric for the cylindrical region is found to be

$$W = J^T J = \begin{pmatrix} c\phi & s\phi & 0 \\ -rs\phi & rc\phi & 0 \\ 0 & 0 & 1 \end{pmatrix} \begin{pmatrix} c\phi & -rs\phi & 0 \\ s\phi & rc\phi & 0 \\ 0 & 0 & 1 \end{pmatrix} = \begin{pmatrix} 1 & 0 & 0 \\ 0 & r^2 & 0 \\ 0 & 0 & 1 \end{pmatrix},$$

and its determinant $\det W = r^2$ such that $\det J = \sqrt{\det W} = r$. While for the spherical region, the Riemannian metric becomes

$$W = J^T J = \begin{pmatrix} 1 & 0 & 0 \\ 0 & R^2 & 0 \\ 0 & 0 & R^2 s^2 \theta \end{pmatrix}.$$

Clearly, the determinant $\det J = \sqrt{\det W} = R^2 \sin \theta$. Similarly, for the 2-torus surface, the Riemannian metric turns out to be

$$W = J^T J = \begin{pmatrix} (R + rc_2)^2 & 0 \\ 0 & r^2 \end{pmatrix}, \tag{2.16}$$

which is always positive-definite as $R > r > 0$.

Note that in the above first two $m = n$ cases, the determinant $\det(J)$ of each Jacobian matrix J is exactly the differential volume conversion between two different coordinate systems in calculus. Namely,

$$dx \wedge dy \wedge dz = r dr \wedge d\phi \wedge dz, \quad \text{and} \quad dx \wedge dy \wedge dz = R^2 \sin \theta dR \wedge d\theta \wedge d\phi.$$

However, in the case of robot dynamics modeling, the ambient Euclidean m-space $\mathbb{R}^m$ has to be more spacious than the n-space in order to completely one-to-one map the n-dimensional C-manifold M^n into $\mathbb{R}^m$. Namely, $m > n$, and $m = n$ would be almost impossible in general because of the fact that the C-manifold, as a dynamic constraint hyper-surface for an n-joint robotic system, is topologically more complex than those of the cylindrical and spherical regions. If one could find such a continuous one-to-one mapping, called an **embedding** as formulated by equation (2.10), then its Jacobian matrix J in equation (2.11) would have to be m by n with $m > n$, a tall matrix, similar to the above torus surface case in equation (2.15).

The Riemannian metric W, based on (2.14), should be an n by n positive-definite symmetric matrix, which would be exactly equal to the physical **inertial matrix** W of the robot if the embedding could also be **isometric** to preserve both the topology and geometry during the mapping. In other words, *the Riemannian metric endowed on a robot C-manifold given by W could exactly match with the inertial matrix of the robot if all the joint positions in $q = (\theta_1 \cdots \theta_n)^T$ would be defined as the local coordinates on M^n and an* **isometric embedding** *would be found to send M^n to $\mathbb{R}^m$.*

Conventionally, to find an inertial matrix W for a robot dynamic system, one has to seek the total kinetic energy K first and then to find W, or using a numerical algorithm to determine it, as will be presented in Chapter 3 for details. If one could directly find such an isometric embedding $z = \zeta(q)$, and through its Jacobian matrix to determine $W = J^T J$, then, it would become an alternative but much short and elegant way to simplify the robot dynamics modeling.

Now, a fundamental question challenges us: what is, in general, the least dimension m of an ambient Euclidean space $\mathbb{R}^m$ such that the C-manifold M^n of an n-joint robot can be immersed or embedded into $\mathbb{R}^m$? To answer this question, let us first go over the mathematical definitions of **the immersion and embedding** between two manifolds [4, 5].

Definition 2.3. A manifold M^n is said to be immersed into a manifold Y^m with $m \geq n$, if there is a smooth mapping $f : M \rightarrow Y$ such that the induced mapping $\bar{f} : T_x(M) \rightarrow T_{f(x)}(Y)$ at each point $x \in M^n$ is a one-to-one mapping of the tangent plane. The mapping f is called an immersion of M^n into Y^m. The immersion f is further called an embedding if f itself is one-to-one.

The American mathematician Hassler Whitney (1907-1989) first showed in 1936 the following interesting immersion/embedding theorem [4, 5, 16, 17]:

Theorem 2.3. (Whitney) – *A smooth and compact manifold M of dimension n can be smoothly immersed in $\mathbb{R}^{2n}$, and can also be smoothly embedded into $\mathbb{R}^{2n+1}$.*

Based on this theorem, any smooth and compact manifold M^n can be embedded into the Euclidean $(2n+1)$-space $\mathbb{R}^{2n+1}$ to preserve its topology. In many cases, however, the necessary dimension m of the ambient Euclidean space can be less than $2n$ (or $2n + 1$) for an immersion (or embedding), depending on the topological structure for a particular manifold M^n to be immersed (or embedded). For example, both the cylindrical region equation and spherical region equation are at least the immersions due to their non-singular Jacobian matrices. If all the local coordinates are restricted within the well-defined domains, they are also the embeddings. Similarly, the 2-torus equation as a mapping $\zeta : T^2 \rightarrow \mathbb{R}^3$ given by (2.9) is also an embedding because its 3 by 2 Jacobian matrix in (2.15) is fully ranked at each point on T^2, and both θ_1 and θ_2 are restricted within the quotient space $\mathbb{R}^1/2\pi\mathbb{Z}^1$. In fact, it can be shown that all the known compact and orientable 2-dimensional manifolds can be embedded into $\mathbb{R}^3$ [4].

In differential geometry, an n-dimensional smooth manifold M^n endowed with a Riemannian metric w_{ij} is called the **Riemannian manifold**, denoted by (M^n, w). Two French mathematicians Elie Joseph Cartan (1869–1951) and Maurice Janet (1888–1983) initiated the study on the isometric embedding, and showed the following well-known theorem in the 1920s:

Theorem 2.4. (Cartan-Janet) – *Any analytic n-dimensional Riemannian manifold M^n can be analytically, locally and isometrically embedded into $\mathbb{R}^{s_n}$.*

In the above theorem,

$$s_n = \frac{n(n+1)}{2}$$

is called the *Janet dimension*, which is actually the number of different elements in the positive-definite symmetric Riemannian metric matrix W, i.e., the number of elements $\{w^i_j\}$ for $i \geq j$, including the elements on the diagonal plus the lower-left corner due to its symmetric nature.

After the Cartan-Janet theorem, the American mathematician John Forbes Nash (1928–2015) also investigated the isometric embedding theories to extended the local and analytic conditions to global and smooth, and showed the well-known **Nash Embedding Theorem** in the 1950s to isometrically embed a smooth and compact M^n globally and smoothly into $\mathbb{R}^m$ with $m = \max\{s_n + 2n, s_n + n + 5\}$.

In 1970, a Russian-French mathematician Mikhail Leonidovich Gromov and two others: Vladimir Rokhlin and Robert Greene independently showed that any smooth Riemannian manifold M^n can be locally, smoothly and isometrically embedded into a much smaller Euclidean space $\mathbb{R}^{s_n+n}$ [18–20]. For instance, if $n = 2$, then $m = s_n = 3$ based on the Cartan-Janet theorem, and $m = 10$ by the Nash Embedding Theorem, while $m = s_n + n = 5$ by Gromov, Rokhlin and Greene. It appears to be recognizable that the dimension m for the ambient Euclidean space could be lower by relaxing some conditions, but none of the isometric embedding theorems in history can tell us how to find it.

Nevertheless, let us just explore by studying a simple but typical example: a planar robot arm with two revolute joints to see how we can expand the ambient Euclidean space in order to meet the embedding and further the isometric embedding requirements. Before this example is introduced in detail, let us first state the following useful theorem for testing a mapping to see if it is qualified as an embedding:

Theorem 2.5. *For a smooth mapping $\zeta : M^n \to \mathbb{R}^m$ with $m \geq n$, it is an embedding if and only if its Jacobian matrix $J = \frac{\partial \zeta}{\partial q}$ is full ranked, i.e., $rank(J) = n$.*

Proof. A qualified embedding requires one-to-one under the mapping condition $m \geq n$. This means that any change of the local coordinates q will make the z coordinates in an ambient Euclidean n-space distinguishable. Since $\dot{z} = J\dot{q}$, if J is full-ranked, based on linear algebra, any nonzero local coordinates change $\dot{q} \neq 0$ cannot lead to $\dot{z} = 0$ so that z is distinguishable.

On the other hand, if $z = \zeta(q)$ is an embedding, it must be one-to-one, this implies that J must be full ranked. Therefore, the sufficient and necessary condition is proven.

$\square$

This theorem can be employed as a mathematical tool to test if the mapping $z = \zeta(q)$ with $m \geq n$ is an embedding.

Now, let us return to the 2-joint planar robot example. As shown on the left of Figure 2.3, if the link lengths are a_1 and a_2 for the two links of the planar 2-joint arm, and the joint angles are θ_1 and θ_2, then based on the robot kinematics modeling algorithm, the tip-point position with respect to the base is given by

$$\begin{cases} x = a_1 c_1 + a_2 c_{12} \\ y = a_1 s_1 + a_2 s_{12}, \end{cases} \tag{2.17}$$

where and hereafter we will always adopt the conventional short notations: $c_i = \cos \theta_i$, $s_i = \sin \theta_i$, $c_{ij} = \cos(\theta_i + \theta_j)$ and $s_{ij} = \sin(\theta_i + \theta_j)$ for $i, j = 1, \cdots, n$.

Let $z^1 = x$ and $z^2 = y$. This tip position vector equation can first be considered as a mapping $\zeta : M^2 \to \mathbb{R}^2$, where M^2 is the compact C-manifold of this planar 2-joint robot arm, and $q = (\theta_1 \ \theta_2)^T$ is the set of local coordinates on M^2. Then, its 2 by 2 Jacobian matrix can be calculated as follows:

$$J = \frac{\partial \zeta}{\partial q} = \begin{pmatrix} -a_1 s_1 - a_2 s_{12} & -a_2 s_{12} \\ a_1 c_1 + a_2 c_{12} & a_2 c_{12} \end{pmatrix},$$

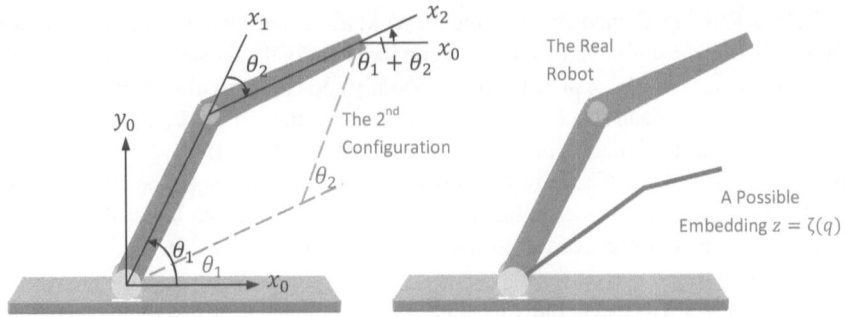

Figure 2.3   A planar RR-type robot arm and its multi-configuration and embedding

which is also the same Jacobian matrix in the robot kinematics. However, since its determinant $\det(J) = a_1 a_2 \sin \theta_2$, which vanishes at $\theta_2 = 0$ or $\pm\pi$, the robot arm has a singularity posture when the two links are lined up. Thus, we may predict that there is a *multi-configuration* issue, i.e., the two distinct configurations $(\theta_1^{(1)}, \theta_2)$ and $(\theta_1^{(2)}, -\theta_2)$ can result in the same tip position (x, y). These two distinct configurations with different values of θ_1 are, in general, symmetric with respect to its singular position at $\theta_2 = 0$ or $\pm\pi$. Such a multi-configuration issue may directly cause the mapping in (2.17) to be disqualified as an embedding.

To remedy the issue, we can add more equations into (2.17) as an effort of expanding the ambient Euclidean space. For such a simple robot arm, intuitively, we may add one more equation $z^3 = h_1(\theta_1 + \theta_2)$ that is actually representing the orientation of link 2, with a constant $h_1 \neq 0$ into the mapping to expand the ambient Euclidean space from $\mathbb{R}^2$ to $\mathbb{R}^3$, i.e.,

$$\begin{cases} z^1 = b_1 c_1 + b_2 c_{12} \\ z^2 = b_1 s_1 + b_2 s_{12} \\ z^3 = h_1(\theta_1 + \theta_2), \end{cases} \tag{2.18}$$

where we purposely replace each link length a_i by more generic coefficient b_i for $i = 1, 2$. Its new Jacobian matrix is now 3 by 2:

$$J = \frac{\partial \zeta}{\partial q} = \begin{pmatrix} -b_1 s_1 - b_2 s_{12} & -b_2 s_{12} \\ b_1 c_1 + b_2 c_{12} & b_2 c_{12} \\ h_1 & h_1 \end{pmatrix},$$

which is full-ranked at each point of M^2. Therefore, the mapping in (2.18) is qualified to be an embedding, but may not be an isometric embedding until we add one more equation $z^4 = h_2 \theta_1$ that represents the orientation of link 1 with a new constant $h_2 \neq 0$.

Thus, a new updated embedding becomes

$$
\begin{cases}
z^1 = b_1 c_1 + b_2 c_{12} \\
z^2 = b_1 s_1 + b_2 s_{12} \\
z^3 = h_1 (\theta_1 + \theta_2) \\
z^4 = h_2 \theta_1 .
\end{cases}
\tag{2.19}
$$

Its Jacobian matrix is expanded to 4 by 2:

$$
J = \frac{\partial \zeta}{\partial q} =
\begin{pmatrix}
-b_1 s_1 - b_2 s_{12} & -b_2 s_{12} \\
b_1 c_1 + b_2 c_{12} & b_2 c_{12} \\
h_1 & h_1 \\
h_2 & 0
\end{pmatrix},
$$

and the Riemannian metric becomes

$$
W = J^T J =
\begin{pmatrix}
b_1^2 + b_2^2 + 2 b_1 b_2 c_2 + h_1^2 + h_2^2 & b_2^2 + b_1 b_2 c_2 + h_1^2 \\
b_2^2 + b_1 b_2 c_2 + h_1^2 & b_2^2 + h_1^2
\end{pmatrix}.
\tag{2.20}
$$

Now, let us compare this Riemannian metric W in (2.20) with the following physical inertial matrix of the robot:

$$
W =
\begin{pmatrix}
w_{11} & m_2 l_2^2 + I_2 + m_2 a_1 l_2 c_2 \\
m_2 l_2^2 + I_2 + m_2 a_1 l_2 c_2 & m_2 l_2^2 + I_2
\end{pmatrix},
\tag{2.21}
$$

where $w_{11} = m_1 l_1^2 + I_1 + m_2 l_2^2 + I_2 + m_2 a_1^2 + 2 m_2 a_1 l_2 c_2$, and m_1 and m_2 are the masses of link 1 and link 2, while l_1 and l_2 are the mass center coordinates of link 1 and link 2, respectively. In addition, I_1 and I_2 are the moments of inertia for the two links. If we **deform** the smooth compact C-manifold M^2 by adjusting all the parameters to set

$$
b_1 = \sqrt{m_2}\, a_1, \; b_2 = \sqrt{m_2}\, l_2, \; h_1 = \sqrt{I_2}, \text{ and } h_2 = \sqrt{m_1 l_1^2 + I_1},
\tag{2.22}
$$

then, the Riemannian metric in (2.20) can be exactly equal to the inertial matrix of the robot in (2.21). This implies that the mapping (2.19) is, indeed, an isometric embedding to preserve both the topology and geometry for the compact C-manifold M^2 of the 2-joint planar robot under the mapping $\zeta : M^2 \to \mathbb{R}^4$.

Table 2.1 summarizes the above discussions on how a mapping can be defined and qualified as an embedding or even an isometric embedding through the example of the 2-revolute-joint planar robot. It can also be observed that for such a 2-joint robot arm, its compact C-manifold M^2 is 2-dimensional, but the ambient Euclidean 2-space $\mathbb{R}^2$ is too small to be fitted. In other words, the mapping (2.17) has a singular point, causing a multi-configuration issue so that it is disqualified to be an embedding. After adding the 3rd equation that represents the orientation of link 2, the new mapping becomes an embedding to preserve the topology of M^2 after it is sitting into $\mathbb{R}^3$. If one wishes the mapping could also preserve the geometry so that its Riemannian metric could match with the inertial matrix of the robot by smoothly deforming the

The Mapping $z = \zeta(q)$	Equation (2.17)	Equation (2.18)	Equation (2.19)
Physical Meaning at the Kinematics Perspective	Tip Position p_0^2 of the Robot	p_0^2 & Orientation of Link 2	p_0^2 & Orientations of Both Links 1 & 2
Jacobian Matrix $J = \frac{\partial \zeta}{\partial q}$	Singular Points	Full-Ranked	Full-Ranked
Ambient Euclidean Space	$\mathbb{R}^2$	$\mathbb{R}^3$	$\mathbb{R}^4$
Qualifications	Not Embedding	Embedding	Isometric Embedding

Table 2.1

Map the C-manifold M^2 of a 2-joint robot into an ambient $\mathbb{R}^n$

C-manifold M^2, one would have to further expand the ambient space to $\mathbb{R}^4$ by adding one more equation that represents the orientation of link 1, which could thoroughly eliminate the multi-configuration issue as well.

As a quantitative analysis, let the dynamic parameters of the 2-joint planar robot be given by

$$m_1 = 1.2 \text{ Kg}, \ m_2 = 0.8 \text{ Kg}, \ a_1 = 0.6 \text{ m}, \ a_2 = 0.5 \text{ m}, \ l_1 = 0.3 \text{ m}, \ l_2 = 0.25 \text{ m},$$

and the moments of inertia follow the standard formula for a solid rod rotating about the center axis, i.e., $I_1 = \frac{1}{12}m_1 a_1^2 = 0.036 \text{ Kg-m}^2$ and $I_2 = \frac{1}{12}m_2 a_2^2 = 0.0167 \text{ Kg-m}^2$. Then, according to the parameter adjustment in (2.22), we obtain

$$b_1 = 0.5367, \ b_2 = 0.2236, \ h_1 = 0.1292, \text{ and } h_2 = 0.3795.$$

Thus, the 2 by 1 tip-point position vector p_0^2 of the robot given by (2.17) and the 4 by 1 isometric embedding $z = \zeta(q)$ in (2.19) have their numerical new looks:

$$p_0^2 = \begin{pmatrix} 0.6c_1 + 0.5c_{12} \\ 0.6s_1 + 0.5s_{12} \end{pmatrix}, \quad \text{and} \quad z = \zeta(q) = \begin{pmatrix} 0.5367c_1 + 0.2236c_{12} \\ 0.5367s_1 + 0.2236s_{12} \\ 0.1292(\theta_1 + \theta_2) \\ 0.3795\theta_1 \end{pmatrix}.$$

Comparing the top two elements of the isometric embedding $z = \zeta(q)$ with the tip position vector p_0^2, we can clearly see that they have the same variable ingredients, but the distinct tip-point positions due to the different coefficients in their equations. In addition, a mapping to be an isometric embedding must at least cover both the position and orientation of the last link for a robot arm plus more lower link orientations to completely eliminate the multi-configuration issues [13–15]. A further discussion with more examples on how to find an isometric embedding for a robotic system will be given in Chapter 4.

2.1.3 DIFFERENTIAL CONNECTIONS AND GEODESIC EQUATIONS

In differential geometry, there is one more generalized directional derivative, called a **covariant derivative** that can be operated on a smooth Riemannian manifold

(M^n, w). Let u and v be two vector fields at a point p of M^n. Inside the neighborhood U of the point p, there is a basis $\{e_i\}$ of the local coordinates $q = (q^1 \cdots q^n)^T$. The covariant derivative $\nabla_v u$ at the point p must meet the following requirements:

1. It is additive in u: $\nabla_v(u_1 + u_2) = \nabla_v u_1 + \nabla_v u_2$;
2. It is linear in v: $\nabla_{ax+by} u = a\nabla_x u + b\nabla_y u$ for two scalar fields a and b, and two vector fields x and y;
3. It follows the product rule: $\nabla_v(au) = a\nabla_v u + (\nabla_v a)u$.

Let the two vector fields $u = \alpha^1 e_1 + \cdots + \alpha^n e_n = \alpha^i e_i$ and $v = \beta^1 e_1 + \cdots + \beta^n e_n = \beta^i e_i$, both of which are projected on the basis $\{e_i\}$ of the local coordinates at the point p in $U \subset M^n$. Note that the above $u = \alpha^i e_i$ and $v = \beta^i e_i$ are in the tensor form that has conventionally omitted the summation symbol. Then,

$$\nabla_v u = \nabla_v \alpha^i e_i = \beta^j \frac{\partial \alpha^i}{\partial q^j} e_i + \alpha^i \nabla_v e_i.$$

It can be observed that the above covariant derivative not only calculates the derivative of the vector field u along the direction of the vector v, but also takes care of the basis frame changes on such a curved non-Euclidean manifold. In order to find the frame changes, let the second term of the above equation be

$$\alpha^i \nabla_v e_i = \Gamma(v)u = \Gamma_i^k(v)\alpha^i e_k$$

for some coefficient $\Gamma(v)$ as a *differential connection 1-form* so that the vector $\Gamma(v)u$ can measure the difference between the frame and its *parallel transport* in the direction v, weighted by the components of u.

If the connection 1-form is written in terms of components, we obtain a more explicit form of the connection coefficient:

$$\Gamma_i^k(v) = \beta^j \Gamma_{ij}^k.$$

This Γ_{ij}^k can measure the k-th component of the difference between e_i and its parallel transport in the direction of e_j. Intuitively, the connection coefficient Γ_{ij}^k, called a **Christoffel symbol**, should be directly related to the Riemannian metric $\{w_j^i\}$ on the manifold M^n. Thus,

$$\alpha^i \nabla_v e_i = \alpha^i \beta^j \Gamma_{ij}^k e_k.$$

On the other hand, since $\alpha^i \nabla_v e_i = \alpha^i \beta^j \nabla_{e_j} e_i$, this implies that

$$\Gamma_{ij}^k e_k = \nabla_{e_j} e_i.$$

Therefore,

$$\nabla_v u = \beta^j \frac{\partial \alpha^i}{\partial q^j} e_i + \alpha^i \nabla_v e_i = \beta^j \frac{\partial \alpha^k}{\partial q^j} e_k + \alpha^i \beta^j \Gamma_{ij}^k e_k = \left(\beta^j \frac{\partial \alpha^k}{\partial q^j} + \alpha^i \beta^j \Gamma_{ij}^k \right) e_k.$$

$$(2.23)$$

If we replace the direction vector field v by one of the basis e_j, then it can be reduced without β^j as

$$\nabla_{e_j} u = \left(\frac{\partial \alpha^k}{\partial q^j} + \alpha^i \Gamma_{ij}^k \right) e_k. \tag{2.24}$$

Obviously, the first term in (2.24) represents the changes of the components of the vector field u, while the second term reflects a "twist" of the local coordinate frame due to the curved manifold. Also, it can be easily seen that the differential connection is symmetric: $\Gamma_{ij}^k = \Gamma_{ji}^k$ [3–6].

In order to determine the Christoffel symbol Γ_{ij}^k, let us first assume that every covariant derivative is under the **Levi-Civita connection**, i.e., as the orthogonal projection of each derivative onto the tangent space of the manifold M^n. Also, consider that the Riemannian manifold M^n is embedded in a Euclidean m-space $\mathbb{R}^m$ $(m > n)$. Then, the tangent space $T_p(M^n)$ at a point p on M^n is spanned by all the column vectors of the Jacobian matrix J of the embedding $z = \zeta(q)$, as given in equation (2.11). In other words, the coordinates frame basis in the tangent space $T_p(M^n)$ at each point p of M^n is

$$e_k = \frac{\partial \zeta}{\partial q^k}, \quad \text{for} \quad k = 1, \cdots, n.$$

On the other hand, based on equation (2.13), the Riemannian metric endowed on M^n should be

$$w_{ij} = \langle e_i, e_j \rangle = e_i \cdot e_j.$$

If we also define the vector field u on the tangent space $T_p(M^n)$, i.e., $u = \alpha^i e_i$, then,

$$\nabla_{e_j} u = \nabla_{e_j} (\alpha^i e_i) = \frac{\partial \alpha^i}{\partial q^j} e_i + \alpha^i \frac{\partial e_i}{\partial q^j}.$$

The second term of the above equation should be equal to the second term of equation (2.24) so that

$$\frac{\partial e_i}{\partial q^j} = \Gamma_{ij}^k e_k.$$

Now let us multiply (inner-product) e_l to both sides, we obtain

$$\left\langle \frac{\partial e_i}{\partial q^j}, e_l \right\rangle = \Gamma_{ij}^k \langle e_k, e_l \rangle = \Gamma_{ij}^k w_{kl}. \tag{2.25}$$

Moreover, by taking derivatives on the Riemannian metrics, we have

$$\frac{\partial w_{ij}}{\partial q^k} = \frac{\partial}{\partial q^k} \langle e_i, e_j \rangle = \left\langle \frac{\partial e_i}{\partial q^k}, e_j \right\rangle + \left\langle e_i, \frac{\partial e_j}{\partial q^k} \right\rangle,$$

$$\frac{\partial w_{jk}}{\partial q^i} = \frac{\partial}{\partial q^i} \langle e_j, e_k \rangle = \left\langle \frac{\partial e_j}{\partial q^i}, e_k \right\rangle + \left\langle e_j, \frac{\partial e_k}{\partial q^i} \right\rangle,$$

$$\frac{\partial w_{ki}}{\partial q^j} = \frac{\partial}{\partial q^j} \langle e_k, e_i \rangle = \left\langle \frac{\partial e_k}{\partial q^j}, e_i \right\rangle + \left\langle e_k, \frac{\partial e_i}{\partial q^j} \right\rangle.$$

Because each partial derivative

$$\frac{\partial e_i}{\partial q^k} = \frac{\partial^2 \zeta}{\partial q^k \partial q^i} = \frac{\partial^2 \zeta}{\partial q^i \partial q^k} = \frac{\partial e_k}{\partial q^i},$$

on the right-hand sides of the above three inner-product equations, there are, though, totally six terms, only three of them are independent. Thus,

$$\frac{\partial w_{jk}}{\partial q^i} + \frac{\partial w_{ki}}{\partial q^j} - \frac{\partial w_{ij}}{\partial q^k} = 2 \left\langle \frac{\partial e_i}{\partial q^j}, e_k \right\rangle.$$

By comparing it with equation (2.25), the Christoffel symbol is finally solved to be

$$\Gamma_{ij}^k = \frac{1}{2} w^{kl} \left(\frac{\partial w_{jl}}{\partial q^i} + \frac{\partial w_{li}}{\partial q^j} - \frac{\partial w_{ij}}{\partial q^l} \right), \tag{2.26}$$

where w^{kl} is the inverse of w_{kl}, i.e., $w^{kl} = \left(w^{-1} \right)_{kl}$.

Now, we apply the concepts and formulas of covariant derivatives to find the shortest path between two distinct points on M^n. If a manifold is Euclidean, then the shortest path is a straight line, a trivial case. However, on a curved non-Euclidean manifold, the shortest path is no longer a straight line, but is a special curved path on M^n that takes off from a point $p \in M^n$. Let $T_p(M^n)$ be the tangent space at the point p. It is conceivable that any curved path on M^n, starting from p, can always be orthogonally projected on the tangent M^n. If the orthogonal projection is a straight line on $T_p(M^n)$, then this curve should be the shortest path on M^n, which is often referred to as a **geodesic curve**, as shown in Figure 2.4. Based on this notion, let us seek a geodesic equation, the solution of which will always be the shortest path at the point p on the manifold M^n.

Let $q = q(t)$ be an arc length as $t \in [a, b]$ for $0 < a < b$. Its derivative $\dot{q}(t) = \frac{dq(t)}{dt}$ should represent the tangent vector to this arc length. The shortest path starting from p at $t = a$ should keep a straight line after it is projected on the tangent plane $T_p(M^n)$. In differential geometry, this action is called a **parallel transport** [3, 6]. More formally, the tangent vector $\dot{q}(t)$ is making change when it is taking off from p. If the difference of the tangent vector $\dot{q}(t + dt) - \dot{q}(t)$ in a small time interval dt can always be perpendicular to the tangent plane $T_p(M^n)$, then this is a parallel transport with respect to the tangent space. Now, the difference of the tangent vectors vector that is projected on $T_p(M^n)$ can be perfectly represented by the covariant derivative $\nabla_v u$ with both the vector fields $u = v = \dot{q} = \dot{q}^i e_i$, i.e., $\nabla_{\dot{q}} \dot{q}$. Then, the parallel transport implies that

$$\nabla_{\dot{q}} \dot{q} = 0.$$

Based on equation (2.23), after setting each $\alpha^i = \dot{q}^i$ and $\beta^j = \dot{q}^j$, we obtain

$$\nabla_{\dot{q}} \dot{q} = \left(\dot{q}^j \frac{\partial \dot{q}^k}{\partial q^j} + \dot{q}^i \dot{q}^j \Gamma_{ij}^k \right) e_k = 0.$$

Since

$$\dot{q}^j \frac{\partial \dot{q}^k}{\partial q^j} = \frac{dq^j}{dt} \frac{\partial}{\partial q^j} \left(\frac{dq^k}{dt} \right) = \frac{d^2 q^k}{dt^2} = \ddot{q}^k,$$

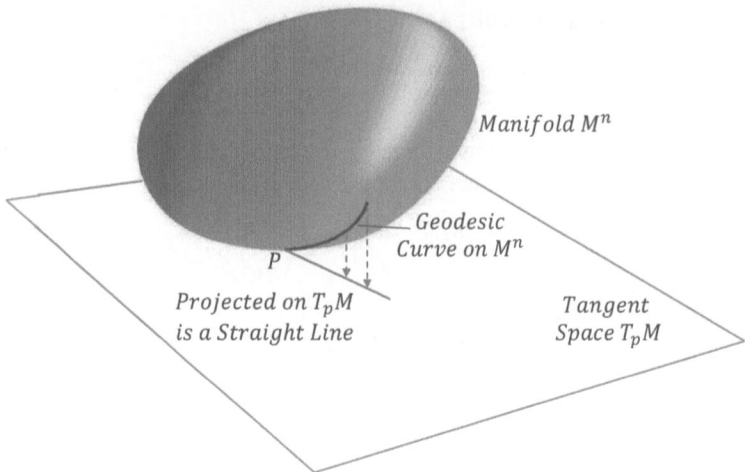

Figure 2.4   A geodesic curve on M^n and its projection on the tangent space T_pM

we achieve the following **geodesic equation** in a tensor form:

$$\ddot{q}^k + \Gamma^k_{ij}\dot{q}^i\dot{q}^j = 0. \qquad (2.27)$$

This is a second-order differential equation and its geodesic solution is the shortest curve $q(t)$ taking off at $q(0)$ with the initial velocity $\dot{q}(0)$.

Recalling the matrix form of the Riemannian metric that is an n by n positive-definite symmetric matrix W in equation (2.14), let us further define the following *derivative matrix*:

$$W_d = \begin{pmatrix} \dot{q}^T \frac{\partial W}{\partial q^1} \\ \vdots \\ \dot{q}^T \frac{\partial W}{\partial q^n} \end{pmatrix}. \qquad (2.28)$$

Since $\dot{W}$ is also symmetric, while W_d is not, $\dot{W} \neq W_d^T$. However, we can show the following useful identity:

$$\dot{W}\dot{q} \equiv W_d^T \dot{q} \qquad (2.29)$$

that holds for any q and $\dot{q}$. In fact, because

$$\dot{W} = \frac{\partial W}{\partial q^1}\dot{q}^1 + \cdots + \frac{\partial W}{\partial q^n}\dot{q}^n,$$

then,

$$\dot{W}\dot{q} = \frac{\partial W}{\partial q^1}\dot{q}\dot{q}^1 + \cdots + \frac{\partial W}{\partial q^n}\dot{q}\dot{q}^n = \left(\frac{\partial W}{\partial q^1}\dot{q} \quad \cdots \quad \frac{\partial W}{\partial q^n}\dot{q} \right)\dot{q}.$$

By comparing the last matrix with the definition of W_d in (2.28) and noticing that W is symmetric, we reach the identity (2.29).

Using both the matrices W and W_d, we can now convert the geodesic equation (2.27) plus the Christoffel symbol equation (2.26) in the tensor form into a matrix form:

$$\ddot{q} + \frac{1}{2}W^{-1}(\dot{W} + W_d^T - W_d)\dot{q} = 0. \tag{2.30}$$

In order to further reduce the computation, let us multiply the Riemannian metric matrix W for both sides of the above matrix form of the geodesic equation in (2.30) to avoid the matrix inverse and also notice the identity (2.29). Namely,

$$W\ddot{q} + \left(W_d^T - \frac{1}{2}W_d\right)\dot{q} = 0. \tag{2.31}$$

This is a simplified version of the geodesic equation in the matrix form. We now use (2.31) to find the geodesic equations for both the 2-dimensional spherical surface S^2 and torus surface T^2. Since we have already derived the 3 by 3 Riemannian metric W for the 3D spherical region in Section 2.1.2, we can now let the radius R be a constant to reduce the dimension down to 2, and immediately get the 2 by 2 Riemannian metric for S^2 by removing the first row and first column. Namely,

$$W = \begin{pmatrix} R^2 & 0 \\ 0 & R^2\sin^2\theta \end{pmatrix}.$$

The local coordinates on S^2 are now (θ, ϕ). Its derivative matrix becomes

$$W_d = \begin{pmatrix} \dot{q}^T \frac{\partial W}{\partial \theta} \\ \dot{q}^T \frac{\partial W}{\partial \phi} \end{pmatrix} = \begin{pmatrix} 0 & R^2\dot{\phi}\sin(2\theta) \\ 0 & 0 \end{pmatrix}.$$

Substituting both the matrices W and W_d into (2.31), we have

$$\begin{pmatrix} R^2 & 0 \\ 0 & R^2\sin^2\theta \end{pmatrix}\begin{pmatrix} \ddot{\theta} \\ \ddot{\phi} \end{pmatrix} + \begin{pmatrix} 0 & -\frac{1}{2}R^2\dot{\phi}\sin(2\theta) \\ R^2\dot{\phi}\sin(2\theta) & 0 \end{pmatrix}\begin{pmatrix} \dot{\theta} \\ \dot{\phi} \end{pmatrix} = 0.$$

Through more symbolical derivations, the geodesic equation for S^2 is found to be

$$\begin{cases} \ddot{\theta} - \dot{\phi}^2\sin\theta\cos\theta = 0 \\ \ddot{\phi} + 2\dot{\theta}\dot{\phi}\cot\theta = 0. \end{cases}$$

If we let the elevation angle $\theta = 90°$, the starting point p on S^2 is just at the equator, then $\cos\theta = 0$ and the geodesic equation is reduced to a special case that both $\ddot{\theta} = 0$ and $\ddot{\phi} = 0$. Their solutions are $\theta(t) = \theta(0) + \dot{\theta}(0)t$ and $\phi(t) = \phi(0) + \dot{\phi}(0)t$. This means that either traveling along the equator, or along the big circle perpendicular to the equator plane will be the shortest path, depending on these initial conditions.

The Riemannian metric W for the T^2 surface has also been derived in equation (2.16) in Section 2.1.2, i.e.,

$$W = J^T J = \begin{pmatrix} (R + rc_2)^2 & 0 \\ 0 & r^2 \end{pmatrix}.$$

The local coordinates on T^2 are now (θ_1, θ_2). The derivative matrix can be calculated by

$$W_d = \begin{pmatrix} \dot{q}^T \frac{\partial W}{\partial \theta_1} \\ \dot{q}^T \frac{\partial W}{\partial \theta_2} \end{pmatrix} = \begin{pmatrix} 0 & 0 \\ -2r\dot{\theta}_1(R+rc_2)s_2 & 0 \end{pmatrix}.$$

Substituting both W and W_d into (2.31), we obtain

$$\begin{pmatrix} (R+rc_2)^2 & 0 \\ 0 & r^2 \end{pmatrix} \begin{pmatrix} \ddot{\theta}_1 \\ \ddot{\theta}_2 \end{pmatrix} + \begin{pmatrix} 0 & -2r\dot{\theta}_1(R+rc_2)s_2 \\ r\dot{\theta}_1(R+rc_2)s_2 & 0 \end{pmatrix} \begin{pmatrix} \dot{\theta}_1 \\ \dot{\theta}_2 \end{pmatrix} = 0.$$

Finally, the geodesic equation for T^2 is determined as

$$\begin{cases} (R+rc_2)\ddot{\theta}_1 - 2rs_2\dot{\theta}_1\dot{\theta}_2 = 0 \\ r\ddot{\theta}_2 + (R+rc_2)s_2\dot{\theta}_1^2 = 0. \end{cases}$$

Similarly, let $\theta_2 = 0$. The starting point p on T^2 in this special case is at the outer major circle on the x-y plane, as shown in Figure 2.2. The geodesic equations are also reduced to a special case $\ddot{\theta}_1 = 0$ and $\ddot{\theta}_2 = 0$ so that their solutions are $\theta_1(t) = \theta_1(0) + \dot{\theta}_1(0)t$ and $\theta_2(t) = \theta_2(0) + \dot{\theta}_2(0)t$. Each one will take off from p to travel along the major or minor circle on the 2-torus surface T^2, and guarantee to be the shortest path.

If the principle of geodesics is applied on the C-manifold M^n of a robotic system with n joints, then its kinetic energy K, as a positive-definite quadratic form of the tangent vectors, is defined as a Riemannian metric endowed on the C-manifold M^n. The geodesic equation (2.30) or (2.31) will match with the force-free Lagrange equation, while the Riemannian metric matrix W in (2.30) or (2.31) is exactly equal to the inertial matrix of the robot dynamic system. This once again demonstrates a beautiful consistency between mathematics and physics. We will have a fairly detailed discussion on the C-manifold of a robotic system in Chapter 4.

2.1.4   DUAL NUMBERS, DUAL VECTORS AND DUAL MATRICES

In mathematics, two real numbers $a, b \in \mathbb{R}$ can be combined with a unit factor j to create a new number $c = a + jb$. In general, this unit factor j is defined by nothing more but the following three different ways:

$$j^2 = \begin{cases} +1 & \text{double number} \\ 0 & \text{dual number} \\ -1 & \text{complex number.} \end{cases}$$

Then, the new numbers $c = a + jb$ can constitute three different number categories, each of which possesses unique algebra and geometry properties that distinguish from each other. In mathematical history, these three types of new numbers have already been studied [21–24], and they can briefly be described in Table 2.2.

$j^2 =$	Name of $c = a + jb$	Magnitude	Inverse	Algebra	Geometry
$+1$	Double Number	$\sqrt{|a^2 - b^2|}$	$\frac{a - jb}{a^2 - b^2}$	Ring	Hyperbolic
0	Dual Number	$|a|$	$\frac{a - jb}{a^2}$	Ring	Parabolic
-1	Complex Number	$\sqrt{a^2 + b^2}$	$\frac{a - jb}{a^2 + b^2}$	Field	Elliptic

Table 2.2
Three types of combined numbers and their mathematical properties

It can be clearly seen that each category of the combined numbers constitutes an additive group, but only the set of complex numbers is qualified to be a multiplicative group and thus, forms an algebraic field, called the **complex field**. The other two: double number and dual number categories can only form a multiplicative semigroup for each because of the inverse loss if $a = b$ for a double number and $a = 0$ for a dual number, as seen in Table 2.2. Therefore, each of them just constitutes an algebraic ring, instead of a field, and is called the **double ring** and **dual ring**, respectively.

It is also noticeable that the third category is the well-known complex field that has been widely used as one of the most popular mathematical tools. In contrast, the double numbers are the least useful set because their algebraic properties are not quite simple nor meaningful. The dual number, however, could be overlooked in history. In fact, it is quite unique and interesting with a relatively simpler algebra, which can be adopted as an effective mathematical tool in 3D space transformations.

The dual ring, though, may lose the inverse if the real part a vanishes, many attractive and excellent algebraic properties owned by the dual ring will make itself useful in the applications to space transformations, especially for the representation of a *line vector*. In order to clearly distinguish a dual number from the others in the next discussions, let j be re-denoted by ε as $a + \varepsilon b$ under $\varepsilon^2 = 0$.

The dual number, first introduced by Clifford in 1873 [21], was further developed by Study in 1901 [22] to represent a dual angle in spatial geometry [23–30]. For two arbitrary straight lines in 3D space, let the shortest distance between them be d and the angle between their directions be α, then a *dual angle* was defined by

$$\hat{\alpha} = \alpha + \varepsilon d$$

under $\varepsilon^2 = 0$. The geometric meaning for this so-defined *dual angle* is just like a screw motion. Namely, using a screw driver to twist a screw by an angle α while pressing and moving the screw down for a vertical distance d.

Similar to every smooth function defined in either the real field or the complex field, a smooth function of dual numbers $f(a + \varepsilon b)$ can also be expanded into a formal Taylor series at the point a, and the expansion will be soon ended due to $\varepsilon^2 = 0$:

$$f(a + \varepsilon b) = f(a) + \varepsilon b f'(a) + \varepsilon^2 \frac{b^2}{2!} f''(a) + \cdots$$

$$= f(a) + \varepsilon b f'(a), \tag{2.32}$$

where $f'(a)$ and $f''(a)$ are the first and second derivatives evaluated at the point a, respectively. Based on such a short finite expansion, it can be easily derived that

$$e^{a+\varepsilon b} = e^a \cdot e^{\varepsilon b} = e^a + \varepsilon b e^a, \tag{2.33}$$

$$\ln(a + \varepsilon b) = \ln a + \varepsilon \frac{b}{a}, \quad \text{for } a > 0, \tag{2.34}$$

$$\sin(\alpha + \varepsilon d) = \sin \alpha + \varepsilon d \cos \alpha, \tag{2.35}$$

and

$$\cos(\alpha + \varepsilon d) = \cos \alpha - \varepsilon d \sin \alpha. \tag{2.36}$$

More significantly, if we replace $a + \varepsilon b$ in (2.32) by $q + \varepsilon \dot{q}$, where $q = (q^1 \cdots q^n)^T \in \mathbb{R}^n$ is an n-dimensional column vector and $\dot{q}$ is its time-derivative vector, while the function itself $f(\cdot)$ could also be a high-dimensional vector or even a matrix (note that in this case, the dimensions in each matrix multiplication must be compatible), then,

$$f(q + \varepsilon \dot{q}) = f(q) + \varepsilon \frac{\partial f(q)}{\partial q} \dot{q} = f(q) + \varepsilon \dot{f}(q), \tag{2.37}$$

where $\dot{f}(q) = \frac{d}{dt} f(q) = \frac{\partial f(q)}{\partial q} \dot{q}$, the time-derivative of the function $f(q)$ based on the derivative chain-rule. Clearly, the dual part of the dual function in (2.37) is $\dot{f}(q)$. Furthermore, if the above variable $q + \varepsilon \dot{q}$ inside the function $f(\cdot)$ is replaced by $q + \varepsilon \delta_i$ for a same dimensional zero vector δ_i except its i-th element being 1 ($1 \leq i \leq n$), i.e., $\delta_i = (0 \cdots 0 \ 1 \ 0 \cdots 0)^T$, then due to (2.37)

$$f(q + \varepsilon \delta_i) = f(q) + \varepsilon \frac{\partial f(q)}{\partial q} \delta_i = f(q) + \varepsilon \frac{\partial f(q)}{\partial q^i}. \tag{2.38}$$

The dual part of the dual function in (2.38) is now a partial derivative of $f(q)$ with respect to q^i and evaluated at q, i.e., $\frac{\partial f(q)}{\partial q^i}$.

These new properties in (2.37) and (2.38) will extend the unique advantage of (2.32) to finding time-derivatives or partial derivatives with respect to some element for large-scale scalar, vector or matrix functions. The advantages will be overwhelming for numerical solutions. As we have all experienced in the numerical computation, to calculate the derivative of a function numerically in computer, it is difficult to achieve a satisfactory high-order approximation without a long memory of previous sampling data. Now, if a computational software package, such as MATLAB^{TM}, can implement an internal subroutine to operate the dual-number calculation, like the complex number calculation that has already been built in MATLAB^{TM}, we can numerically determine any kind of derivatives of a large-scale function accurately and instantaneously at a point of interest in computer via equations (2.32), (2.37) and (2.38) without any symbolical pre-derivation.

For example, let

$$f(t) = e^{-\sigma t} \sin \omega t$$

for two constants σ and ω. Then, directly taking derivative yields

$$f'(t) = -\sigma e^{-\sigma t} \sin \omega t + \omega e^{-\sigma t} \cos \omega t.$$

If we replace t by $\hat{t} = t + \varepsilon$ with a unity dual part $b = 1$, then,

$$f(\hat{t}) = e^{-\sigma(t+\varepsilon)} \sin \omega(t + \varepsilon) = e^{-\sigma t - \varepsilon \sigma} \sin(\omega t + \varepsilon \omega).$$

By applying (2.33) and (2.35), we obtain

$$f(t + \varepsilon) = (e^{-\sigma t} - \varepsilon \sigma e^{-\sigma t})(\sin \omega t + \varepsilon \omega \cos \omega t)$$

$$= e^{-\sigma t} \sin \omega t + \varepsilon(-\sigma e^{-\sigma t} \sin \omega t + \omega e^{-\sigma t} \cos \omega t).$$

Its real part is the original function $f(t)$ and its dual part is exactly equal to the above $f'(t)$.

The second example is the following scalar function of two variables $\theta_1(t)$ and $\theta_2(t)$:

$$f(q) = 3 \sin^2 \theta_1 \cos \theta_2 - 4 \cos^2 \theta_2,$$

and q is a 2 by 1 vector as an implicit function of t: $q(t) = (\theta_1(t) \ \theta_2(t))^T$. In order to find its total time-derivative $\dot{f}(q)$ at a certain time instant, the regular way is to directly take time-derivative on $f(q)$ so that

$$\dot{f}(q) = 3 \sin(2\theta_1) \cos \theta_2 \dot{\theta}_1 - 3 \sin^2 \theta_1 \sin \theta_2 \dot{\theta}_2 + 4 \sin(2\theta_2) \dot{\theta}_2.$$

If at a time instant, $q = (1.2 \ -0.8)^T$ in radians, and $\dot{q} = (-1.5 \ 0.5)^T$ in rad/sec., then $f(q) = -0.1259$ and $\dot{f}(q) = -3.1821$. We now use the dual function under each θ_i is replaced by $\hat{\theta}_i = \theta_i + \varepsilon \dot{\theta}_i$ for $i = 1, 2$, i.e.,

$$f(q + \varepsilon \dot{q}) = 3 \sin^2(\theta_1 + \varepsilon \dot{\theta}_1) \cos(\theta_2 + \varepsilon \dot{\theta}_2) - 4 \cos^2(\theta_2 + \varepsilon \dot{\theta}_2)$$

$$= 3 \sin^2(1.2 - \varepsilon 1.5) \cos(-0.8 + \varepsilon 0.5) - 4 \cos^2(-0.8 + \varepsilon 0.5).$$

Then, by using (2.35) and (2.36) and through dual number manipulations, we find that

$$f(\hat{q}) = f(q + \varepsilon \dot{q}) = f(q) + \varepsilon \dot{f}(q) = -0.1259 - \varepsilon 3.1821,$$

where both the real and dual parts agree with the direct computations. Comparing the dual number approach with the direct computation, the former can substitute the given numbers as soon as each θ_i is replaced by $\theta_i + \varepsilon \dot{\theta}_i$, while the latter must complete the symbolical derivation thoroughly before the numbers can be plugged in. Therefore, the dual number approach has an exclusive numerical algorithmic advantage, especially for a large-scale matrix function that would be almost unmanageable for symbolical derivations, but could be numerically handled by the dual number advantage, provided that all the dual number computations could be implemented in MATLAB^{TM} programming.

The third example is a 3 by 3 matrix R as a function of three angular variables $q = (\theta_1 \ \theta_2 \ \theta_3)^T$. Given

$$R = \begin{pmatrix} c_1 c_{23} & -s_1 & c_1 s_{23} \\ s_1 c_{23} & c_1 & s_1 s_{23} \\ -s_{23} & 0 & c_{23} \end{pmatrix},$$

where $c_1 = \cos \theta_1$, $s_1 = \sin \theta_1$, $c_{23} = \cos(\theta_2 + \theta_3)$, and $s_{23} = \sin(\theta_2 + \theta_3)$. We want to find the partial derivation of R with respect to θ_2. For a direct computation,

$$\frac{\partial R}{\partial \theta_2} = \begin{pmatrix} -c_1 s_{23} & 0 & c_1 c_{23} \\ -s_1 s_{23} & 0 & s_1 c_{23} \\ -c_{23} & 0 & -s_{23} \end{pmatrix}.$$

If using the dual number approach, in this particular case, we need only to replace θ_2 by $\hat{\theta}_2 = \theta_2 + \varepsilon$ so as to replace s_{23} by $\hat{s}_{23} = \sin(\theta_2 + \theta_3 + \varepsilon) = s_{23} + \varepsilon c_{23}$ and replace c_{23} by $\hat{c}_{23} = \cos(\theta_2 + \theta_3 + \varepsilon) = c_{23} - \varepsilon s_{23}$. Through a few more calculation, we find that

$$\hat{R} = \begin{pmatrix} c_1 c_{23} & -s_1 & c_1 s_{23} \\ s_1 c_{23} & c_1 & s_1 s_{23} \\ -s_{23} & 0 & c_{23} \end{pmatrix} + \varepsilon \begin{pmatrix} -c_1 s_{23} & 0 & c_1 c_{23} \\ -s_1 s_{23} & 0 & s_1 c_{23} \\ -c_{23} & 0 & -s_{23} \end{pmatrix}.$$

Evidently, the real part matches with the original matrix R and the dual part exactly agrees with the above partial derivative of R with respect to θ_2.

As have discussed in Section 2.1.3, we will handle a much larger Riemannian metric matrix W for the C-manifold of a multi-joint robotic system in the next dynamics modeling. In order to determine its geodesic equation or its dynamic equation, we have to calculate the time-derivative $\dot{W}$ or the derivative matrix W_d given in equation (2.28). If the robotic system is large-scaled, such as a dual-arm robot or a humanoid robot, which often has more than 30 joints. In this case, the metric matrix W will be larger than 30 by 30, and its derivatives have to be calculated in a numerical manner, instead of the symbolical derivations. The dual number approach may showcase a great potential to make a breakthrough in exploring numerical solutions to the large-scale derivative matrix computation.

It can be seen that from the above examples, the dual number definition can be easily extended to forming a dual vector and a dual matrix as follows:

$$\hat{a} = \begin{pmatrix} a_1 + \varepsilon b_1 \\ a_2 + \varepsilon b_2 \\ a_3 + \varepsilon b_3 \end{pmatrix} = \begin{pmatrix} a_1 \\ a_2 \\ a_3 \end{pmatrix} + \varepsilon \begin{pmatrix} b_1 \\ b_2 \\ b_3 \end{pmatrix} = a + \varepsilon b,$$

or

$$\hat{A} = A + \varepsilon B$$

for two same dimensional matrices A and B. It will soon be discovered that the high-dimensional extensions of the dual number definition possess very unique and elegant properties over the complex numbers and double numbers, and they deserve to be further explored and studied to find more applications.

First, the inner product between two dual vectors is the same as the scalar product between two real vectors. Let $\hat{a} = a + \varepsilon b$ and $\hat{c} = c + \varepsilon d$ with each of a, b, c and $d \in \mathbb{R}^3$. Then, the inner (dot) product between $\hat{a}$ and $\hat{c}$ is similar to the dual number multiplication rule:

$$\hat{a}^T \hat{c} = a^T c + \varepsilon (b^T c + a^T d).$$

In order to perform a cross (vector) product $\hat{a} \times \hat{c}$, let the skew-symmetric matrix of a 3 by 1 dual vector $\hat{a}$ be defined by

$$S(\hat{a}) = \begin{pmatrix} 0 & -a_3 - \varepsilon b_3 & a_2 + \varepsilon b_2 \\ a_3 + \varepsilon b_3 & 0 & -a_1 - \varepsilon b_1 \\ -a_2 - \varepsilon b_2 & a_1 + \varepsilon b_1 & 0 \end{pmatrix} = S(a) + \varepsilon S(b).$$

Then,

$$\hat{a} \times \hat{c} = S(\hat{a})\hat{c} = S(a)c + \varepsilon (S(b)c + S(a)d) = a \times c + \varepsilon (b \times c + a \times d). \quad (2.39)$$

It can also be observed that $S(\hat{a})^T = -S(\hat{a})$, and both the real part and dual part of the dual skew-symmetric matrix $S(\hat{a})$ are skew-symmetric as well, but over the real field, i.e.,

$$S(\hat{a})^T = S(a)^T + \varepsilon S(b)^T = -S(a) - \varepsilon S(b) = -S(\hat{a}).$$

Furthermore, let the Lie bracket of commutation $[S(\hat{a}), S(\hat{c})]$, as a dual cross-product operator, be applied on an arbitrary dual vector $\hat{u} = u + \varepsilon v$ with both u, $v \in \mathbb{R}^3$. Then,

$$[S(\hat{a}), S(\hat{c})]\hat{u} = S(\hat{a})S(\hat{c})\hat{u} - S(\hat{c})S(\hat{a})\hat{u} = \hat{a} \times (\hat{c} \times \hat{u}) - \hat{c} \times (\hat{a} \times \hat{u}),$$

which yields a difference between two triple dual vector cross products. Based on (2.39) and the triple-product equation (2.1) in the dual vector version:

$$\hat{a} \times (\hat{c} \times \hat{u}) = (\hat{a}^T \hat{u})\hat{c} - (\hat{a}^T \hat{c})\hat{u},$$

it can be easily shown that

$$[S(\hat{a}), S(\hat{c})] = S(\hat{a} \times \hat{c})$$

is also a 3 by 3 dual skew-symmetric matrix as a dual vector cross product operator $(\hat{a} \times \hat{c})\times$. Therefore, all the 3 by 3 dual skew-symmetric matrices under the Lie bracket of commutation form a Lie algebra $so(3, \mathbb{D})$ over the dual ring [12, 13].

Likewise, the product between two dual matrices $\hat{A} = A + \varepsilon B$ and $\hat{C} = C + \varepsilon D$ with four n by n real square matrices A, B, C and D is similar to the dual vector products:

$$\hat{A}\hat{C} = AC + \varepsilon (BC + AD),$$

and also

$$\hat{A}\hat{a} = Aa + \varepsilon (Ab + Ba)$$

for a dual vector $\hat{a} = a + \varepsilon b$.

If we wish to invert a dual square matrix $\hat{A} = A + \varepsilon B$, we have to first assume that the real part A must be nonsingular. Then, the inverse of $\hat{A}$ can be determined by

$$\hat{A}^{-1} = (A + \varepsilon B)^{-1} = [A(I + \varepsilon A^{-1}B)]^{-1} = (I + \varepsilon A^{-1}B)^{-1}A^{-1}.$$

Since $(I + \varepsilon A^{-1}B)(I - \varepsilon A^{-1}B) = I$ due to the uniquely simple algebra owned by the dual ring,

$$\hat{A}^{-1} = (I - \varepsilon A^{-1}B)A^{-1} = A^{-1} - \varepsilon A^{-1}BA^{-1}, \tag{2.40}$$

which can also be verified by

$$\hat{A}^{-1}\hat{A} = (A^{-1} - \varepsilon A^{-1}BA^{-1})(A + \varepsilon B) = I + \varepsilon O = I,$$

where I and O are the n by n real identity and zero matrix, respectively, and so is its commutation $\hat{A}\hat{A}^{-1} = I$.

In order to find $\det(\hat{A})$ for an n by n dual square matrix $\hat{A}$, let us first expand the exponential function of εM with any n by n matrix M into its Taylor series, which will be soon ended due to $\varepsilon^2 = 0$, i.e.,

$$\exp(\varepsilon M) = I + \varepsilon M. \tag{2.41}$$

This implies that if A is nonsingular, $\exp(\varepsilon A^{-1}B) = I + \varepsilon A^{-1}B$ such that

$$\det(\hat{A}) = \det(A + \varepsilon B) = \det(A) \cdot \det(I + \varepsilon A^{-1}B) = \det(A) \cdot \det(e^{\varepsilon A^{-1}B}).$$

Due to the following well-known identity:

$$\det(e^A) = e^{\text{tr}(A)}$$

for any square matrix A, the above determinant can be reduced to

$$\det(\hat{A}) = \det(A + \varepsilon B) = \det(A) \cdot e^{\varepsilon \text{tr}(A^{-1}B)} = \det(A) \cdot (1 + \varepsilon \text{tr}(A^{-1}B)). \tag{2.42}$$

This concludes that the determinant of a dual matrix $\hat{A}$ is a dual scalar, whose real part is just the determinant of the real part of $\hat{A}$ while the dual part is $\det(A)\text{tr}(A^{-1}B)$. Obviously, equation (2.42) can only be used under $\det(A) \neq 0$.

For example, let a 3 by 3 dual matrix be

$$\hat{A} = A + \varepsilon B = \begin{pmatrix} 5 + \varepsilon 3 & 2 + \varepsilon 4 & -\varepsilon 2 \\ -1 & 2 - \varepsilon & 1 \\ -2 + \varepsilon 4 & 3 - \varepsilon 2 & -4 + \varepsilon 2 \end{pmatrix}$$

$$= \begin{pmatrix} 5 & 2 & 0 \\ -1 & 2 & 1 \\ -2 & 3 & -4 \end{pmatrix} + \varepsilon \begin{pmatrix} 3 & 4 & -2 \\ 0 & -1 & 0 \\ 4 & -2 & 2 \end{pmatrix}.$$

Then,

$$A^{-1} = \begin{pmatrix} 0.1642 & -0.1194 & -0.0299 \\ 0.0896 & 0.2985 & 0.0746 \\ -0.0149 & 0.2836 & -0.1791 \end{pmatrix},$$

and

$$A^{-1}BA^{-1} = \begin{pmatrix} 0.1419 & 0.0949 & 0.1207 \\ 0.0855 & -0.1029 & -0.0183 \\ -0.1187 & 0.0022 & 0.0826 \end{pmatrix}.$$

Hence, based on (2.40), the inverse of the dual matrix $\hat{A}$ can be evaluated by

$$\hat{A}^{-1} = A^{-1} - \varepsilon A^{-1}BA^{-1} =$$

$$\begin{pmatrix} 0.1642 & -0.1194 & -0.0299 \\ 0.0896 & 0.2985 & 0.0746 \\ -0.0149 & 0.2836 & -0.1791 \end{pmatrix} - \varepsilon \begin{pmatrix} 0.1419 & 0.0949 & 0.1207 \\ 0.0855 & -0.1029 & -0.0183 \\ -0.1187 & 0.0022 & 0.0826 \end{pmatrix}.$$

To find the determinant of $\hat{A}$, since $\det(A) = -67 \neq 0$, and $\mathrm{tr}(A^{-1}B) = -0.0448$,

$$\det(\hat{A}) = \det(A) \cdot (1 + \varepsilon \mathrm{tr}(A^{-1}B)) = -67 + \varepsilon 3.$$

This result can also be verified by a direct computation:

$$\det(\hat{A}) = (5 + \varepsilon 3) \begin{vmatrix} 2 - \varepsilon & 1 \\ 3 - \varepsilon 2 & -4 + \varepsilon 2 \end{vmatrix} + \begin{vmatrix} 2 + \varepsilon 4 & -\varepsilon 2 \\ 3 - \varepsilon 2 & -4 + \varepsilon 2 \end{vmatrix}$$

$$+ (-2 + \varepsilon 4) \begin{vmatrix} 2 + \varepsilon 4 & -\varepsilon 2 \\ 2 - \varepsilon & 1 \end{vmatrix} = (5 + \varepsilon 3)(-11 + \varepsilon 10)$$

$$-8 - \varepsilon 12 + (-2 + \varepsilon 4)(2 + \varepsilon 8) = -67 + \varepsilon 3.$$

After we introduced the dual vectors, dual matrices and their attributes and formulations, we now turn our study to focusing on more specific and useful concepts.

Definition 2.4. A 3-dimensional dual vector $\hat{a} = a + \varepsilon b$ is said to be a *dual screw* if $a^T b = 0$. $\hat{a}$ is also called a *unit dual screw* if it is a dual screw and $\|\hat{a}\|^2 = \hat{a}^T \hat{a} = 1$.

Based on the above definition, since $\hat{a}^T \hat{a} = a^T a + \varepsilon 2a^T b$, a dual screw must have its dual norm equal to its real norm, i.e., $\hat{a}^T \hat{a} = \|\hat{a}\|^2 = a^T a = \|a\|^2$. Also, its dual part b is orthogonal to the real part a. In fact, this dual screw definition suggests that a dual part for a dual vector be so created that it can be the best to represent a *line vector* that consists of a vector plus its moment. For instance, if a force f is applied to rotate a rigid body, the torque τ is commonly defined as $\tau = r \times f$, where r is a radial vector tailed at the pivoting point and its arrow touches the force line. Clearly, the force vector f and torque vector τ are always perpendicular to each other. If one wishes to upgrade the force vector f to be a *dual force vector* $\hat{f}$, then its real part must be f itself, while the best choice of its dual part should be the torque τ to represent a moment so that $\hat{f} = f + \varepsilon \tau$. Clearly, this dual force vector $\hat{f}$ is a dual screw. If $\|f\| = 1$, it becomes a unit dual screw. Under the same reason, since a linear velocity v and an angular velocity ω obey $v = \omega \times r$ with a radial vector r, then the *dual velocity vector* can be defined by $\hat{\omega} = \omega + \varepsilon v$. Obviously, it is a dual screw and is also a unit dual screw if $\|\omega\| = 1$.

Definition 2.5. A 3 by 3 dual square matrix $\hat{R} = R + \varepsilon S$ is called a *special orthogonal dual matrix* if $\hat{R} \in SO(3, \mathbb{D})$, where

$$SO(3, \mathbb{D}) = \{\hat{R} \in \mathbb{D}^{3 \times 3} \mid \hat{R}^T \hat{R} = \hat{R} \hat{R}^T = I \ and \ \det(\hat{R}) = +1\}$$

is the special orthogonal group over the dual ring.

In robotics applications, we often use a rotation matrix R to uniquely represent the orientation of a frame, and a 3 by 1 (point) vector p to determine the position of the origin of the frame with respect to a reference. Although each of the three coordinate axes of the frame can be described by the corresponding column r_i of the rotation matrix R for $i = 1, 2, 3$, each of which is just a point vector with a free-moving tail point. It still requires a moment vector to trace the tail point of each r_i. Now, let a moment vector s_i be defined, like the torque τ or linear velocity v, by $s_i = p \times r_i$ for $i = 1, 2, 3$. Thus, a complete dual vector to describe the i-th axis of the frame becomes

$$\hat{r}_i = r_i + \varepsilon s_i = r_i + \varepsilon(p \times r_i). \tag{2.43}$$

By augmenting all the three dual vectors together in order, a 3 by 3 dual rotation matrix is formed to be

$$\hat{R} = R + \varepsilon S \tag{2.44}$$

that can uniquely represent both the orientation and position of the frame with respect to the reference. In other words, each axis of the frame is now treated as a line vector, instead of a point vector.

Since each column r_i of the rotation matrix R is a unit vector and the moment vector $s_i = p \times r_i$ is always perpendicular to r_i, i.e., $r_i^T r_i = 1$ and $r_i^T s_i = 0$ for $i = 1, 2, 3$, each dual column vector $\hat{r}_i$ in $\hat{R}$ is obviously a unit dual screw, and the dual rotation matrix $\hat{R}$ in (2.44) is a special orthogonal dual matrix. Therefore, we have shown the following lemma:

Lemma 2.1. *If a 3 by 3 dual matrix $\hat{R} = R + \varepsilon S \in SO(3, \mathbb{D})$, then each column or each row of $\hat{R}$ is a unit dual screw.*

Through the above analysis of the dual rotation matrix formation, we can show the following theorem:

Theorem 2.6. *A 3 by 3 dual matrix $\hat{R} = R + \varepsilon S$ is a special orthogonal dual matrix, i.e., $\hat{R} \in SO(3, \mathbb{D})$ if and only if $R \in SO(3, \mathbb{R})$ and SR^T is a 3 by 3 real skew-symmetric matrix, i.e., $SR^T \in so(3, \mathbb{R})$.*

Proof. For the "if" part, since $\hat{R} \hat{R}^T = RR^T + \varepsilon(RS^T + SR^T)$, and SR^T is skew-symmetric, $(SR^T)^T = RS^T = -SR^T$ so that $RS^T + SR^T = O$, we have $\hat{R} \hat{R}^T = \hat{R}^T \hat{R} = RR^T = I$. On the other hand, because $\mathrm{tr}(SR^T) = 0$, which also implies that $\mathrm{tr}(R^T S) = \mathrm{tr}(SR^T) = 0$, based on (2.42), $\det(\hat{R}) = \det(R)(1 + \varepsilon \mathrm{tr}(R^T S)) = \det(R) = +1$. Therefore, $\hat{R} \in SO(3, \mathbb{D})$.

Now, for the "only if" part, $\hat{R} \hat{R}^T = RR^T + \varepsilon(RS^T + SR^T) = I$ implies that $RR^T = I$ and $RS^T + SR^T = O$. Thus, SR^T is skew-symmetric. Moreover, $\mathrm{tr}(R^T S) = \mathrm{tr}(SR^T) = 0$, this also implies that $\det(R) = \det(\hat{R}) = +1$. Therefore, $R \in SO(3, \mathbb{R})$. $\qquad\square$

While one can use a direct computation to determine if a given dual matrix is a special orthogonal dual matrix, the above theorem provides an alternative way to perform the test. For example, if a dual matrix is given by

$$\hat{A} = \begin{pmatrix} 1 & -\varepsilon & \varepsilon3 \\ -\varepsilon3 & \varepsilon2 & 1 \\ -\varepsilon & -1 & \varepsilon2 \end{pmatrix} = \begin{pmatrix} 1 & 0 & 0 \\ 0 & 0 & 1 \\ 0 & -1 & 0 \end{pmatrix} + \varepsilon \begin{pmatrix} 0 & -1 & 3 \\ -3 & 2 & 0 \\ -1 & 0 & 2 \end{pmatrix}.$$

The direct computation way is to check $\hat{A}^T \hat{A}$ to see if it is a real identity as well as $\det(\hat{A}) = +1$. However, we here adopt the method from Theorem 2.6. It can be easily tested that the real part is, indeed, a matrix $R \in SO(3,\mathbb{R})$, while the product between its dual part S and the real part transpose R^T becomes

$$SR^T = \begin{pmatrix} 0 & 3 & 1 \\ -3 & 0 & -2 \\ -1 & 2 & 0 \end{pmatrix} = P,$$

which is a skew-symmetric matrix, denoted by P. Thus, the dual matrix $\hat{A} = R + \varepsilon S$ is a special orthogonal dual matrix. The resulting skew-symmetric matrix

$$P = p\times = \begin{pmatrix} 2 \\ 1 \\ -3 \end{pmatrix} \times$$

is the cross-product operator of the vector p, which can represent a position.

To further investigate the connection between the Lie algebra $so(3,\mathbb{D})$ and the Lie group $SO(3,\mathbb{D})$, both of which are over the dual ring, let us test the exponential mapping. Suppose that a 3 by 3 dual skew-symmetric matrix is $S(\hat{a}) = S(a) + \varepsilon S(b)$, which is the cross-product operator of the dual vector $\hat{a} = a + \varepsilon b$. However, according to the well-known Baker-Campbell-Hausdorff (BCH) formula in mathematics [31, 32], for any two square matrices A and B,

$$\exp(A) \cdot \exp(B) = \exp\left(A + B + \frac{1}{2}[A,B] + \frac{1}{12}[A,[A,B]] - \frac{1}{12}[B,[A,B]] - \cdots\right),$$

which is not equal to $\exp(A+B)$ unless A and B are commutable, i.e., $[A,B] = 0$. Therefore, in general,

$$\exp(S(\hat{a})) = \exp(S(a) + \varepsilon S(b)) \neq \exp(S(a)) \cdot \exp(\varepsilon S(b)),$$

and the exponential mapping process encounters a barrier. We have to narrow down the dual skew-symmetric matrices $S(\hat{a})$ to be in more special case. Let $K = k\times$ be a 3 by 3 unit skew-symmetric matrix that is the cross-product operator for a unit vector k under $\|k\| = 1$. We now define a special dual skew-symmetric matrix $S(\hat{a}) = (\phi + \varepsilon\rho)K = \phi K + \varepsilon\rho K$. Their real part and dual part are commutable now so that

$$\exp(S(\hat{a})) = \exp(\phi K + \varepsilon\rho K) = \exp(\phi K) \cdot \exp(\varepsilon\rho K). \tag{2.45}$$

By recalling equation (2.2), $\exp(\phi K) = I + \sin\phi K + (1 - \cos\phi)K^2 = R \in SO(3,\mathbb{R})$, and based on (2.41),

$$\exp(\varepsilon\rho K) = I + \varepsilon\rho K.$$

Thus,

$$\exp(S(\hat{a})) = R(I + \varepsilon\rho K) = R + \varepsilon\rho RK = R + \varepsilon S,$$

and its dual part becomes

$$S = \rho RK = \rho(I + \sin\phi K + (1 - \cos\phi)K^2)K = \rho K(\cos\phi I + \sin\phi K).$$

Then, the skew-symmetric matrix $P = SR^T$ that corresponds to a position vector turns out to be

$$P = \rho K(\cos\phi I + \sin\phi K)(I - \sin\phi K + (1 - \cos\phi)K^2) = \rho K.$$

This result reveals a fact that through the exponential mapping from the Lie algebra $so(3,\mathbb{D})$ to the Lie group $SO(3,\mathbb{D})$ under the dual angle about the same axis, we achieve a simultaneous transformation, as a partial connection, of the rotation about the axis k by an angle ϕ and the translation along the same axis k by a distance ρ. This simultaneous transformation can now be uniquely represented by a 3 by 3 dual rotation matrix $\hat{R} \in SO(3,\mathbb{D})$.

In fact, since for the rotating axis k,

$$Rk = (I + \sin\phi K + (1 - \cos\phi)K^2)k = k,$$

due to $Kk = k \times k = 0$, by comparing $Rk = k$ with the standard eigenvalue and eigenvector definition: $Ax = \lambda x$, *the unit vector k is actually the eigenvector of this particular rotation matrix R corresponding to the eigenvalue $\lambda = +1$.* This fact also implies that if one defines a different k, then the corresponding rotation matrix $R = I + \sin\phi K + (1 - \cos\phi)K^2$ will be different as well. Thus, let this rotation matrix be denoted as R_0^k to more specifically reflect the association with the axis k. Actually, the unit vector k, as an axis of both the rotation and translation for the dual angle $\phi + \varepsilon\rho$, can be arbitrary, and the exponential mapping in (2.45) is always valid.

Definition 2.6. A dual number $\hat{\theta} = \theta + \varepsilon d$ is called a *dual angle* if its real part θ is a rotating angle about an axis, and its dual part d is a translating distance along the exactly same axis of the rotation.

This formal definition of a dual angle $\hat{\theta} = \theta + \varepsilon d$ allows us to generalize and extend the single dual rotation angle $\phi + \varepsilon\rho$ to multiple dual rotation angles about their individual axes.

Theorem 2.7. *Let $R \in SO(3,\mathbb{R})$ be a 3 by 3 real rotation matrix that is a function of multiple rotating angles $\theta_1, \cdots, \theta_n$. If any number of these angles θ_i's are replaced by their corresponding dual angles $\hat{\theta}_i = \theta_i + \varepsilon d_i$, the real rotation matrix R will be converted to a dual rotation matrix $\hat{R} = R + \varepsilon S \in SO(3,\mathbb{D})$. Then, the skew-symmetric matrix $SR^T = P = p\times$ is a cross-product operator of the position vector p that exactly represents a resultant translation due to all the translating distances d_i's along their respective axes.*

Proof. Without loss of generality, let the given rotation matrix R represent the orientation of a frame a so that $R = R_b^a$ with respect to the base frame b. If the first angle θ_1 inside R_b^a is rotated about an axis $z_{1(a)}$ that is projected on frame a, we can define a vector k_1 as the unit vector of $z_{1(a)}$ and let θ_1 be replaced by the dual angle $\theta_1 + \varepsilon d_1$. Then, the exponential mapping $\exp((\theta_1 + \varepsilon d_1)K)$, similar to (2.45), will generate a dual rotation matrix

$$\hat{R}_a^{k_1} = R_a^{k_1} + \varepsilon S_a^{k_1} = (I + \sin\theta_1 K_1 + (1 - \cos\theta_1)K_1^2)(I + \varepsilon d_1 K_1),$$

and $S_a^{k_1}(R_a^{k_1})^T = d_1 K_1$ indicates a translation by a distance d_1 along k_1.

Now, consider to project the new frame k_1 back to frame a in order to be ready for the next rotation and translation. Thus, the process of the first rotation and translation plus projecting from frame k_1 back to frame a can be formulated by

$$\hat{R}_a^{k_1} R_{k_1}^a = (R_a^{k_1} + \varepsilon S_a^{k_1})R_{k_1}^a = I + \varepsilon P_1,$$

where $S_a^{k_1} R_{k_1}^a = P_1 = p_1 \times$ should be a skew-symmetric matrix and the vector p_1 represents the first translation referred to frame a. By the same token, the second rotation and translation plus projecting from frame k_2 back to frame a can be determined by

$$\hat{R}_a^{k_2} R_{k_2}^a = (R_a^{k_2} + \varepsilon S_a^{k_2})R_{k_2}^a = I + \varepsilon P_2.$$

Therefore, the process of n steps of such successive rotations and translations plus projecting from each frame k_i back to frame a turns out to be

$$(I + \varepsilon P_1)\cdots(I + \varepsilon P_n) = \prod_{i=1}^{n}(I + \varepsilon P_i).$$

Because $(I + \varepsilon P_1)(I + \varepsilon P_2) = I + \varepsilon(P_1 + P_2)$, we obtain

$$\prod_{i=1}^{n}(I + \varepsilon P_i) = I + \varepsilon \sum_{i=1}^{n} P_i = I + \varepsilon(p_1 + \cdots + p_n) \times.$$

Clearly, the sum of all P_i's will still be a 3 by 3 skew-symmetric matrix as a cross-product operator of the resultant translation vector due to all the translating distances d_i's along their respective axes.   □

Let us now test the above theorem on the same rotation matrix R as the previous example:

$$R = \begin{pmatrix} c_1 c_{23} & -s_1 & c_1 s_{23} \\ s_1 c_{23} & c_1 & s_1 s_{23} \\ -s_{23} & 0 & c_{23} \end{pmatrix}.$$

Once the angle θ_1 is replaced by its dual angle $\theta_1 + \varepsilon d_1$, and the angle θ_2 is replaced by its dual angle $\theta_2 + \varepsilon d_2$, a new dual rotation matrix is created as

$$\hat{R} = R + \varepsilon S = R + \varepsilon \begin{pmatrix} -d_1 s_1 c_{23} - d_2 c_1 s_{23} & -d_1 c_1 & -d_1 s_1 s_{23} + d_2 c_1 c_{23} \\ d_1 c_1 c_{23} - d_2 s_1 s_{23} & -d_1 s_1 & d_1 c_1 s_{23} + d_2 s_1 c_{23} \\ -d_2 c_{23} & 0 & -d_2 s_{23} \end{pmatrix}.$$

Then, let the dual part S be post-multiplied by the real part R^T. The resultant translation vector due to the distances d_1 and d_2 along their individual axes can be found as follows:

$$P = SR^T = \begin{pmatrix} 0 & -d_1 & d_2 c_1 \\ d_1 & 0 & d_2 s_1 \\ -d_2 c_1 & -d_2 s_1 & 0 \end{pmatrix}.$$

Finally, $p = (-d_2 s_1 \; d_2 c_1 \; d_1)^T$. If the given R is referred to the base, so is the position vector p.

In summary, the exponential mapping of the dual skew-symmetric matrix under the dual angle $\exp((\phi + \varepsilon\rho)K) = \exp(\phi K) \cdot \exp(\varepsilon\rho K)$ can unify the rotation and translation in one form, where the first factor $\exp(\phi K) = R$ uniquely represents a rotation while the second factor $\exp(\varepsilon\rho K) = I + \varepsilon\rho K$ represents a translation. Since for two different unit skew-symmetric matrices $K_1 \neq K_2$, they are not commutable in general, or $[K_1, K_2] \neq 0$ so that according to the BCH formula,

$$\exp(\phi_1 K_1) \cdot \exp(\phi_2 K_2) \neq \exp(\phi_1 K_1 + \phi_2 K_2).$$

This implies that two different rotations are not commutable, or *any successive rotations are sensitive to their order changes*. In contrast, for two different distances $\rho_1 \neq \rho_2$, $[\varepsilon\rho_1 K_1, \varepsilon\rho_2 K_2] = \varepsilon^2 \rho_1 \rho_2 (K_1 K_2 - K_2 K_1) = 0$ due to $\varepsilon^2 = 0$, though $K_1 K_2 - K_2 K_1 = [K_1, K_2] \neq 0$. This means that based on the BCH formula,

$$\exp(\varepsilon\rho_1 K_1) \cdot \exp(\varepsilon\rho_2 K_2) = \exp(\varepsilon\rho_1 K_1 + \varepsilon\rho_2 K_2).$$

In other words, *any successive translations are invariant of their order changes*. These conclusions are completely consistent to the theory of kinematics.

2.2   ROBOT KINEMATICS: THEORIES AND REPRESENTATIONS

2.2.1   UNIQUE REPRESENTATIONS OF POSITION AND ORIENTATION

To uniquely represent a position or a translation, we commonly use a position vector $p \in \mathbb{R}^3$. This can also be converted to a 3 by 3 skew-symmetric matrix as its cross-product operator $P = p \times \; \in so(3)$. Although all the skew-symmetric matrices are disqualified to constitute a group under the multiplication, they can form a Lie algebra under the Lie bracket of commutation, i.e., $[P_1, P_2] = P_1 P_2 - P_2 P_1$. Nevertheless, all such 3 by 3 skew-symmetric matrices P's form an *additive group*. This means that $P = P_1 + P_2 = p_1 \times + p_2 \times = (p_1 + p_2) \times$ is still skew-symmetric and uniquely represents a cross-product operator of the resultant vector $p_1 + p_2 = p$.

However, when two skew-symmetric matrices are added together, $P_1 + P_2$, they must have been projected on a common reference frame. In order to change the projection frame, for a regular position vector p_a^i that is the position of frame i and currently projected on frame a, we can pre-multiply it by an orthogonal matrix $R_b^a \in SO(3)$, then $R_b^a p_a^i = p_b^i$ becomes the same position vector but projected on frame b. In contrast, in order to change the skew-symmetric matrix P_a^i that is currently projected on frame a, we need an orthogonal transformation to project it on frame b, i.e.,

$$P_b^i = R_b^a P_a^i (R_b^a)^T = R_b^a P_a^i R_a^b.$$

As a counterpart, to uniquely represent an orientation or a rotation, there is no any suitable 3 by 1 vector to be qualified. Only the 3 by 3 rotation matrix $R \in SO(3)$ is the best and most secure way to do so. Now, in order to unify the representation between the position and orientation, or between the translation and rotation, as one can apparently see, the former is either a 3 by 1 vector, or a 3 by 3 skew-symmetric matrix, while the latter is a 3 by 3 rotation matrix, and they are mixed in format.

The most effective unification of uniquely representing a 6 degrees of freedom (d.o.f.) rigid motion is called a **4 by 4 homogeneous transformation**, which augments the 3 by 3 rotation matrix and the 3 by 1 position vector together in the following form:

$$H = \begin{pmatrix} R & p \\ 0_3 & 1 \end{pmatrix},$$

where 0_3 is the 1 by 3 zero covector, and the 3 by 3 rotation matrix is given by

$$R = \begin{pmatrix} r_{11} & r_{12} & r_{13} \\ r_{21} & r_{22} & r_{23} \\ r_{31} & r_{32} & r_{33} \end{pmatrix} \in SO(3),$$

while the 3 by 1 position vector is given by

$$p = \begin{pmatrix} p_1 \\ p_2 \\ p_3 \end{pmatrix} \in \mathbb{R}^3.$$

Then, the 4 by 4 augmented square matrix is constructed by

$$H = \begin{pmatrix} r_{11} & r_{12} & r_{13} & p_1 \\ r_{21} & r_{22} & r_{23} & p_2 \\ r_{31} & r_{32} & r_{33} & p_3 \\ 0 & 0 & 0 & 1 \end{pmatrix}. \tag{2.46}$$

If a 3-dimensional vector is to be rotated and translated simultaneously, it must be first temporarily redefined as a 4 by 1 vector by adding 1 at the bottom:

$$v = \begin{pmatrix} p_1 \\ p_2 \\ p_3 \\ 1 \end{pmatrix}.$$

Then, it can be pre-multiplied by a 4 by 4 homogeneous transformation matrix

$$v_2 = H v_1 \tag{2.47}$$

to perform the unified operation of both rotation and translation.

Furthermore, a homogeneous transformation matrix H^i_j with super- and sub-scripts can also be used to uniquely represent both the relative position and orientation of frame i with respect to frame j. The uniqueness, consistency and security of such a unification has made this homogeneous transformation be the most popular and best approach thus far to representing every 6 d.o.f. rigid motion. Clearly, all such 4 by 4 homogeneous transformation matrices constitute a multiplicative **special Euclidean group** $SE(3) = \mathbb{R}^3 \times SO(3)$ [10, 11].

To invert a 4 by 4 homogeneous transformation that is given by

$$H^i_j = \begin{pmatrix} R^i_j & p^i_j \\ 0_3 & 1 \end{pmatrix}, \tag{2.48}$$

one may take advantage of the orthogonality of the rotation matrix $R^i_j \in SO(3)$ at the upper-left 3 by 3 corner of the above H^i_j. In fact, it can be easily verified that

$$(H^i_j)^{-1} = H^j_i = \begin{pmatrix} (R^i_j)^T & -(R^i_j)^T p^i_j \\ 0_3 & 1 \end{pmatrix} = \begin{pmatrix} R^j_i & -R^j_i p^i_j \\ 0_3 & 1 \end{pmatrix}. \tag{2.49}$$

Furthermore, in order to perform a series of successive 6 d.o.f. transformations, it is also convenient to operate multiplications between the homogeneous transformation matrices in certain order. For instance, if both the position and orientation of frame 1 with respect to the base frame 0 is given by a 4 by 4 homogeneous transformation matrix H^1_0, and both the position and orientation of frame 2 referred to frame 1 is H^2_1, then $H^2_0 = H^1_0 H^2_1$ gives both the position and orientation of frame 2 with respect to the base. On the other hand, if H^2_0 is given, then we can find $H^1_0 = H^2_0 H^1_2 = H^2_0 (H^1_2)^{-1}$. Therefore, each multiplication must follow a *common diagonal-index cancellation rule*, such as $H^2_0 H^1_2$, to cancel the common index 2 on the diagonal and result in H^1_0.

Another effective and unique representation for the 6 d.o.f. motion is the special orthogonal dual matrix $\hat{R} \in SO(3, \mathbb{D})$. As we have introduced and studied in fairly details in the last section, any real rotation matrix $R \in SO(3, \mathbb{R})$ can be converted to a special orthogonal dual matrix $\hat{R} = R + \varepsilon S$ by replacing each rotating angle θ_i with its dual angle $\hat{\theta}_i = \theta_i + \varepsilon d_i$, where d_i the moving distance along the same axis as the rotating angle θ_i. Then, the position vector $p \in \mathbb{R}^3$ is implicitly contained in the skew-symmetric matrix $P = p\times = SR^T$. If MATLAB^{TM} could build-in a subroutine for the dual-number computation, similar to the already built-in complex number computation, then, the dual number approach would show a significant numerical advantage. Now, we make a detailed comparative study between the homogeneous transformations and dual rotation matrices, both of which are the best approaches, thus far, to unifying the representations for position and orientation as well as for translation and rotation, and they are summarized in Table 2.3.

According to the comparison, the dual rotation matrix representation, indeed, has a number of advantages over its counterpart homogeneous transformation. But the explicitness of position vector p in a homogeneous transformation matrix could be the major upside to beat the dual rotation matrix representation. Since the dual

Comparison Aspects	Homogeneous Transformation	Dual Rotation Matrix
Orientation R	Explicit	Explicit
Position p	Explicit	Implicit
Inverting	Average Using (2.49)	Very Easy $\hat{R}^T = \hat{R}^{-1}$
Computational Complexity	Average Both Symbolical and Numerical	Lower If Dual Number Computation Can Be Built-in
Jacobian Formation	Need More Computations	Straightforward

Table 2.3

A comparison between homogeneous transformation and dual rotation matrix

number computation has not been built-in MATLAB^{TM} yet, the homogeneous transformation is still the most popular way to represent both the position and orientation simultaneously. However, to construct a Jacobian matrix in robot kinematics, the dual rotation matrix formulation will be overwhelmingly easier than its counterpart, as it will be convinced in the later section.

2.2.2 THE ROTATION SPEED AND ANGULAR VELOCITY

In order to investigate the orientation change rate for a coordinate frame, described by a real rotation matrix $R \in SO(3)$, let us first take time-derivative on the identity $RR^T = I$ to result in

$$\dot{R}R^T + R\dot{R}^T = O,$$

where O is the 3 by 3 zero matrix. This shows that $\dot{R}R^T$ is a skew-symmetric matrix.

On the other hand, let a 3 by 1 vector be defined as $\xi = \phi k$ to conformally represent the orientation of a frame, where k is the unit vector and ϕ is a rotating angle about the axis k. Actually, this conformal definition was to attempt "uniquely" representing the orientation of a frame, which, of course, was unsuccessful due to the failure of the required one-to-one correspondence. Based on equation (2.2), $R = I + \sin\phi K + (1 - \cos\phi)K^2$ and also $R^T = I - \sin\phi K + (1 - \cos\phi)K^2$, one can solve both ϕ and K from a given R by the following recursion:

$$\phi = \cos^{-1}\left(\frac{\operatorname{tr}(R) - 1}{2}\right), \quad \text{and} \quad K = \frac{R - R^T}{2\sin\phi}. \tag{2.50}$$

Then, the conformal orientation vector $\xi = \phi k$, or its skew-symmetric matrix $S(\xi) = \phi K$ can be determined by (2.50) in the following form:

$$S(\xi) = \phi K = \frac{\phi}{2\sin\phi}(R - R^T).$$

Now, taking time-derivative for ξ yields

$$\dot{\xi} = \dot{\phi}k + \phi\dot{k}.$$

The first term $\dot{\phi}k$ is actually the traditional definition of a 3 by 1 angular velocity $\omega \triangleq \dot{\phi}k$, which reflects the rotating angle time rate about a temporarily fixed axis k. Whereas the second term $\phi\dot{k}$ takes care about the rotating axis direction change $\dot{k}$, which is obviously perpendicular to k due to the unit vector k. Thus,

$$\omega = \dot{\phi}k = \dot{\xi} - \phi\dot{k}. \tag{2.51}$$

This traditional definition of the angular velocity causes ω unable to be a total time-derivative of some angular position vector $\mu \in \mathbb{R}^3$. Namely, one cannot find a vector μ such that $\dot{\mu} = \omega$ in general [13].

Let $\Omega = S(\omega) = \omega\times$ be the 3 by 3 skew-symmetric matrix as a cross-product operator of ω. We can easily show that

$$\Omega = \dot{R}R^T. \tag{2.52}$$

In fact, if the unit vector k is temporarily fixed, then based on equation (2.2) and its time-derivative,

$$\dot{R}R^T = (\cos\phi\,\dot{\phi}K + \sin\phi\,\dot{\phi}K^2)(I - \sin\phi K + (1 - \cos\phi)K^2) = \dot{\phi}K = \Omega.$$

This equation provides a direct linkage between the traditional angular velocity ω and the time-derivative of a rotation matrix R.

However, if we take off the condition of fixed rotating axis k, then, the complete $\dot{R}$ should be

$$\dot{R} = \cos\phi\,\dot{\phi}K + \sin\phi\,\dot{\phi}K^2 + \sin\phi\dot{K} + (1 - \cos\phi)(\dot{K}K + K\dot{K}),$$

which includes the terms of $\dot{K}$. By noticing the two identities $K^3 = -K$ and $K\dot{K}K = O$ (this can be proven by a direct computation with a fact that K is a unity skew-symmetric matrix), through a lengthy symbolical derivation, we obtain

$$\dot{R}R^T = \dot{\phi}K + \sin\phi\dot{K} + (1 - \cos\phi)[K, \dot{K}],$$

where $[K, \dot{K}] = K\dot{K} - \dot{K}K$ is the Lie bracket of commutation. Because K and $\dot{K}$ are not commutable in general, this result with both $\dot{\phi}$ and $\dot{K}$ being considered looks quite complicated. In classical mechanics, the consideration of rotating axis change is covered in the linear velocity definitions, i.e., $v = v_k + \omega \times p$, while an angular velocity is just defined by $\omega = \dot{\phi}k$ or $\Omega = \dot{\phi}K$, where v is called the *absolute velocity* and v_k is called the *relative velocity* along a fixed axis k.

Let us now turn to explore the linkage equation (2.52) over the dual ring, i.e., what does $\dot{\hat{R}}\hat{R}^T$ turn out? Since $\hat{R} = R + \varepsilon S$ and $\dot{\hat{R}} = \dot{R} + \varepsilon\dot{S}$, we have

$$\dot{\hat{R}}\hat{R}^T = (\dot{R} + \varepsilon\dot{S})(R^T + \varepsilon S^T) = \dot{R}R^T + \varepsilon(\dot{R}S^T + \dot{S}R^T).$$

Because the position skew-symmetric matrix is $P = SR^T$, $\dot{P} = \dot{S}R^T + S\dot{R}^T$. Substituting $\dot{P} - S\dot{R}^T = \dot{S}R^T$ into the above equation and noticing (2.52) yield

$$\dot{\hat{R}}\hat{R}^T = \Omega + \varepsilon(\dot{P} - [\Omega, P]) = \Omega + \varepsilon(\dot{P} - (\omega \times p) \times).$$

On the other hand, a position vector can also be considered as $p = \rho k$, where ρ is a moving distance along the unit axis k. Then,

$$\dot{p} = \dot{\rho}k + \rho\dot{k}.$$

Similar to the previous time-derivative of the orientation vector $\dot{\xi}$, and also analog to the definition of $\omega = \dot{\phi}k = \dot{\xi} - \phi\dot{k}$, the above $\dot{\rho}k = \dot{p} - \rho\dot{k}$ is also considering only the translating distance along a fixed axis k. This is the main reason of causing the dual part of $\dot{\hat{R}}\hat{R}^T$ to be $\dot{P} - [\Omega, P]$, instead of $\dot{P}$. In fact, the above explanation can also be verified by the homogeneous transformation counterpart in the following form:

$$\dot{H}H^{-1} = \begin{pmatrix} \dot{R} & \dot{p} \\ 0_3 & 0 \end{pmatrix} \begin{pmatrix} R^T & -R^T p \\ 0_3 & 1 \end{pmatrix} = \begin{pmatrix} \Omega & \dot{p} - \Omega p \\ 0_3 & 0 \end{pmatrix},$$

where the upper right corner is $\dot{p} - \Omega p = \dot{p} - \omega \times p$, instead of $\dot{p}$, which is exactly consistent to the outcome in the above dual ring exploration.

One of the greatest mathematicians in history, Leonhard Euler (1707–1783), born in Switzerland, first described the orientation of a coordinate frame by three angles, called the **Euler angles**, with respect to a fixed reference frame. Because the rotation about each axis of a coordinate system can be written by the following *three elementary rotation matrices*:

$$R(x, \alpha) = \begin{pmatrix} 1 & 0 & 0 \\ 0 & \cos\alpha & -\sin\alpha \\ 0 & \sin\alpha & \cos\alpha \end{pmatrix}, \quad R(y, \beta) = \begin{pmatrix} \cos\beta & 0 & \sin\beta \\ 0 & 1 & 0 \\ -\sin\beta & 0 & \cos\beta \end{pmatrix},$$

and

$$R(z, \gamma) = \begin{pmatrix} \cos\gamma & -\sin\gamma & 0 \\ \sin\gamma & \cos\gamma & 0 \\ 0 & 0 & 1 \end{pmatrix}, \tag{2.53}$$

based on Euler, the orientation of a frame after any rotations can be determined by a combination of three independent elementary rotations.

For example, one of the well-known rotations is so-called *Roll-Pitch-Yaw*, which is a *ZYX* rotation:

1. First, rotate an original frame 0 about its z-coordinate axis by an angle γ to create a new frame a;
2. Second, rotate frame a about the new y-coordinate axis by an angle β to reach the second new frame b;
3. Then, rotate frame b about the new x-coordinate axis by an angle α to arrive at the destinating frame c.

The final orientation of the destinating frame c with respect to frame 0 is given by the product

$$R_0^c = R(z, \gamma)R(y, \beta)R(x, \alpha),$$

and traditionally call γ a roll angle, β a pitch angle, and α a yaw angle. Obviously, R_0^c is a function of the three Euler angles to respond the 3-dimensional Lie group $SO(3)$.

Another typical rotation with three Euler angles is a ZYZ rotation:

1. First, rotate an original frame 0 about its z-coordinate axis by an angle α to create a new frame a;
2. Second, rotate frame a about the new y-coordinate axis by an angle β to reach the second new frame b;
3. Third, rotate frame b about the new z-coordinate axis by an angle γ to arrive at the destinating frame c.

Then, the final destinating orientation with respect to frame 0 becomes:

$$R_0^c = R(z, \alpha)R(y, \beta)R(z, \gamma). \tag{2.54}$$

Actually, many more combinations of rotations with three Euler angles can be defined to represent the orientation of frame c referred to the base, such as ZXZ, XYX, YZY, XYZ, YZX, etc. In other words, for any frame, its orientation with respect to the base can always be determined by a set of three Euler angles. However, between the orientation of a frame and a set of the three Euler angles, they are not in one-to-one correspondence. In other words, given a combination, then the destinating orientation R_0^c can be unique, but if given an orientation R, the solutions for the three Euler angles are not unique, even if the combination of Euler angles is fixed. This fact once again shows the complication of rotation.

Let us now take the ZYZ rotation as an example. According to (2.54),

$$R_0^c = \begin{pmatrix} \cos\alpha & -\sin\alpha & 0 \\ \sin\alpha & \cos\alpha & 0 \\ 0 & 0 & 1 \end{pmatrix} \begin{pmatrix} \cos\beta & 0 & \sin\beta \\ 0 & 1 & 0 \\ -\sin\beta & 0 & \cos\beta \end{pmatrix} \begin{pmatrix} \cos\gamma & -\sin\gamma & 0 \\ \sin\gamma & \cos\gamma & 0 \\ 0 & 0 & 1 \end{pmatrix},$$

which is a function of the three Euler angles α, β and γ. If we wish to solve for these three Euler angles from a given R_0^c in the ZYZ combination, then, the solutions are not unique, as can be seen in the following algorithm:

Algorithm 2.1. Given

$$R_0^c = \begin{pmatrix} r_{11} & r_{12} & r_{13} \\ r_{21} & r_{22} & r_{23} \\ r_{31} & r_{32} & r_{33} \end{pmatrix}.$$

First, let $\beta = \text{atan2}\left(\sqrt{r_{31}^2 + r_{32}^2}, r_{33}\right).$

If $\sin \beta \neq 0$, then

$$
\begin{aligned}
\alpha &= \text{atan2}(r_{23}/\sin \beta, \, r_{13}/\sin \beta), \\
\gamma &= \text{atan2}(r_{32}/\sin \beta, \, -r_{31}/\sin \beta).
\end{aligned}
$$

If $\beta = 0$, then

$$
\begin{aligned}
\alpha &= 0, \\
\gamma &= \text{atan2}(-r_{12}, \, r_{11}).
\end{aligned}
$$

If $\beta = 180°$, then

$$
\begin{aligned}
\alpha &= 0, \\
\gamma &= \text{atan2}(r_{12}, \, -r_{11}).
\end{aligned}
$$

In this algorithm, atan2(y, x) is the four-quadrant arc-tangent function of the two dummy elements x and y in MATLAB^{TM}.

Moreover, when one formulates a series of three successive elementary rotations in a certain order, it is essential to know not only the current rotation is about which coordinate axis, but also this axis is on the old (original) coordinate frame or the new frame. Since each elementary rotation matrix given in (2.53) only indicates which coordinate axis, for instance,

$$
R(z, \gamma) = \begin{pmatrix} \cos \gamma & -\sin \gamma & 0 \\ \sin \gamma & \cos \gamma & 0 \\ 0 & 0 & 1 \end{pmatrix}
$$

that is rotating about the z-axis of a frame. However, rotating about the z-axis of the new coordinate frame, or of the old frame will have to follow the exact same matrix $R(z, \gamma)$. Therefore, it is crucial to answer how to distinguish a rotation that is about a certain axis of the old frame from the same axis of the new frame in their successive rotation formula?

In general, if the current rotation is about a certain axis of the new frame, then the rotation matrix should be right-multiplied to the previously resulting matrix. For instance, let the orientation of an original frame a be R_0^a that is referred to frame 0, and to be rotated about its x-axis by an angle α to reach a new orientation R_0^b. Thus, $R_0^b = R_0^a R(x, \alpha)$. If R_0^b is further rotated about the new z-axis by β, then $R_0^c = R_0^b R(z, \beta) = R_0^a R(x, \alpha) R(z, \beta)$. In contrast, if R_0^b is further rotated about the old z-axis, i.e., the z-axis of frame a by the same angle β, then, $R_0^c = R_0^a R(z, \beta) R(x, \alpha)$, i.e., $R(z, \beta)$ must be left-multiplied to $R(x, \alpha)$.

In fact, when R_0^b is desired to be rotated about the old z-axis, since frame a has already been a "history", we need to recover it before R_0^b can be rotated about the z-axis of frame a. This recovery is actually the *orthogonal transformation* in linear algebra. Namely, the orientation of frame b after frame a is recovered becomes

$$
\tilde{R}_0^b = R_0^a R(z, \beta) R(x, \alpha) R^T(z, \beta).
$$

Then, the final orientation should be

$$R_0^c = \tilde{R}_0^b R(z,\beta) = R_0^a R(z,\beta)R(x,\alpha),$$

instead of $R_0^c = R_0^b R(z,\beta) = R_0^a R(x,\alpha)R(z,\beta)$. Therefore, **the resultant orientation after rotating about a coordinate axis of the new frame or old frame is fully distinguished by the right-multiplication or left-multiplication of the rotation matrix, respectively.** Clearly, the above Roll-Pitch-Yaw, or the ZYX three Euler angles rotation (each rotated about the new axis) should be equivalent to the XYZ rotation, where each rotation is about the axis of the old frame, instead of the new one [13].

Finally, we turn to explore the direct relation between an angular velocity ω and a rotation speed in terms of three Euler angles. Let us take the ZYZ rotation given by (2.54) as an illustrative example. Its time-derivative becomes

$$\dot{R}_0^c = \dot{R}(z,\alpha)R(y,\beta)R(z,\gamma) + R(z,\alpha)\dot{R}(y,\beta)R(z,\gamma) + R(z,\alpha)R(y,\beta)\dot{R}(z,\gamma).$$

Since

$$(R_0^c)^T = R_c^0 = R(z,\gamma)^T R(y,\beta)^T R(z,\alpha)^T,$$

according to (2.52),

$$\dot{R}_0^c R_c^0 = \Omega = \dot{R}(z,\alpha)R(z,\alpha)^T + R(z,\alpha)\dot{R}(y,\beta)R(y,\beta)^T R(z,\alpha)^T +$$

$$+R(z,\alpha)R(y,\beta)\dot{R}(z,\gamma)R(z,\gamma)^T R(y,\beta)^T R(z,\alpha)^T.$$

Because

$$\dot{R}(z,\alpha)R(z,\alpha)^T = \begin{pmatrix} 0 & -1 & 0 \\ 1 & 0 & 0 \\ 0 & 0 & 0 \end{pmatrix}\dot{\alpha}, \quad \dot{R}(y,\beta)R(y,\beta)^T = \begin{pmatrix} 0 & 0 & 1 \\ 0 & 0 & 0 \\ -1 & 0 & 0 \end{pmatrix}\dot{\beta},$$

and

$$\dot{R}(z,\gamma)R(z,\gamma)^T = \begin{pmatrix} 0 & -1 & 0 \\ 1 & 0 & 0 \\ 0 & 0 & 0 \end{pmatrix}\dot{\gamma},$$

we have

$$R(z,\alpha)\dot{R}(y,\beta)R(y,\beta)^T R(z,\alpha)^T =$$

$$\begin{pmatrix} \cos\alpha & -\sin\alpha & 0 \\ \sin\alpha & \cos\alpha & 0 \\ 0 & 0 & 1 \end{pmatrix}\begin{pmatrix} 0 & 0 & 1 \\ 0 & 0 & 0 \\ -1 & 0 & 0 \end{pmatrix}\begin{pmatrix} \cos\alpha & \sin\alpha & 0 \\ -\sin\alpha & \cos\alpha & 0 \\ 0 & 0 & 1 \end{pmatrix}\dot{\beta}$$

$$= \begin{pmatrix} 0 & 0 & \cos\alpha \\ 0 & 0 & \sin\alpha \\ -\cos\alpha & -\sin\alpha & 0 \end{pmatrix}\dot{\beta}.$$

Similarly, since

$$R(z,\alpha)R(y,\beta) = \begin{pmatrix} \cos\alpha\cos\beta & -\sin\alpha & \cos\alpha\sin\beta \\ \sin\alpha\cos\beta & \cos\alpha & \sin\alpha\sin\beta \\ -\sin\beta & 0 & \cos\beta \end{pmatrix},$$

we obtain

$$R(z,\alpha)R(y,\beta)\dot{R}(z,\gamma)R(z,\gamma)^T R(y,\beta)^T R(z,\alpha)^T$$

$$= \begin{pmatrix} 0 & -\cos\beta & \sin\alpha\sin\beta \\ \cos\beta & 0 & -\cos\alpha\sin\beta \\ -\sin\alpha\sin\beta & \cos\alpha\sin\beta & 0 \end{pmatrix}\dot{\gamma}.$$

Substituting all the derivations into the above equation containing Ω yields

$$\Omega = \begin{pmatrix} 0 & -\dot{\alpha}-\cos\beta\dot{\gamma} & \cos\alpha\dot{\beta}+\sin\alpha\sin\beta\dot{\gamma} \\ \dot{\alpha}+\cos\beta\dot{\gamma} & 0 & \sin\alpha\dot{\beta}-\cos\alpha\sin\beta\dot{\gamma} \\ -\cos\alpha\dot{\beta}-\sin\alpha\sin\beta\dot{\gamma} & -\sin\alpha\dot{\beta}+\cos\alpha\sin\beta\dot{\gamma} & 0 \end{pmatrix}.$$

Therefore,

$$\omega = \begin{pmatrix} -\sin\alpha\dot{\beta}+\cos\alpha\sin\beta\dot{\gamma} \\ \cos\alpha\dot{\beta}+\sin\alpha\sin\beta\dot{\gamma} \\ \dot{\alpha}+\cos\beta\dot{\gamma} \end{pmatrix} = \begin{pmatrix} 0 & -\sin\alpha & \cos\alpha\sin\beta \\ 0 & \cos\alpha & \sin\alpha\sin\beta \\ 1 & 0 & \cos\beta \end{pmatrix}\begin{pmatrix} \dot{\alpha} \\ \dot{\beta} \\ \dot{\gamma} \end{pmatrix}.$$

It is now apparent to see that the column vector $(\dot{\alpha}\ \dot{\beta}\ \dot{\gamma})^T$ is a perfect total time-derivative of the 3-dimensional vector $(\alpha\ \beta\ \gamma)^T$ that is formed by the three Euler angles. However, it has been left-multiplied by a 3 by 3 matrix that is a function of the three Euler angles, which destroys the total time-derivative property for the angular velocity ω. This once again convinces us that the differential 1-form ωdt is not *exact*, i.e., one cannot find a 3-dimensional vector μ such that $d\mu = \omega dt$ in general [13].

2.2.3 THE DENAVIT-HARTENBERG (D-H) CONVENTION

In 1955, two engineering scientists Jacques Denavit and Richard Hartenberg introduced a convention in order to standardize the coordinate frame assignment for spatial linkages. Actually, there were a number of different conventions to describe mechanical linkages, such as the Hayati–Roberts line representation, etc., but the method of the Denavit-Hartenberg line coordinates was the best and most popular convention to uniquely represent the mechanical structures. A serial-chain robot manipulator is a typical mechanical linkage system that can be uniquely represented by the Denavit-Hartenberg (D-H) convention. The major advantageous features of this convention can be outlined as follows:

1. Each joint/link representation is referred to only two axes z and x of the link coordinate frame;

2. Only four parameters: joint angle θ, joint offset d, twist angle α and link length a are defined to uniquely describe each joint linkage;
3. Among the four parameters at each joint/link level, the pair of θ and d can form a dual angle $\hat{\theta} = \theta + \varepsilon d$ about/along the z-axis, while the pair of α and a can form another dual angle $\hat{\alpha} = \alpha + \varepsilon a$ about/along the x-axis;
4. Each link is assigned to have a fixed coordinate frame attached on, which will be convenient to define every dynamic parameter for the link.

A detailed implementation of the D-H convention on a given serial-chain robotic system can be step-by-step procedurized as follows:

1. First, identify all the moving joints for the given robot, and determine all the z-axes in order, starting from the z_0-axis for joint 1, z_1 for joint 2, and so forth;
2. Second, determine each x_i-axis as a common normal vector of the two adjacent z_{i-1} and z_i-axes, and this x_i-axis must also attach (touch) both the adjacent z_{i-1} and z_i-axes;
3. Each joint angle θ_i is defined as a rotating angle about the z_{i-1}-axis from the x_{i-1}-axis to the x_i-axis;
4. Each joint offset d_i is defined as a displacement along the z_{i-1}-axis from the x_{i-1}-axis to the x_i-axis;
5. Each twist angle α_i is defined as a rotating angle about the x_i-axis from the z_{i-1}-axis to the z_i-axis;
6. Each link length a_i is defined as a displacement along the x_i-axis from the z_{i-1}-axis to the z_i-axis.

After following all the steps of the procedure, z_0 will be the z-axis of the base coordinate system, and frame i will be attached on link i for $i = 1, \cdots, n$. All the four parameters are signed values, i.e., they can be positive or negative. If the adjacent z-axes are parallel to each other, one can define any common normal but must attach (touch) both the two parallel adjacent z-axes. Among the four parameters at each joint/link level, only one and nothing more than one of them is a joint variable, and the others are constant parameters. In general, θ_i is a variable if joint i is revolute, or d_i is a variable if joint i is prismatic. Each twist angle and each link length are often the constant parameters to describe the mechanical structure twist between two adjacent joints. If one encounters a spherical joint, since its topology is a 2-sphere S^2, it must be treated as two levels of joint/link to define two independent sets of four D-H parameters. If it is a universal joint, whose topology is $SO(3)$, one must treat it as three levels of joint/link to generate three independent sets of four D-H parameters [11, 13].

Once we generate all the sets of four D-H parameters, it is time to create a D-H table. Through the D-H table, we can write a *one-step homogeneous transformation matrix* A_{i-1}^i, or a *one-step dual rotation matrix* $\hat{R}_{i-1}^i$ at each level of joint/link. Since the four D-H parameters at the i-th level of joint/link are either about/along the z_{i-1}-axis or about/along the x_i-axis, a general form of each one-step homogeneous transformation should be

$$A^i_{i-1} = \begin{pmatrix} c\theta_i & -s\theta_i & 0 & 0 \\ s\theta_i & c\theta_i & 0 & 0 \\ 0 & 0 & 1 & d_i \\ 0 & 0 & 0 & 1 \end{pmatrix} \begin{pmatrix} 1 & 0 & 0 & a_i \\ 0 & c\alpha_i & -s\alpha_i & 0 \\ 0 & s\alpha_i & c\alpha & 0 \\ 0 & 0 & 0 & 1 \end{pmatrix}, \tag{2.55}$$

or a general form of each one-step dual rotation matrix should be

$$\hat{R}^i_{i-1} = \begin{pmatrix} c\hat{\theta}_i & -s\hat{\theta}_i & 0 \\ s\hat{\theta}_i & c\hat{\theta}_i & 0 \\ 0 & 0 & 1 \end{pmatrix} \begin{pmatrix} 1 & 0 & 0 \\ 0 & c\hat{\alpha}_i & -s\hat{\alpha}_i \\ 0 & s\hat{\alpha}_i & c\hat{\alpha}_i \end{pmatrix}, \tag{2.56}$$

where the dual angles are given by $\hat{\theta}_i = \theta_i + \varepsilon d_i$ and $\hat{\alpha}_i = \alpha_i + \varepsilon a_i$.

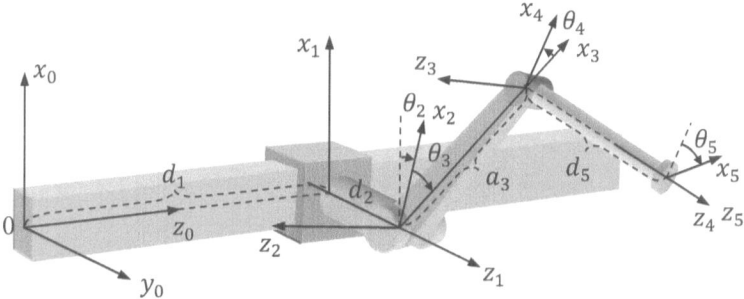

Figure 2.5 A (4+1)-Joint Redundant Robot Manipulator

Example 2.1. A 4-joint robot arm sitting on a linear track (beam) forms a so-called (4+1)-joint robot manipulator, as shown in Figure 2.5. According to the D-H convention procedure, let us first determine all the z-axes, starting from the z_0-axis, which is also the z-axis of the fixed base, and keep continuing until the z_5-axis of the tool frame is assigned. Then, find all the x-axes as the common normal vectors of every pair of the adjacent z-axes. After following the D-H convention, we can find the D-H table of this (4+1)-joint robot in Table 2.4. The D-H table can also be re-listed in terms of all the dual angles, as depicted in Table 2.5.

Now, based on (2.55), all the one-step homogeneous transformation matrices for the (4+1)-joint robot can be determined as

$$A^1_0 = \begin{pmatrix} 1 & 0 & 0 & 0 \\ 0 & 1 & 0 & 0 \\ 0 & 0 & 1 & d_1 \\ 0 & 0 & 0 & 1 \end{pmatrix} \begin{pmatrix} 1 & 0 & 0 & 0 \\ 0 & 0 & 1 & 0 \\ 0 & -1 & 0 & 0 \\ 0 & 0 & 0 & 1 \end{pmatrix} = \begin{pmatrix} 1 & 0 & 0 & 0 \\ 0 & 0 & 1 & 0 \\ 0 & -1 & 0 & d_1 \\ 0 & 0 & 0 & 1 \end{pmatrix},$$

$$A^2_1 = \begin{pmatrix} c_2 & 0 & -s_2 & 0 \\ s_2 & 0 & c_2 & 0 \\ 0 & -1 & 0 & d_2 \\ 0 & 0 & 0 & 1 \end{pmatrix}, \quad A^3_2 = \begin{pmatrix} c_3 & -s_3 & 0 & a_3 c_3 \\ s_3 & c_3 & 0 & a_3 s_3 \\ 0 & 0 & 1 & 0 \\ 0 & 0 & 0 & 1 \end{pmatrix},$$

Joint Angle θ_i	Joint Offset d_i	Twist Angle α_i	Link Length a_i
$\theta_1 = 0$	d_1	$-90°$	0
θ_2	d_2	$-90°$	0
θ_3	0	0	a_3
θ_4	0	$90°$	0
θ_5	d_5	0	0

Table 2.4

The D-H table for the (4+1)-joint robot manipulator (1)

Dual Joint Angle $\hat{\theta}_i = \theta_i + \varepsilon d_i$	Dual Twist Angle $\hat{\alpha}_i = \alpha_i + \varepsilon a_i$
εd_1	$-90°$
$\theta_2 + \varepsilon d_2$	$-90°$
θ_3	εa_3
θ_4	$90°$
$\theta_5 + \varepsilon d_5$	0

Table 2.5

The D-H table for the (4+1)-joint robot manipulator (2)

$$A_3^4 = \begin{pmatrix} c_4 & 0 & s_4 & 0 \\ s_4 & 0 & -c_4 & 0 \\ 0 & 1 & 0 & 0 \\ 0 & 0 & 0 & 1 \end{pmatrix}, \quad A_4^5 = \begin{pmatrix} c_5 & -s_5 & 0 & 0 \\ s_5 & c_5 & 0 & 0 \\ 0 & 0 & 1 & d_5 \\ 0 & 0 & 0 & 1 \end{pmatrix}.$$

If one adopts the dual number approach, based on (2.56), all the one-step dual rotation matrices for the robot can be formulated by

$$\hat{R}_0^1 = \begin{pmatrix} 1 & -\varepsilon d_1 & 0 \\ \varepsilon d_1 & 1 & 0 \\ 0 & 0 & 1 \end{pmatrix} \begin{pmatrix} 1 & 0 & 0 \\ 0 & 0 & 1 \\ 0 & -1 & 0 \end{pmatrix} = \begin{pmatrix} 1 & 0 & -\varepsilon d_1 \\ \varepsilon d_1 & 0 & 1 \\ 0 & -1 & 0 \end{pmatrix},$$

$$\hat{R}_1^2 = \begin{pmatrix} c_2 - \varepsilon d_2 s_2 & 0 & -s_2 - \varepsilon d_2 c_2 \\ s_2 + \varepsilon d_2 c_2 & 0 & c_2 - \varepsilon d_2 s_2 \\ 0 & -1 & 0 \end{pmatrix}, \quad \hat{R}_2^3 = \begin{pmatrix} c_3 & -s_3 & \varepsilon a_3 s_3 \\ s_3 & c_3 & -\varepsilon a_3 c_3 \\ 0 & \varepsilon a_3 & 1 \end{pmatrix},$$

$$\hat{R}_3^4 = \begin{pmatrix} c_4 & 0 & s_4 \\ s_4 & 0 & -c_4 \\ 0 & 1 & 0 \end{pmatrix}, \quad \hat{R}_4^5 = \begin{pmatrix} c_5 - \varepsilon d_5 s_5 & -s_5 - \varepsilon d_5 c_5 & 0 \\ s_5 + \varepsilon d_5 c_5 & c_5 - \varepsilon d_5 s_5 & 0 \\ 0 & 0 & 1 \end{pmatrix}.$$

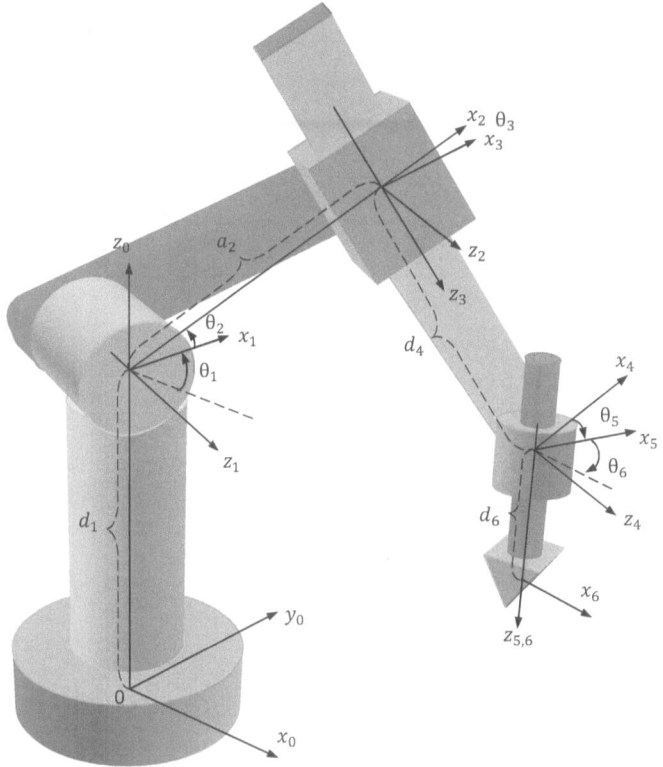

Figure 2.6 A 6-Joint Robot Manipulator

Example 2.2. A 6-joint robot manipulator is given in Figure 2.6. According to all the z-axes and x-axes assignments, its D-H tables for both the homogeneous transformation method and the dual number approach can be listed in Table 2.6 and Table 2.7, respectively.

Then, based on Table 2.6, all the one-step homogeneous transformation matrices can be determined by

$$
A_0^1 = \begin{pmatrix} c_1 & 0 & s_1 & 0 \\ s_1 & 0 & -c_1 & 0 \\ 0 & 1 & 0 & d_1 \\ 0 & 0 & 0 & 1 \end{pmatrix}, \quad
A_1^2 = \begin{pmatrix} c_2 & -s_2 & 0 & a_2 c_2 \\ s_2 & c_2 & 0 & a_2 s_2 \\ 0 & 0 & 1 & 0 \\ 0 & 0 & 0 & 1 \end{pmatrix},
$$

$$
A_2^3 = \begin{pmatrix} c_3 & 0 & s_3 & 0 \\ s_3 & 0 & -c_3 & 0 \\ 0 & 1 & 0 & 0 \\ 0 & 0 & 0 & 1 \end{pmatrix}, \quad
A_3^4 = \begin{pmatrix} 1 & 0 & 0 & 0 \\ 0 & 0 & 1 & 0 \\ 0 & -1 & 0 & d_4 \\ 0 & 0 & 0 & 1 \end{pmatrix},
$$

Joint Angle θ_i	Joint Offset d_i	Twist Angle α_i	Link Length a_i
θ_1	d_1	90°	0
θ_2	0	0	a_2
θ_3	0	90°	0
$\theta_4 = 0$	d_4	−90°	0
θ_5	0	90°	0
θ_6	d_6	0	0

Table 2.6

The D-H table for the 6-joint robot manipulator (1)

Dual Joint Angle $\hat{\theta}_i = \theta_i + \varepsilon d_i$	Dual Twist Angle $\hat{\alpha}_i = \alpha_i + \varepsilon a_i$
$\theta_1 + \varepsilon d_1$	90°
θ_2	εa_2
θ_3	90°
εd_4	−90°
θ_5	90°
$\theta_6 + \varepsilon d_6$	0

Table 2.7

The D-H table for the 6-joint robot manipulator (2)

$$A_4^5 = \begin{pmatrix} c_5 & 0 & s_5 & 0 \\ s_5 & 0 & -c_5 & 0 \\ 0 & 1 & 0 & 0 \\ 0 & 0 & 0 & 1 \end{pmatrix}, \quad A_5^6 = \begin{pmatrix} c_6 & -s_6 & 0 & 0 \\ s_6 & c_6 & 0 & 0 \\ 0 & 0 & 1 & d_6 \\ 0 & 0 & 0 & 1 \end{pmatrix}.$$

By following Table 2.7, all the one-step dual rotation matrices can also be found as

$$\hat{R}_0^1 = \begin{pmatrix} c_1 - \varepsilon d_1 s_1 & 0 & s_1 + \varepsilon d_1 c_1 \\ s_1 + \varepsilon d_1 c_1 & 0 & -c_1 + \varepsilon d_1 s_1 \\ 0 & 1 & 0 \end{pmatrix}, \quad \hat{R}_1^2 = \begin{pmatrix} c_2 & -s_2 & \varepsilon a_2 s_2 \\ s_2 & c_2 & -\varepsilon a_2 c_2 \\ 0 & \varepsilon a_2 & 1 \end{pmatrix},$$

$$\hat{R}_2^3 = \begin{pmatrix} c_3 & 0 & s_3 \\ s_3 & 0 & -c_3 \\ 0 & 1 & 0 \end{pmatrix}, \quad \hat{R}_3^4 = \begin{pmatrix} 1 & 0 & -\varepsilon d_4 \\ \varepsilon d_4 & 0 & 1 \\ 0 & -1 & 0 \end{pmatrix},$$

$$\hat{R}_4^5 = \begin{pmatrix} c_5 & 0 & s_5 \\ s_5 & 0 & -c_5 \\ 0 & 1 & 0 \end{pmatrix}, \quad \hat{R}_5^6 = \begin{pmatrix} c_6 - \varepsilon d_6 s_6 & -s_6 - \varepsilon d_6 c_6 & 0 \\ s_6 + \varepsilon d_6 c_6 & c_6 - \varepsilon d_6 s_6 & 0 \\ 0 & 0 & 1 \end{pmatrix}.$$

The above two examples are both typical as the open serial-chain robots. In fact, the D-H convention for each joint/link coordinate frame assignment and procedure is also applicable to many open hybrid serial-parallel-chain robotic systems.

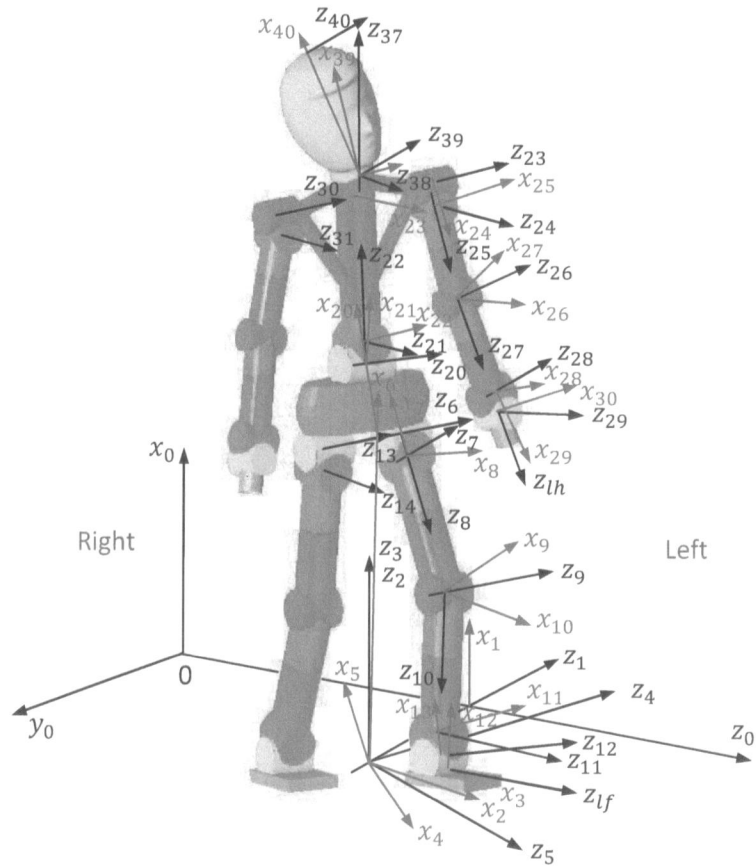

Figure 2.7 A 40-joint humanoid robot

Example 2.3. This example is to illustrate how the D-H convention can be applied for an open hybrid-chain robot, typically for a humanoid robot, as shown in Figure 2.7. Starting from the base frame (x_0, y_0, z_0), it first needs 6 virtual joints to represent the 6 d.o.f. global motion of the humanoid robot: three for translation, and three for rotation. Then, the frame assignment immediately goes to the pelvis, called the *H-triangle*. Through the H-triangle, the frame assignment procedure will split into four paths and one for each of the four limbs, such as the left leg, right leg, left arm and the right arm, as described in Table 2.8.

θ_i	d_i	α_i	a_i	$\theta_i(home)$	Global Motion Joint Names
0	d_1	90°	0	0	Translation along z_0
90°	d_2	90°	0	90°	Translation along z_1
0	d_3	0	0	0	Translation along z_2
θ_4	0	−90°	0	0	Body Spin about z_3
θ_5	0	−90°	0	−90°	Fall Over about z_4
θ_6	0	90°	a_6	0	Fall Aside about z_5

θ_i	d_i	α_i	a_i	$\theta_i(home)$	The Left Leg Joint Names
θ_7	d_7	−90°	$-a_7$	0	Left Hip Flexion/Extension
θ_8	0	90°	0	−90°	Left Hip Abduction/Adduction
θ_9	d_9	90°	0	90°	Left Hip Medial/Lateral
θ_{10}	0	−90°	0	0	Left Knee Flexion/Extension
θ_{11}	d_{11}	−90°	0	−90°	Left Knee Medial/Lateral
θ_{12}	0	90°	$-a_{12}$	90°	Left Ankle Dorsiflexion/plantar
θ_{13}	0	−90°	$-a_{13}$	0	Left Toe Flexion/Extension

θ_i	d_i	α_i	a_i	$\theta_i(home)$	The Right Leg Joint Names
θ_{14}	$-d_{14}$	−90°	$-a_{14}$	0	Right Hip Flexion/Extension
θ_{15}	0	90°	0	−90°	Right Hip Abduction/Adduction
θ_{16}	d_{16}	90°	0	90°	Right Hip Medial/Lateral
θ_{17}	0	−90°	0	0	Right Knee Flexion/Extension
θ_{18}	d_{18}	−90°	0	−90°	Right Knee Medial/Lateral
θ_{19}	0	90°	$-a_{19}$	90°	Right Ankle Dorsiflexion/plantar
θ_{20}	0	−90°	$-a_{20}$	0	Right Toe Flexion/Extension

θ_i	d_i	α_i	a_i	$\theta_i(home)$	The Waist Joint Names
θ_{21}	0	−90°	a_{21}	0	Waist Flexion/Extension
θ_{22}	0	−90°	0	−90°	Waist Medial/Lateral
θ_{23}	d_{23}	−90°	0	−90°	Waist Rotation

θ_i	d_i	α_i	a_i	$\theta_i(home)$	The Left Arm Joint Names
θ_{24}	d_{24}	90°	a_{24}	90°	Left Shoulder Flexion/Extension
θ_{25}	0	90°	0	90°	Left Shoulder Abduction/Adduction
θ_{26}	d_{26}	90°	0	90°	Left Shoulder Medial/Lateral
θ_{27}	0	−90°	0	0	Left Elbow Flexion/Extension
θ_{28}	d_{28}	90°	0	0	Left Elbow Pronation/Supination
θ_{29}	0	90°	a_{29}	90°	Left Wrist Flexion/Extension
θ_{30}	0	90°	0	90°	Left Wrist Radial/Ulnar

θ_i	d_i	α_i	a_i	$\theta_i(home)$	The Right Arm Joint Names
θ_{31}	$-d_{31}$	90°	a_{31}	90°	Right Shoulder Flexion/Extension
θ_{32}	0	90°	0	90°	Right Shoulder Abduction/Adduction
θ_{33}	d_{33}	90°	0	90°	Right Shoulder Medial/Lateral
θ_{34}	0	−90°	0	0	Right Elbow Flexion/Extension
θ_{35}	d_{35}	90°	0	0	Right Elbow Pronation/Supination
θ_{36}	0	90°	a_{36}	90°	Right Wrist Flexion/Extension
θ_{37}	0	90°	0	90°	Right Wrist Radial/Ulnar

θ_i	d_i	α_i	a_i	$\theta_i(home)$	The Head Joint Names
θ_{38}	d_{38}	90°	0	90°	Head Turn
θ_{39}	0	90°	0	90°	Head Nod
θ_{40}	0	0	a_{40}	0	Head Tilt

Table 2.8
The D-H table for the 40-joint humanoid robot

From all the specific D-H tables in Table 2.8, one can clearly see that between the left leg and right leg as well as between the left arm and right arm, the coordinate frames are purposely made in the same assignment so that the D-H tables between the left and right can almost be the same. This will ensure their one-step homogeneous transformations to have the same formulation, allowing to ease the symbolical derivations and programming. In the D-H tables, the "Home" position of each joint is also listed to show the initial "at ease" posture when one simulates or animates this humanoid robot in computer.

2.2.4   CARTESIAN MOTION VS. DIFFERENTIAL MOTION

For an n-joint robot manipulator, once each of n one-step homogeneous transformations A_{i-1}^i's has been found as a function of either the i-th joint angle θ_i if it is revolute or the joint offset d_i if it is prismatic, we can determine the homogeneous transformation A_0^n to represent both the position and orientation of the last frame n with respect to the base frame 0 as a function of all the n joint variables. If we want frame n to match with a desired set of position and orientation, which can be uniquely described by a numerical homogeneous transformation H_0^n, then let $A_0^n = H_0^n$. This match will generate totally six independent equations to reflect a 6 d.o.f. kinematic motion of the robot.

However, most of the equations are trigonometric with non-unique solutions in nature. We can solve those trigonometric equations either symbolically, or numerically to determine all the n joint variables. This process of solving all the joint variables in terms of a given H_0^n is referred to as **inverse kinematics**, or IK in abbreviation. The motion driven by the inverse kinematics at each sampling point is called the **Cartesian motion**. Clearly, if all the joint variables are given, one can compute A_0^n to determine both the position and orientation of the last frame of the robot, which is called the **forward kinematics**, or FK.

As a counterpart of the Cartesian motion, one can also drive the robot by updating its joint variables at each sampling point in terms of the differential increment $\Delta\theta_i$ or Δd_i of each joint variable θ_i or d_i. This process is called the **differential motion**. If a forward kinematics of the robot could be determined by a vector equation $y = f(q)$, where y could represent the 6 d.o.f. positions and orientations, and q is a joint position vector, then,

$$\dot{y} = \frac{\partial f}{\partial q}\dot{q} = J\dot{q}.$$

If the Jacobian matrix J is a non-singular square matrix, we can find a differential increment $dq = J^{-1}dy$ at each sampling point to update the joint variables $q(t+dt) = q(t) + dq = q(t) + J^{-1}dy$ under the first-order approximation.

However, due to the fact that we are unable to find a 3 by 1 orientation vector $\xi \in \mathbb{R}^3$ to uniquely represent the orientation or rotation of a coordinate frame in reality, such an ideal Jacobian matrix J cannot be found by using the partial derivative $\partial y/\partial q$. We have to find an alternative way to construct a transformation matrix in tangent space, similar to the mathematical Jacobian matrix J as a bridge between the joint velocity and Cartesian velocity.

Let a linear velocity v be augmented by an angular velocity ω to form a so-called **twist** vector:

$$V = \begin{pmatrix} v \\ \omega \end{pmatrix},$$

which is a 6-dimensional vector in the tangent space of the Euclidean topological space $SE(3) = \mathbb{R}^3 \times SO(3)$ to represent the 6 d.o.f. motion velocity for a rigid body. Since the angular velocity for the last n-th frame of a robot manipulator with all n revolute joints is contributed by each joint rotation $\dot{\theta}_i$ about the z_{i-1}-axis according to the D-H convention, each of such contribution should be

$$\omega_{i(n)} = \dot{\theta}_i r_n^{i-1},$$

where ω_i contributed by the i-th joint rotation is projected on the n-th frame. While the linear velocity $v_{i(n)}$ due to the i-th joint rotation is given by

$$v_{i(n)} = \omega_{i(n)} \times p_{i-1(n)}^n = (p_n^{i-1} \times r_n^{i-1})\dot{\theta}_i.$$

Thus, the total twist vector of frame n for the robot becomes

$$V_{(n)} = \begin{pmatrix} v_{(n)} \\ \omega_{(n)} \end{pmatrix} = \begin{pmatrix} \sum_{i=1}^n (p_n^{i-1} \times r_n^{i-1})\dot{\theta}_i \\ \sum_{i=1}^n r_n^{i-1}\dot{\theta}_i \end{pmatrix}.$$

Let us now define a *kinematic Jacobian matrix* that is 6 by n-dimensional:

$$J_{(n)} = \begin{pmatrix} p_n^0 \times r_n^0 & \cdots & p_n^{n-1} \times r_n^{n-1} \\ r_n^0 & \cdots & r_n^{n-1} \end{pmatrix}, \tag{2.57}$$

where r_n^0 and r_n^{n-1} are the last (3rd) columns of R_n^0 and R_n^{n-1}, respectively. If $\dot{q} = (\dot{\theta}_1 \cdots \dot{\theta}_n)^T$ is a joint velocity vector, then,

$$V_{(n)} = \begin{pmatrix} v_{(n)} \\ \omega_{(n)} \end{pmatrix} = J_{(n)}\dot{q}. \tag{2.58}$$

In this kinematic Jacobian equation, the matrix $J_{(n)}$ is not a mathematical Jacobian matrix due to the angular velocity $\omega_{(n)}$ is not a total time-derivative of any vector μ in general, but we traditionally keep calling $J_{(n)}$ a Jacobian matrix. Also, in equation (2.58), the twist vector $V_{(n)}$ and the Jacobian matrix $J_{(n)}$ must have the same projection frame. If one needs to change the projection frame, the Jacobian matrix should be pre-multiplied by a 6 by 6 augmented rotation matrix:

$$J_{(k)} = \begin{pmatrix} R_k^j & O \\ O & R_k^j \end{pmatrix} J_{(j)},$$

for any $j,k = 0,1,\cdots,n$. If the i-th joint is prismatic, due to no contribution to the angular velocity $\omega_{(n)}$ by a prismatic joint, the i-th column $\begin{pmatrix} p_n^{i-1} \times r_n^{i-1} \\ r_n^{i-1} \end{pmatrix}$ of the Jacobian matrix $J_{(n)}$ in (2.57) should be replaced by $\begin{pmatrix} r_n^{i-1} \\ 0_3 \end{pmatrix}$ with the 3 by 1 zero vector 0_3.

For a general dual rotation matrix $\hat{R} = R + \varepsilon S$, $SR^T = P$, a skew-symmetric matrix for the position vector so that $S = PR$. This means that each column of S is equal to the position vector p cross-product with the corresponding column of R to represent a moment vector. Thus, the i-th column $p_n^{i-1} \times r_n^{i-1}$ in the Jacobian matrix is actually equal to s_n^{i-1}, the last column of the dual part S_n^{i-1} of the dual rotation matrix $\hat{R}_n^{i-1}$.

Therefore, if the dual number operation is adopted, the formation of Jacobian matrix in (2.57) becomes more straightforward without any calculation of cross-products:

$$J_{(n)} = \begin{pmatrix} s_n^0 & \cdots & s_n^{n-1} \\ r_n^0 & \cdots & r_n^{n-1} \end{pmatrix}. \tag{2.59}$$

Obviously, if the i-th joint is prismatic, then the i-th column $\begin{pmatrix} s_n^{i-1} \\ r_n^{i-1} \end{pmatrix}$ of the Jacobian matrix $J_{(n)}$ in (2.59) should be replaced by $\begin{pmatrix} r_n^{i-1} \\ 0_3 \end{pmatrix}$ with the 3 by 1 zero vector 0_3.

It can be further observed that the 6 by n Jacobian matrix $J_{(n)}$ in (2.57) can be divided into two equal-size portions:

$$J_{(n)} = \begin{pmatrix} J_{(n)}(v) \\ J_{(n)}(\omega) \end{pmatrix},$$

and each is 3 by n and they are given by

$$J_{(n)}(v) = (p_n^0 \times r_n^0 \quad \cdots \quad p_n^{n-1} \times r_n^{n-1}), \quad J_{(n)}(\omega) = (r_n^0 \quad \cdots \quad r_n^{n-1}). \tag{2.60}$$

Then, a 3 by n *dual Jacobian matrix* can be formed by $\hat{J}_{(n)} = J_{(n)}(\omega) + \varepsilon J_{(n)}(v)$.

Clearly, the angular velocity portion $J_{(n)}(\omega)$ has no chance to be a mathematical Jacobian, but the linear velocity portion $J_{(n)}(v)$ is always possible as long as it is projected onto the base, i.e., $R_0^n J_{(n)}(v) = J_{(0)}$. Let us now show a following theorem:

Theorem 2.8. *If $p_0^n \in \mathbb{R}^3$ is a position vector projected onto the fixed base for an n-joint robot arm, and its Jacobian matrix $J_{(n)}(v)$ for the linear velocity portion is formed by*

$$J_{(n)}(v) = (p_n^0 \times r_n^0 \quad \cdots \quad p_n^{n-1} \times r_n^{n-1}) = (s_n^0 \quad \cdots \quad s_n^{n-1}),$$

then,

$$J_{(0)}(v) = R_0^n J_{(n)}(v) = \frac{\partial p_0^n}{\partial q}. \tag{2.61}$$

Proof. The position vector p_0^n that is a function of all the joint variables should be tailed at the origin of the base frame 0, and arrow-pointing to the origin of frame n. If the i-th joint is revolute, and its variable θ_i is replaced by a dual angle with the unity length shift, i.e., $\hat{\theta}_i = \theta_i + \varepsilon$, then the real part p_0^n of the new dual position vector $\hat{p}_0^n$ should have only a parallel shift by one unit. As a point vector, it does not have any change. However, the dual part is imposed by a new moment $p_0^{i-1} \times r_0^{i-1} = s_0^{i-1}$, because θ_i is always rotated about the z_{i-1}-axis, whose unit vector is r_0^{i-1} based on the D-H convention. Therefore, $\hat{p}_0^n = p_0^n + \varepsilon s_0^{i-1}$. Now, by applying the dual function property given in equation (2.38),

$$\hat{p}_0^n(\theta_i + \varepsilon) = p_0^n + \varepsilon s_0^{i-1} = p_0^n + \varepsilon \frac{\partial p_0^n}{\partial \theta_i}.$$

On the other hand, if the i-th joint is prismatic, the variable d_i is along z_{i-1}-axis so that it will contribute additional $d_i r_0^{i-1}$ shift to the position. Now, d_i is replaced by $d_i + \varepsilon$ with the unity dual part, resulting in a dual shift $(d_i + \varepsilon) r_0^{i-1}$ with the dual part r_0^{i-1} as the new moment. Once again, by applying the dual function property given in equation (2.38),

$$\hat{p}_0^n(d_i + \varepsilon) = p_0^n + \varepsilon r_0^{i-1} = p_0^n + \varepsilon \frac{\partial p_0^n}{\partial d_i}.$$

With the same reason, the above two equations for either revolute or prismatic joints should be valid for all $i = 1, \cdots, n$. Based on equation (2.59), we finally show equation (2.61). □

This theorem reveals that to determine the Jacobian matrix for the linear velocity portion $J_{(0)}(v)$ of an n-joint robot, an alternative way is to directly take partial derivative on the position vector p_0^n with respect to the joint position vector q, provided that p_0^n is referred to and projected onto the fixed base. This direct way can avoid manipulating all the cross-products, and guarantee that the resultant Jacobian matrix has already been projected on the base. However, the Jacobian matrix for the angular velocity portion $J_{(0)}(\omega)$ is not qualified to be found by taking partial derivatives so that it has to be formed by augmenting all the last (3rd) columns $r_n^0, \cdots, r_n^{n-1}$ of the rotation matrices $R_n^0, \cdots, R_n^{n-1}$, respectively, as shown in equation (2.60).

Let us look back at the (4+1)-joint robot arm in Example 2.1, and forget the tool disc spinning by setting $\theta_5 = 0$. It becomes a 4-joint robot. Through all the one-step homogeneous transformation matrices, we can derive that

$$A_0^5 = A_0^1 A_1^2 A_2^3 A_3^4 A_4^5 = \begin{pmatrix} c_2 c_{34} & -s_2 & c_2 s_{34} & d_5 c_2 s_{34} + a_3 c_2 c_3 \\ -s_{34} & 0 & c_{34} & d_2 + d_5 c_{34} - a_3 s_3 \\ -s_2 c_{34} & -c_2 & -s_2 s_{34} & d_1 - d_5 s_2 s_{34} - a_3 s_2 c_3 \\ 0 & 0 & 0 & 1 \end{pmatrix}.$$

The last column is just the position vector p_0^5. According to Theorem 2.8 and equation (2.61), by noticing that the joint position vector except θ_5 is $q = (d_1 \ \theta_2 \ \theta_3 \ \theta_4)^T$, taking time-derivative for p_0^5 yields

$$J_{(0)} = \frac{\partial p_0^5}{\partial q} = \begin{pmatrix} 0 & -d_5 s_2 s_{34} - a_3 s_2 c_3 & d_5 c_2 c_{34} - a_3 c_2 s_3 & d_5 c_2 c_{34} \\ 0 & 0 & -d_5 s_{34} - a_3 c_3 & -d_5 s_{34} \\ 1 & -d_5 c_2 s_{34} - a_3 c_2 c_3 & -d_5 s_2 c_{34} + a_3 s_2 s_3 & -d_5 s_2 c_{34} \end{pmatrix}. \quad (2.62)$$

This is a 3 by 4 Jacobian matrix, which means that the robot has 4 joints except θ_5 while the output is only considering the 3D position, and thus it is a redundant robot.

However, if deriving the Jacobian matrix $J_{(0)}$ by using $J_{(n)}(v)$ in (2.60) before pre-multiplying it with R_0^n to project the resulting Jacobian matrix onto the base, the calculation is conceivably much more time-consuming than the above direct way.

Once we determine the Jacobian matrix $J_{(0)}$ and it is non-singular, the differential motion can be implemented to update the joint positions at each sampling point. One can use the first-order approximation:

$$q(t + \Delta t) = q(t) + \dot{q}\Delta t = q(t) + J_{(0)}^{-1}V_{(0)}\Delta t,$$

or the second-order approximation:

$$q(t + \Delta t) = q(t) + J_{(0)}^{-1}V_{(0)}\Delta t + \frac{1}{2}J_{(0)}^{-1}(\dot{V}_{(0)} - \dot{J}_{(0)}J_{(0)}^{-1}V_{(0)})\Delta t^2,$$

or calling the following Runge-Kutta algorithm to get the 4-th order accuracy:

Algorithm 2.2. (Runge-Kutta) - To solve $\dot{x} = f(x)$ for $x \in \mathbb{R}^n$ with $x(0) = x_0$, first, let

$$z_1 = f(x_k)\Delta t, \quad z_2 = f(x_k + \frac{z_1}{2})\Delta t,$$

$$z_3 = f(x_k + \frac{z_2}{2})\Delta t, \quad z_4 = f(x_k + z_3)\Delta t.$$

Then,

$$x_{k+1} = x_k + \frac{1}{6}(z_1 + 2z_2 + 2z_3 + z_4), \quad k = 0, 1, \cdots.$$

Using the above Runge-Kutta numerical algorithm to solve for $\dot{q} = J_{(0)}^{-1}V_{(0)}$, we can achieve quite satisfactory motion tracking results, but the cost we have to pay is to run each update four times of iteration in computing z_1 through z_4.

A typical example is to show the simulation results for a classic 6-joint Stanford robot that is drawing a sine wave on the board in Figure 2.8. The tracking errors at a side view show a significant contrast between the first-order approximation on the left and the Runge-Kutta 4th order algorithm on the right of Figure 2.9. The sine wave deviation from the board occurs in the direction from the board to the robot, which, though, does not interfere very much the sine wave drawing at the front view. In addition to the tracking errors, the simulation also requires to keep the orientation of the pen gripped by the robot end-effector being always perpendicular to the board. However, the left one of Figure 2.9 with the first-order approximation shows the pen orientation no longer kept normal to the board after the two periods of sine wave drawing, while the right one using the Runge-Kutta algorithm performs very well to keep the normal direction.

The Cartesian motion is still the most popular and dominant way to kinematically control robot manipulators in industrial applications. The accuracy is guaranteed by executing the IK algorithm once at each sampling point. However, once the number of joints and the degree of redundancy are significantly increased, like a humanoid robot, the Cartesian motion with the IK algorithm may not be an efficient way any more. Instead, the differential motion with the Jacobian equation may take over the major motion control for large number of joints or highly redundant robotic systems.

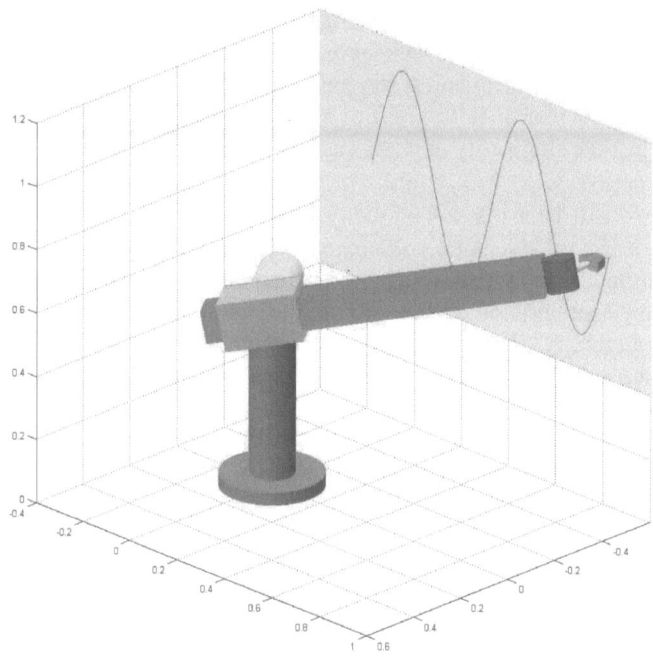

Figure 2.8   A 6-Joint Stanford robot drawing a sine wave on the board

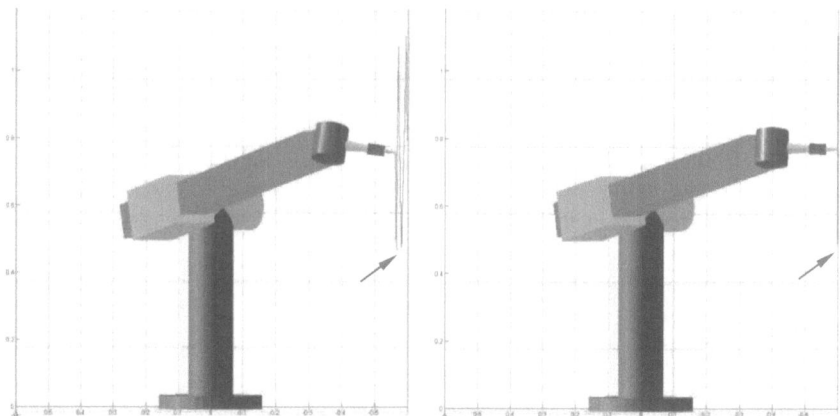

Figure 2.9   Comparison of the drawing errors: using the first-order approximation on the left, while using the Runge-Kutta algorithm on the right

2.2.5   KINEMATIC SINGULARITY AND REDUNDANCY

A **kinematic singularity** issue occurs if one or more d.o.f. are lost. Mathematically, some IK equations to solve for joint positions encounter a solution divergence. The best way to test the singularity is checking $\det(J_{(k)})$ to see if it is nearly zero or not, where the Jacobian matrix $J_{(k)}$ can be projected onto any coordinate frame k, and its determinant is always invariant. If the joint number for a robot is greater than six, the total d.o.f. number in $\mathbb{R}^3$ space, i.e., $n > 6$, then the Jacobian becomes a short matrix. To test the singularity in $n > 6$ cases, one has to check $\det(J_{(k)}J_{(k)}^T)$ to see if it vanishes or not.

It will be seen later that a singularity issue not only damages the kinematics, but also destroys the dynamics and control of the entire system. It is a dangerous issue, but difficult to avoid, because it is often unpredictable. The singularity will also cause a multi-configuration issue, as shown in the 2-link planar robot arm example, depicted in Figure 2.3. Let us formally define the multi-configuration phenomenon, and then state a theorem on the close relationship to the singularity.

Definition 2.7. A robot has a **multi-configuration** case if the joint position q has more than one independent solutions under a fixed Cartesian position and/or orientation output of the last robot coordinate frame.

Theorem 2.9. *For an n-joint robot in $m = n$ d.o.f. motion, it has a multi-configuration case if and only if the robot kinematics has singularity.*

Proof. Since $m = n$, the number of joints is equal to the number of d.o.f., the kinematic Jacobian J of the robot is an n by n square matrix. If it has a singular point, then, $\det(J) = 0$ at that point. Since the Jacobian equation is given by $J\dot{q} = V$ for a twist $V = \begin{pmatrix} v \\ \omega \end{pmatrix}$. Let $V = 0$, which means that the last frame of the robot has no position/orientation changes at all. Then, based on linear algebra, for this homogeneous Jacobian equation $J\dot{q} = 0$, $\dot{q}$ has only a trivial solution, i.e., $\dot{q} = 0$ if J is non-singular. Now, at the singular point, J is singular so that $\dot{q}$ can have a non-trivial solution, and $\pm$ of this nonzero solution or multiplied by any constant number will still make $J\dot{q} = 0$ valid. This implies that the robot configuration can move under $V = 0$.

In the reverse direction, if the robot has multi-configuration with respect to a fixed position/orientation output, $\dot{q}$ can be nonzero at some points under $V = 0$, which occurs only at a singular point. $\qquad\square$

As discussed in the previous sections, the multi-configuration will bring difficulty and complication in finding a qualified embedding to represent the C-manifold of a robot in a Euclidean space. We will continue to study and address this issue in Chapter 4.

In order to reduce or even eliminate the chance of singularity, one may build a robot with $n > 6$, called a **redundant robot**. Actually, a redundant robot may not be the case of $n > 6$. If a required number of d.o.f. is m, the robot arm can be called a redundant robot if its number of joint $n > m$. For instance, a planar robot having 4 joints, but it has only 3 d.o.f. of movement: two for translation and one for rotation on the 2D plane so that $n = 4 > m = 3$, and thus this planar arm is a redundant robot.

Let m be the number of rows of a Jacobian matrix J, and n be the number of columns of J. A redundant robot has a short Jacobian matrix due to $n > m$. Its Jacobian equation $V = J\dot{q}$ will have an infinite number of solutions according to linear algebra, i.e.,

$$\dot{q} = J^+ V + (I - J^+ J)\eta, \tag{2.63}$$

where $J^+ = J^T (JJ^T)^{-1}$ is a **pseudo-inverse**, or called a **Moore-Penrose inverse** of the short Jacobian matrix J, and η is an arbitrary n-dimensional vector [33–35].

The American mathematician Eliakim Hastings Moore (1862–1932) first generalized the matrix inverse and described the pseudo-inverse concept in 1920, and then the English mathematician Sir Roger Penrose (1931-) independently reintroduced and redeveloped the concepts and properties of the generalized inverses in 1955. The motivation of proposing a pseudo-inverse for a matrix is to find its unique representation. According to their mathematical theories, a pseudo-inverse A^+ for any kind of matrix A must obey the following four conditions simultaneously:

$$AA^+A = A, \ A^+AA^+ = A^+, \ (A^+A)^T = A^+A, \ \text{and} \ (AA^+)^T = AA^+. \tag{2.64}$$

In general cases, if A is m by n, no matter $m > n$ or $m < n$ or $m = n$, its rank is $k < \min(m, n)$, we can always make a so-called *maximum-rank decomposition*, i.e., $A = BC$, where B is m by k and C is k by n with rank(B) =rank$(C) = k$. Then, a pseudo-inverse candidate of the matrix A can be defined as

$$A^+ = C^T (CC^T)^{-1} (B^T B)^{-1} B^T. \tag{2.65}$$

We now test each of the above four conditions in (2.64) for both $A = BC$ and the candidate of A^+. It is evident that all the four conditions hold. Therefore, A^+ defined by (2.65) is the pseudo-inverse of the matrix A. Although the maximum-rank decomposition $A = BC$ may not be unique for B and C, it has been proven that the definition of A^+ given in (2.65) is always unique for a given matrix A. More specifically, there are a few special cases to be further emphasized:

1. If $m < n$ and rank$(A) = m$, A is a short full-ranked matrix. Then B should be m by m while C is m by n with both full-ranked. Since A^+ in equation (2.65) is always unique, no matter what B and C are chosen as long as $A = BC$ is the maximum-rank decomposition, let $B = I$, the m by m identity such that $C = A$. Based on (2.65), $A^+ = C^T (CC^T)^{-1} = A^T (AA^T)^{-1}$, as a special case for the short full-ranked matrix A.
2. If now $m > n$ and rank$(A) = n$, A is a tall full-ranked matrix. Then B should be m by n and C is n by n with both full-ranked. Likewise, let $C = I$, the n by n identity such that $B = A$. Based on (2.65), $A^+ = (B^T B)^{-1} B^T = (A^T A)^{-1} A^T$, as another special case for the tall full-ranked matrix A.
3. If $m = n$, A is a square matrix but rank$(A) = k < m = n$ as a singular matrix. Let the maximum-rank decomposition be $A = BC$ with an m by k full-ranked matrix B and a k by $n = m$ full-ranked matrix C. Then, the pseudo-inverse A^+ just follows equation (2.65).

4. If now $m = n$ and the square matrix A is non-singular, obviously, $A^+ = A^{-1}$, a trivial case.

In most robotic application cases, the Jacobian matrix for a redundant robot is in the first special case so that $J^+ = J^T(JJ^T)^{-1}$.

As a numerical example, let

$$A = \begin{pmatrix} 2 & -1 & 2 & 1 \\ -1 & 0 & -2 & 1 \\ 1 & -1 & 0 & 2 \end{pmatrix},$$

which has $\text{rank}(A) = 2 < 3 = m$. One of its maximum-rank decomposition $A = BC$ is given by

$$B = \begin{pmatrix} 1 & -1 \\ 0 & 1 \\ 1 & 0 \end{pmatrix}, \quad \text{and} \quad C = \begin{pmatrix} 1 & -1 & 0 & 2 \\ -1 & 0 & -2 & 1 \end{pmatrix},$$

while the other one for $A = BC$ can be found as

$$B = \begin{pmatrix} -0.8165 & 0 \\ 0.4082 & -0.7071 \\ -0.4082 & -0.7071 \end{pmatrix}, \quad \text{and}$$

$$C = \begin{pmatrix} -2.4495 & 1.2247 & -2.4495 & -1.2247 \\ 0 & 0.7071 & 1.4142 & -2.1213 \end{pmatrix}.$$

Then, based on (2.65), both the different maximum-rank decomposition cases reach to the exact same pseudo-inverse:

$$A^+ = C^T(CC^T)^{-1}(B^TB)^{-1}B^T = \begin{pmatrix} 0.1333 & -0.0667 & 0.0667 \\ -0.0667 & -0.0381 & -0.1048 \\ 0.1333 & -0.2095 & -0.0762 \\ 0.0667 & 0.1810 & 0.2476 \end{pmatrix}.$$

One can verify numerically that all the four conditions in (2.64) hold. Therefore A^+ is the pseudo-inverse of the 3 by 4 lower-ranked matrix A. MATLAB^{TM} has an internal function pinv($\cdot$) that can numerically calculate the pseudo-inverse of a given matrix ($\cdot$), and the result of A^+ in the above example exactly agrees with the answer of calling pinv(A) in MATLAB^{TM}.

To solve the Jacobian equation $V = J\dot{q}$ for a redundant robot with the number of joints $n > m$, where m is the required d.o.f., equation (2.63) is the best form of solution to the differential motion planning. The general solution $\dot{q}$ has been divided into two terms: the first one J^+V is called a *rank solution*, while the second one $(I - J^+J)\eta$ is called a *null solution*. In fact, for the entire n-dimensional joint space, its tangent space is divided into two subspaces: one is the m-dimensional **rank space** R^m and the other one is the $r = (n - m)$-dimensional **null space** N^r, and they are orthogonal to each other.

Because the pseudo-inverse $J^+ = J^T (JJ^T)^{-1}$ of the Jacobian matrix in a redundant robot case is always unique, and it also satisfies the four conditions along with J itself, this will allow us to gain more physical insight into the kinematic redundancy with motion control applications.

1. $JJ^+ = I$, the m by m identity, but $J^+J \neq I$. Instead, the n by n matrix $P_r \triangleq J^+J = J^T (JJ^T)^{-1}J$ with its rank$(J^+J) = m < n$ can play a **projector** role in projecting any n-dimensional vector into the rank space R^m, i.e., for any $x \in \mathbb{R}^n$, $P_r x = J^+Jx = y \in \mathsf{R}^m$. Since $P_r^2 = J^+J \cdot J^+J = J^+J = P_r$, it is *idempotent* so that $P_r = J^+J$ is qualified to be a projector.

2. As a counterpart, $P_n \triangleq I - J^+J = I - P_r$ with its rank$(I - J^+J) = r = n - m$ also plays a projector role but in projecting any n-dimensional vector into the null space N^r, i.e., $P_n x = (I - J^+J)x = (I - P_r)x = z \in \mathsf{N}^r$. Similarly, since $P_n^2 = (I - P_r)^2 = I^2 - 2P_r + P_r^2 = I - P_r = P_n$, it is also idempotent so that P_n is a qualified projector, too.

3. The rank solution J^+V and the null solution $(I - J^+J)\eta$ are always orthogonal to each other. In fact, it can be easily shown that the dot-product of the two solutions:
$$\eta^T (I - J^+J)^T \cdot J^+V = \eta^T (J^+ - J^+JJ^+)V = 0,$$
due to the four conditions for the pseudo-inverse J^+ given in (2.64). Therefore, using the null solution to adjust the instantaneous posture for a redundant robot will not interfere the main task operation that is governed by the rank solution.

4. For any non-zero vector $z \in \mathbb{R}^n$, $z \in \mathsf{N}^r$ if and only if $Jz = 0$. It also implies that if J is a tall or square full-ranked matrix, i.e., $n \leq m$ with rank$(J) = n$, the Jacobian matrix J will have no null space at all.

5. For the entire IK solution $\dot{q}$ in (2.63), if turning off the null solution by setting $\eta = 0$, obviously, the rank solution J^+V becomes the **minimum-norm solution** for $\dot{q}$.

It is now conceivable that the rank solution J^+V of the redundant robot IK equation (2.63) takes care of the main task operation, while the null solution $(I - J^+J)\eta$ is to control the instantaneous posture of a redundant robot to perform one or more desired subtasks. Due to the orthogonality between the two types of solution, the subtasks performance will not interfere the main task operation. Now, the question is how to define and implement the desired subtasks to be optimized?

Actually, any subtask to be performed by adjusting the redundant robot posture can always be described by a scalar **potential function** $p(q)$ as a function of all the joint positions. Then,
$$\dot{p} = \frac{\partial p(q)}{\partial q}\dot{q} = \eta^T \dot{q},$$
where the column vector η is the *gradient vector* of the potential function $p(q)$. Consider that $p(q) = c$ with a constant c represents a super-surface in $\mathbb{R}^n$. It is well-known that the fastest direction to increase or decrease the function value c is along

its gradient vector $\eta = \frac{\partial p(q)}{\partial q}$. This is often called a **gradient approach**. Now, for the null solution $\dot{q} = (I - J^+ J)\eta$ of a redundant robot, if η is the gradient vector of a potential function $p(q)$, then

$$\dot{p}(q) = \eta^T \dot{q} = \eta^T (I - J^+ J)\eta,$$

which is a semi-positive-definite scalar because $(I - J^+ J) = (I - J^+ J)(I - J^+ J)^T$ is a null space projector P_n and $\text{rank}(I - J^+ J) = r < n$. In other words, if the null solution is formed as $k(I - J^+ J)\eta$ with the gradient vector η of $p(q)$ and a gain constant k, then, the potential function $p(q)$ value will be monotonically increasing if $k > 0$, or monotonically decreasing if $k < 0$ in the fastest direction.

Therefore, the key to control the posture of the redundant robot is to define an effective potential function $p(q)$. For example, if $p(q) = \det(JJ^T)$, it may control the posture to stay away from a singular configuration. Since $\det(JJ^T)$ requires a lengthy symbolical derivation, to find its gradient vector η is often impractical. We may redefine a potential function in a more local fashion. Suppose that we knew $\theta_4 = 0$ as the only singular point of the robot. Then, let $p(q) = \sin^2 \theta_4$. Its gradient becomes much simpler and also very effective to perform a singularity avoidance subtask.

One can also define a potential function $p(q)$ to perform a subtask of collision avoidance. In the next section, we will develop a new potential function $p(q) = \text{tr}(JJ^T)$ as a trace of the positive-definite matrix JJ^T to minimize the norm of static joint torque in pursuing the best posture for a humanoid robot in carrying out a heavy-loaded task [13].

2.3 ROBOT STATICS AND APPLICATIONS

2.3.1 TWIST, WRENCH AND STATICS OF ROBOTIC SYSTEMS

As introduced in the last section, the augmentation between a 3-dimensional linear velocity v and a 3-dimensional angular velocity ω can form a 6-dimensional **twist** vector $V = \begin{pmatrix} v \\ \omega \end{pmatrix}$, provided that both of them are projected on the same coordinate frame. Similarly, a 3-dimensional force vector f and a 3-dimensional moment vector m can also be augmented to form a 6-dimensional **wrench** vector $F = \begin{pmatrix} f \\ m \end{pmatrix}$ as long as both f and m are projected on the same frame. The dot-product between a twist and a wrench under the same frame projection will become $F^T V = f^T v + m^T \omega$, which is the total power P, including both the translational and rotational powers.

On the other hand, if each joint of a robot is acted with a joint torque τ_i, and the joint velocity is $\dot{q}_i$, then the total power P can also be determined in the joint space by $P = \tau^T \dot{q}$, where $\tau = (\tau_1 \ \cdots \ \tau_n)^T$ and $\dot{q} = (\dot{q}_1 \ \cdots \ \dot{q}_n)^T$. Due to the principle of energy conservation, $F^T V = P = \tau^T \dot{q}$. By substituting the Jacobian equation $V = J\dot{q}$, as given in (2.58), into this power conservation equation, and noticing that this is for any arbitrary $\dot{q}$, we obtain

$$\tau = J^T F = J^T \begin{pmatrix} f \\ m \end{pmatrix}. \tag{2.66}$$

This is a well-known **robot statics** equation, where both the Jacobian matrix J and the wrench F must be projected onto the same coordinate frame. In other words, if we know that the end-effector of a robot is acted by a static force f and a static moment m, both with respect to the last frame (the end-effector) of the robot, then the static **joint torque distribution** τ can be determined by (2.66).

It is also conceivable that for a certain wrench F acting on the robot end-effector, the static joint torque distribution τ will be different if the Jacobian matrix J of the robot varies. This convinces us that the Jacobian matrix as a function of all the joint positions determines the instantaneous posture of the robot. In particular, if the robot is a humanoid with a large number of joints, intuitively like a real human, a different posture will have a different joint torque distribution even if the two hands are acted with the same wrench.

2.3.2 STATIC JOINT TORQUE DISTRIBUTIONS

Once the number of joints is getting large for a humanoid robot, the static joint torque vector τ becomes high-dimensional. We often plot a bar chart to visualize the joint torque distribution. For example, a 40-joint humanoid robot has been successfully simulated and graphically animated in MATLAB^{TM}, and Figure 2.7 just shows a typical posture of the robot at a time instant of walking. If the left hand is acted with a force $f_{lh} = (1\ 2\ 3)^T$, the right hand is acted by a force $f_{rh} = (-2\ 0\ -1)^T$, while the top of the head is also acted with a force $f_{hd} = (1\ 1\ 1)^T$, all with the zero moments, then, the statics equation (2.66) generates a bar chart of the 20-joint upper body torque distribution, as shown in Figure 2.10. The first three are the joints of the humanoid waist, the next seven are the joints of the left arm, and they are followed by another group of seven joints for the right arm. The last three are the head/neck joint torques.

The bar chart of joint torque distribution will be quite useful for the analysis of joint overweight incidents due to an awkward posture for a humanoid robot to perform a heavy-loaded task. However, how can we determine such a large-scale Jacobian matrix J? Let us take this 40-joint humanoid robot model as an example to illustrate the formation of its up to 36 by 40 large-scale global Jacobian matrix.

Since a Jacobian matrix J is defined as a transformation of tangent space between the joint velocity and Cartesian velocity, i.e., $V = J\dot{q}$, we need first to identify how many "end-effectors" a humanoid robot possesses and where they are. In general, the humanoid robot often performs a task by two hands, plus two feet standing or walking on the ground. Sometimes its H-triangle and head are also desired to be controllable. Therefore, up to six "end-effectors" are identified with each required to specify a twist $V_i = \begin{pmatrix} v_i \\ \omega_i \end{pmatrix}$ for $i = 1, \cdots, 6$. In other words, the left-hand side of the Jacobian equation $V = J\dot{q}$ should be a global twist augmented by all the six twists

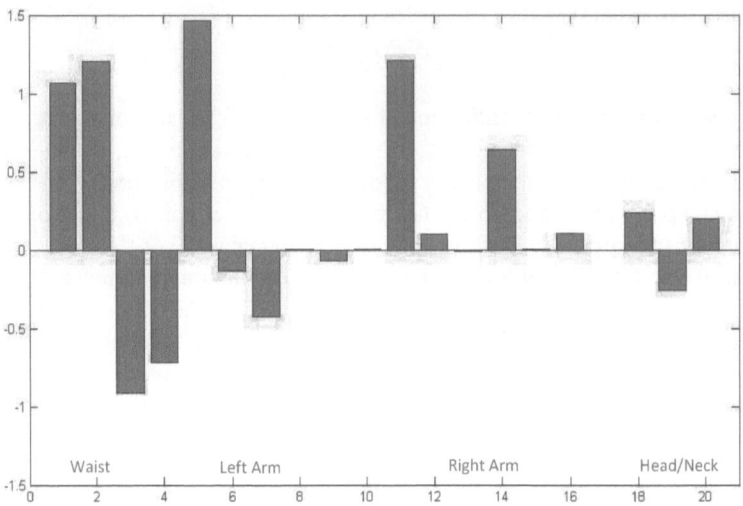

Figure 2.10 A joint torque distribution bar chart for the upper body of the humanoid robot

V_i's. Because each individual V_i is 6-dimensional, the global twist V should be up to 36-dimensional. If the humanoid robot has 40 joints, the global Jacobian matrix J will be up to 36 by 40 dimensional.

Let V_{ht} be the twist vector for the H-triangle of the robot, V_{lh}, V_{rh} and V_{hd} be the twist vectors for the left hand, right hand and the head, respectively. Similarly, let V_{lf} and V_{rf} be the twist vectors for the left leg and right leg, respectively. Then, the global twist vector becomes $V = (V_{ht}^T \ V_{lf}^T \ V_{rf}^T \ V_{lh}^T \ V_{rh}^T \ V_{hd}^T)^T$. Also, according to the frame assignment and the D-H tables for the humanoid robot, as depicted in Figure 2.7 and Table 2.8, the 40-dimensional joint position vector is $q = (d_1 \ d_2 \ d_3 \ \theta_4 \ \cdots \ \theta_{40})^T$.

Let J_{ht} be the 6 by 6 local Jacobian matrix for the H-triangle on frame 6 with respect to the base frame 0. Let J_{lf} and J_{rf} be the 6 by 13 local Jacobian matrices for the left foot on frame lf and right foot on frame rf, respectively, both with respect to the base.

Also, let J_{lh} and J_{rh} be the 6 by 16 local Jacobian matrices for the left hand on frame lh and right hand on frame rh, respectively, with respect to the base, as well as let J_{hd} be the 6 by 12 local Jacobian matrix for the head at frame 40 with respect to the base. Then, the global Jacobian matrix can be constructed as follows:

$$J = \begin{pmatrix} J_{ht} & O & O & O & O & O & O \\ J_{lf}(:,1:6) & J_{lf}(:,7:13) & O & O & O & O & O \\ J_{rf}(:,1:6) & O & J_{rf}(:,7:13) & O & O & O & O \\ J_{lh}(:,1:6) & O & O & J_{lh}(:,7:9) & J_{lh}(:,10:16) & O & O \\ J_{rh}(:,1:6) & O & O & J_{rh}(:,7:9) & O & J_{rh}(:,10:16) & O \\ J_{hd}(:,1:6) & O & O & J_{hd}(:,7:9) & O & O & J_{hd}(:,10:12) \end{pmatrix}.$$

Each Jacobian block is constructed by following equation (2.57) or (2.59) and all the blocks must be projected onto the common frame. When each Jacobian block is formed by starting from the base and ending up to the corresponding humanoid robot end-effector, the joint path may be too long, which could cause an unstable issue when the humanoid robot is running in differential motion. Therefore, an alternative way is to shorten the joint path by setting the H-triangle of the robot on frame 6 as a moving base so that the global Jacobian matrix J splits into three regional Jacobian matrices: one for the H-triangle J_{ht}, one for the lower body J_{leg} and one for the upper body J_{arm}. The 6 by 6 Jacobian matrix J_{ht} is the same as the block in the above global Jacobian matrix J. The lower body regional Jacobian matrix J_{leg} has a size of 12 by 14 to cover the two legs and refer to the H-triangle, while the upper body regional Jacobian matrix J_{arm} is 18 by 20 dimensional to cover the waist, two arms and head. Obviously, the 6 by 6 Jacobian matrix J_{ht} corresponds to the first six joints, while the lower body Jacobian matrix becomes

$$J_{leg} = \begin{pmatrix} J_{lf}(:,7:13) & O \\ O & J_{rf}(:,7:13) \end{pmatrix}$$

that corresponds to joint 7 through joint 20. The upper body Jacobian is the following 18 by 20 matrix:

$$J_{arm} = \begin{pmatrix} J_{lh}(:,7:9) & J_{lh}(:,10:16) & O & O \\ J_{rh}(:,7:9) & O & J_{rh}(:,10:16) & O \\ J_{hd}(:,7:9) & O & O & J_{hd}(:,10:12) \end{pmatrix},$$

which is corresponding to joint 21 through joint 40.

In all the above local matrices, we adopt a common MATLAB^{TM} notation. For instance, $J(:,7:9)$ means a block that covers all the rows but only takes the 7th column through the 9th column from the matrix J. As we can see so far, the Jacobian matrix J for a robot is quite useful and significant not only in performing a trajectory-tracking task in differential motion, but also to determine and show a joint torque distribution against any Cartesian wrenches.

2.3.3 MANIPULABILITY AND POSTURE OPTIMIZATION

As discussed in the previous section, the singularity is one of the toughest issues involved in both robot kinematics and dynamics, which is often unpredictable. One of the solutions to avoid or even eliminate the singularity occurrence is to use a redundant robot with the number of joints n greater than the desired number of d.o.f. m, i.e., $n > m$ to gain $r = n - m$ degrees of redundancy, or gain a non-trivial null space N^r. The concept of *manipulability*, defined by $\det(JJ^T)$, was created a few decades ago as a measure of how good posture a robot can configure. It is also a fantastic idea to define a potential function $p(q) = \det(JJ^T)$ and allow the redundant robot to perform an effective subtask for singularity avoidance. However, the idea is theoretically legitimate, but is practically difficult to derive its gradient vector η. As aforementioned in the last section, one often defines a much simpler potential function in a more local fashion to allow the redundant robot staying away from one or more isolated singular points.

Suppose that the Jacobian matrix J of a redundant robot is m by n-dimensional with $n > m$. We can always find its m positive eigenvalues $\lambda_1, \cdots, \lambda_m$ for the positive-definite matrix JJ^T. Then, based on linear algebra,

$$\det(JJ^T) = \prod_{i=1}^{m} \lambda_i.$$

On the other hand, the trace of the positive-definite matrix JJ^T satisfies

$$\text{tr}(JJ^T) = \sum_{i=1}^{m} \lambda_i.$$

Recall the following well-known basic inequality:

$$\frac{a_1 + \cdots + a_m}{m} \geq \sqrt[m]{a_1 \cdots a_m}$$

for m positive real numbers $a_1 > 0$, $\cdots$, $a_m > 0$. This means that the arithmetic mean is always greater than the geometric mean, and they are equal if and only if $a_1 = \cdots = a_m$. Applying this inequality to the redundant robot case, we have

$$\frac{\text{tr}(JJ^T)}{m} \geq \sqrt[m]{\det(JJ^T)}, \tag{2.67}$$

and they are close to being equal if and only if all the positive eigenvalues of JJ^T are squeezed to be equal to each other, i.e., $\lambda_1 = \cdots = \lambda_m$. This implies that increasing the manipulability $\det(JJ^T)$ is equivalent to decreasing $\text{tr}(JJ^T)$. However, to derive the gradient vector η for $p(q) = \text{tr}(JJ^T)$ is much easier than that for $p(q) = \det(JJ^T)$. Therefore, defining $p(q) = \text{tr}(JJ^T)$ as a potential function is more feasible than $p(q) = \det(JJ^T)$ [13].

Furthermore, according to the robot statics in (2.66), the inner product $\tau^T \tau = F^T(JJ^T)F$. This means that

$$\|\tau\|^2 = \tau_1^2 + \cdots + \tau_n^2 = F^T(JJ^T)F.$$

On the other hand, according to the **Rayleigh Quotient Inequality** in linear algebra [33, 34],

$$\lambda_{min} \leq \frac{F^T(JJ^T)F}{F^T F} \leq \lambda_{max},$$

where λ_{min} and λ_{max} are the minimum and maximum eigenvalues of the positive-definite matrix JJ^T, respectively. Then,

$$\lambda_{min} \leq \frac{\|\tau\|^2}{\|F\|^2} \leq \lambda_{max}.$$

Therefore, minimizing $p(q) = \text{tr}(JJ^T)$ is in the direction of squeezing the distance between λ_{min} and λ_{max}, which is also in the same direction to minimize the norm of

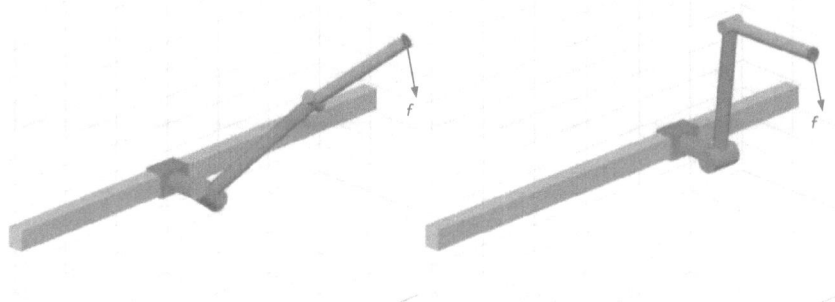

Figure 2.11 The (4+1)-joint robot to find the best posture under a force acting on its tool center point (TCP)

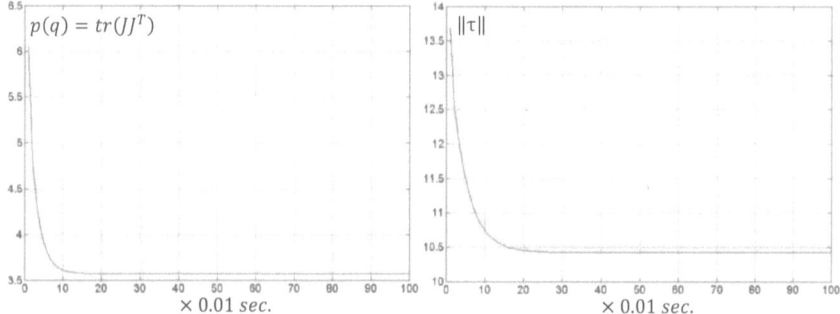

Figure 2.12 Minimization of the potential function $p(q) = \text{tr}(JJ^T)$ and norm of joint torque $\|\tau\|$

the joint torque so as to approach to more even torque distribution. In other words, *the best posture for a redundant robot is the posture that can have the minimum norm of joint torque and further approach to an even joint torque distribution against a certain external wrench F.*

Let us use the previous (4+1)-joint redundant robot, as shown in Figure 2.5, to illustrate how to effectively control the null solution for adjusting the posture without interfering the main task operation. Let the tool disc spinning angle $\theta_5 = 0$ again. This $n = 4$-joint robot arm possesses a single degree of redundancy, i.e., $r = 1$ against the $m = 3$ d.o.f. of translation. In the $r = 1$-dimensional null space N^1, we adopt the potential function $p(q) = \text{tr}(JJ^T)$ to minimize the norm $\|\tau\|$ of the joint torque vector τ if the tool center point (TCP) is acted by a force $f = (-5\ \ 4\ \ -4)^T$ Newtons with respect to the base frame 0, as seen in Figure 2.5. Let us rewrite its symbolical Jacobian matrix $J_{(0)}$ given in (2.62) here again:

$$J_{(0)} = \frac{\partial p_0^5}{\partial q} = \begin{pmatrix} 0 & -d_5 s_2 s_{34} - a_3 s_2 c_3 & d_5 c_2 c_{34} - a_3 c_2 s_3 & d_5 c_2 c_{34} \\ 0 & 0 & -d_5 s_{34} - a_3 c_3 & -d_5 s_{34} \\ 1 & -d_5 c_2 s_{34} - a_3 c_2 c_3 & -d_5 s_2 c_{34} + a_3 s_2 s_3 & -d_5 s_2 c_{34} \end{pmatrix}.$$

It can be observed that to find the trace $\mathrm{tr}(JJ^T)$, one has to first derive the norm square $\|h_i\|^2$ of each row h_i for the non-square short Jacobian matrix J, and then to add them together, i.e., $\mathrm{tr}(JJ^T) = \|h_1\|^2 + \cdots + \|h_m\|^2$ if J has m rows. Now, for the purpose of finding the gradient vector $\eta = \frac{\partial p(q)}{\partial q}$, let us create a column vector y that lists all the variable elements from J but skips over every constant element. Namely, from the above 3 by 4 Jacobian matrix,

$$
y = \begin{pmatrix}
-d_5 s_2 s_{34} - a_3 s_2 c_3 \\
d_5 c_2 c_{34} - a_3 c_2 s_3 \\
d_5 c_2 c_{34} \\
-d_5 s_{34} - a_3 c_3 \\
-d_5 s_{34} \\
-d_5 c_2 s_{34} - a_3 c_2 c_3 \\
-d_5 s_2 c_{34} + a_3 s_2 s_3 \\
-d_5 s_2 c_{34}
\end{pmatrix},
$$

which has kicked out those 0 or 1 constants. Then, $y^T y$ represents the sum of all norm squares of h_i's without constants. Thus, the gradient vector

$$
\eta = \frac{\partial \mathrm{tr}(JJ^T)}{\partial q} = \frac{\partial y^T y}{\partial q} = 2 \left(\frac{\partial y}{\partial q} \right)^T y.
$$

Once we have a symbolical column vector y, we can directly derive its mathematical Jacobian matrix $\frac{\partial y}{\partial q}$ with patience. Then, let MATLAB^{TM} multiply it to the column vector y to achieve a numerical gradient vector η at each sample point.

Figure 2.11 depicts an animation study of the (4+1)-joint robot. The left picture is the initial awkward posture at $t = 0$ with the force f acted at the TCP, where $d_1 = 0$, $\theta_2 = -39°$, $\theta_3 = -30°$, and $\theta_4 = 80°$. The right picture shows the final nearly best posture less than a second of the null space control:

$$
\dot{q} = (I - J^+ J) k \eta
$$

with the gain constant $k = -5$. As can be evidently seen in Figure 2.12, the left plot is the potential function value $p(q) = \mathrm{tr}(JJ^T)$ vs. time, and the right plot is the norm of the joint torque $\|\tau\|$ vs. time. Both of them show a successful and effective minimization process.

REFERENCES

1. Gilmore, R., (1974) Lie Groups, Lie Algebras, and Some of Their Applications. John Wiley, New York.
2. Karger, A. and Novak, J., (1985) Space Kinematics and Lie Groups. Gordon and Breach Science, New York.
3. Dubrovin, B., Fomenko, A. and Novikov, S., (1992) Modern Geometry – Methods and Applications, Part I - The Geometry of Surfaces, Transformation Groups, and Fields. Springer-Verlag, New York.

4. Dubrovin, B., Fomenko, A., Novikov, S., (1992) Modern Geometry – Methods and Applications, Part II - The Geometry and Topology of Manifolds. Springer-Verlag, New York.

5. Bredon, G., (1993) Topology and Geometry. Springer-Verlag, New York.

6. Berger, M., Gostiaux, B., (1988) Differential Geometry: Manifolds, Curves, and Surfaces. Springer-Verlag, New York.

7. Fomenko, A., Manturov, O. and Trofimov, V., (1998) Tensor and Vector Analysis: Geometry, Mechanics and Physics. Mathematics, CRC Press.

8. Danielson, D., (2003) Vectors and Tensors in Engineering and Physics, 2nd Edition. Westview (Perseus).

9. Brockett, R., (1984) Robotic Manipulators and the Product of Exponentials Formula. In: Fuhrman, P. (ed) Mathematical Theory of Networks and Systems, Springer-Verlag, New York.

10. Spong, M. and Vidyasagar, M., (1989) Robot Dynamics and Control. John Wiley & Sons, New York.

11. Murray, R., Li, Z. and Sastry, S., (1994) A Mathematical Introduction to Robotic Manipulation. CRC Press, Boca Raton, London, New York.

12. Gu, Edward Y.L. and Luh, J., (1987) Dual-Number Transformation and Its Applications to Robotics. *IEEE Journal of Robotics and Automation*, Vol. RA-3, No.6, December, pp. 615-623.

13. Gu, Edward Y.L., (2013) A Journey from Robot to Digital Human. Springer, Heidelberg, New York.

14. Gu, Edward Y.L., (2000) Configuration Manifolds and Their Applications to Robot Dynamic Modeling and Control. *IEEE Transactions on Robotics and Automation*, Vol.16, No.5, October, pp. 517-527.

15. Gu, Edward Y.L., (2000) A Configuration Manifold Embedding Model for Dynamic Control of Redundant Robots. *International Journal of Robotics Research*, Vol.19, No.3, March, pp. 289-304.

16. Whitney, H., (1936) Differentiable Manifolds. *Ann. of Math.*, No.2, Vol.37, pp. 645-680.

17. Warner, F., (1983) Foundations of Differentiable Manifolds and Lie Groups. Springer-Verlag, New York.

18. Nash, J., (1956) The Embedding Problem for Riemannian Manifolds. *Ann. of Math.*, No.2, Vol.63, pp. 20-63.

19. Greene, R., (1970) Isometric Embeddings of Riemannian and Pseudo-Riemannian Manifolds. *Memories of the American Mathematical Society*, No.97, iii+63.

20. Gromov, M., (1986) Partial Differential Relations. Springer-Verlag, Berlin, New York.

21. Clifford, W., (1873) Preliminary Sketch of Bi-Quaternions. *Proc. London Math. Soc.*, vol.4, pp. 381-395.

22. Study, E., (1903) Geometrie der Dynamen, Leipzig, Germany.

23. Yang, A., (1974) Calculus of Screws. In: Spillers, W. (ed) Basic of Design Theory. New York: North-Holland/American Elsevier, pp. 266-281.

24. Yaglom, I., (1968) Complex Numbers in Geometry. New York: Academic.

25. Rooney, J., (1978) A Comparison of Representations of General Spatial Screw Displacement. *Envision, Planning (England)*, Series B, Vol.5, pp. 45-88.

26. Rooney, J., (1978) On the Three Types of Complex Number and Planar Transformation. *Envision, Planning (England)*, Series B, Vol.5, pp. 89-99.

27. Rooney, J., (1975) On the Principle of Transference. *Inst. Mechanical Engineering Proceedings of 4th World Congress*, Newcastle-upon-Tyne, England, Sept., pp. 1089-1094.

28. Pennock, G. and Yang, A., (1985) Application of Dual-Number Matrices to the Inverse Kinematics Problem of Robot Manipulators. *ASME Trans. Journal of Mechanisms, Transmissions, Automation in Design*, Vol.107, June, pp. 201-208.

29. McCarthy, J., (1986) Dual Orthogonal Matrices in Manipulator Kinematics. *International Journal of Robotics Research*, Vol.5, pp. 45-51.

30. Duffy, J., (1980) Analysis of Mechanisms and Robot Manipulators, Halstead Press, London.

31. Jacobson, N., (1966) Lie Algebras. John Wiley and Sons, New York.

32. Bakhturin, Y., (2001) Campbell–Hausdorff Formula. In: Hazewinkel, M. (ed) Encyclopedia of Mathematics, Springer, Heidelberg, New York.

33. Bellman, R., (1960) Introduction to Matrix Analysis. McGraw-Hill Book Inc., New York.

34. Horn, R., Johnson, C., (1985) Matrix Analysis. Cambridge University Press, New York.

35. Boullion, T., Odell, P., (1971) Generalized Inverse Matrices. John Wiley and Sons, New York.

3 Robot Dynamics Modeling

3.1 THE HISTORY OF ROBOT DYNAMIC FORMULATIONS

Dynamics is a study of motion response to force and/or torque. Thus, kinematics is a topological structure of dynamics for every dynamic system, while dynamics determines geometry of the motion constraints. The architecture of dynamics is actually a linear combination of the kinematic structure by dynamic parameters. To gain more insights into the dynamics of robotic systems, we need more mathematical and physical perceptions on the relationship between kinematics and dynamics before developing advanced dynamic models, formulations and new computational algorithms for control of robotic systems.

Any kind of robots is commonly modeled as a multi-rigid-body system, and thus the robot dynamics modeling is a typical application of the multi-rigid-body dynamics to robotic systems. Modeling the robot dynamics includes two directions of exploration: the forward dynamics or direct dynamics, and the inverse dynamics. The forward dynamics is to find the motion response to the given force and/or torque, while the inverse dynamics is to determine the force and/or torque, due to which a specified motion can be generated. Therefore, the control design for robotic systems often requires to know the inverse dynamic model.

Dynamics modeling and control of nonlinear and highly coupled multi-body systems, such as a robotic system, have attracted extensive researches and developments in the past decades. All the investigations and explorations have closely followed the advanced theories of mathematics and physics [1–3]. Since the early 1980s, the robot dynamics modeling and control strategies have become the hottest research and development areas drawing a great deal of attention.

A robot manipulator is recognized to be one of the most typical multi-body and highly coupled mechanical systems. Over the past four decades, the Newton-Euler recursive algorithm and the Lagrangian formulation were the two main streams in the robot dynamics research [4–7]. Since the later 1980s, a variety of alternative formulations and control algorithms have been developed to improve the modeling quality, computational speed, control accuracy and robustness as well as the theoretical generalizability, and the breadth and depth of physical insight into robot kinematics and dynamics. Many of them were based on more advanced mathematical theories than the classical Lagrangian formulation. Brockett first utilized the Lie group, Lie algebra and their exponential maps to deal with robot kinematic models and made some extensions to new dynamic models [8–10]. Many other researchers followed to develop their new ideas more broadly and specifically. The representative and influential work includes seeking an isometry to linearize the dynamic equations [11], finding a special output function to factorize the robot inertial matrix and then to "flatten" the metric surfaces [12, 13], applying classical groups to develop recursive dynamic algorithms [14], using a spatial operator algebra to simplify dynamic formulations [15, 16], analysis and design of redundant robots under isotropy conditions [17, 18], exploring a canonical transformation for curvature-vanishing [19] and using a Lie group and product-of-exponential formulation to reduce dynamic models via an extended inertial matrix factorization [20, 21]. More Lie algebraic representation methods with geometrical formulations can be found in [6, 22]. Such a large volume of the extensive studies have brought much profound interpretations and perceptions with advanced mathematical theories into robotics research.

To further understand and answer what is the essence behind an n-th order or an n-body dynamic system, such as an open serial-chain robot arm with n joints, or a closed parallel-chain robotic system, let us first introduce a concept of **configuration manifold**, or **C-Manifold** that is an n-dimensional differentiable (smooth) generalized surface as a unique dynamic motion constraint for the robotic system. More precisely, every dynamic motion trajectory of the robotic system is constrained on this invisible C-manifold. In other words, such a C-manifold, denoted by M^n, becomes the identity of the system. Similar to the personality of a human, each type of robotic systems has own individual C-manifold that distinguishes itself from the others [23–25].

Furthermore, the C-manifold for a robotic system has its unique topological structure that can be directly determined by the kinematics of the robot. The C-manifold can also be endowed at each point with an arc-length that is related to the kinetic energy of the motion as a Riemannian metric to "personalize" its geometry. In an overall mathematical perspective, a space is called a manifold if it is globally non-Euclidean, but it can be locally Euclidean. In more abstract mathematical language, a small open neighborhood around each point on the manifold can be homeomorphic to an open region in a Euclidean space of the same dimension. Namely, every small open non-Euclidean neighborhood on the manifold can be topologically equivalent to the same dimensional open region of the Euclidean space. A coordinate system can then be defined in the small neighborhood that is topologically equivalent to the Euclidean space, and it is commonly called the **local coordinates** at each point of the manifold.

Under such a mathematical interpretation, it is now reasonable to define the n-tuple of joint positions in $q = (q_1 \cdots q_n)^T$ for an n-joint robotic system as a set of the local coordinates on its C-manifold M^n. Although all the joint variables in q may not be orthogonal to each other, they are all linearly independent and thus qualified to form a basis of the locally homeomorphic Euclidean space. In fact, there are many ordinary examples of local coordinates to represent non-Euclidean manifolds in mathematics, such as the well-known spherical coordinates and cylindrical coordinates in calculus, etc.

3.2 THE ASSUMPTION OF RIGID BODY AND RIGID MOTION

3.2.1 THE RIGID BODY AND RIGID MOTION

In the literature [1–3, 6], a rigid body is commonly defined as follows:

Definition 3.1. A **rigid body** is a system of point masses, which is constrained by holonomic relations expressed by the fact that the distance between any two points is constant:

$$\|x_i - x_j\| = r_{ij} = const.$$

Thus, a rigid body is a theoretical model where any two distinct points on the body (or object) have a fixed time-invariant distance. Otherwise, the object is called a deformable body. In the areas of robotics research, we often assume that each link of a robot manipulator is a rigid body, unless it will indicate otherwise. A rigid body is relatively easy to be investigated and modeled in both kinematics and dynamics.

The motion of a rigid body is referred to as a **rigid motion** in the 6-dimensional $SE(3) = \mathbb{R}^3 \times SO(3)$ Euclidean space. The French mathematician Michel Chasles (1793–1880) first discovered and showed the following well-known theorem:

Theorem 3.1. (Chasles) – *The most general rigid body displacement can be produced by a translation along a line followed (or preceded) by a rotation about an axis parallel to that line.*

Based on the Chasles Theorem in classical mechanics [1–3], any rigid motion is a screw displacement about/along some screw axis. A 3 by 3 dual rotation matrix $\hat{R} = R + \varepsilon S \in SO(3, \mathbb{D})$ that was introduced and studied in detail in Chapter 2 is just the most typical combination of both translation and rotation through the dual screw motion. Therefore, *the Chasles Theorem suggests that any rigid motion be completely decomposed without coupling into a translation of the body mass center (or the center of gravity CG) and a rotation of the body about an axis passing through the mass center.* Note that the mass center of a rigid body can be the same as the *centroid* that is the geometrical center of the body if the mass of the rigid body is uniformly distributed, or the mass density is a constant everywhere.

Before we model a robot dynamic system, a number of key concepts from classical mechanics must be carefully reviewed and revisited. First, for a rigid body with its mass center c, the 3D linear velocity v at any point of the rigid body with respect to the fixed base is determined by

$$v = v_c + \omega \times r, \tag{3.1}$$

where v_c is the linear velocity of the mass center referred to the base, ω is the angular velocity of the rigid body projected onto the base and r is a radial vector that is tailed at the mass center and arrow-pointing to the point of interest and projected onto the base as well. This equation once again convinces that *any motion of a rigid body can be fully decoupled into a translation of its mass center and a rotation about the mass center.*

To mathematically justify equation (3.1), let $p_0^i \in \mathbb{R}^3$ be a position vector for point i other than the mass center of the rigid body with respect to the base. If $p_0^c \in \mathbb{R}^3$ is the mass center position vector that is also referred to the base, then

$$p_0^i = p_0^c + R_0^c r_c^i,$$

where R_0^c represents the orientation of frame c with respect to the base to project the radial vector r_c^i onto the base. After every vector is projected onto the common base, taking time-derivatives for both sides can yield

$$\dot{p}_0^i = \dot{p}_0^c + R_0^c \dot{r}_c^i + \dot{R}_0^c r_c^i.$$

Since the distance between any two distinct points in a rigid body is invariant, the radial vector r_c^i is always fixed as seen in the rigid body frame c so that $\dot{r}_c^i = 0$. In addition, according to equation (2.52), $\dot{R}_0^c R_c^0 = \Omega = \omega \times$. Due to $\dot{p}_0^i = v$ and $\dot{p}_0^c = v_c$, we finally reach the agreement with equation (3.1):

$$v = \dot{p}_0^i = \dot{p}_0^c + \dot{R}_0^c R_c^0 R_0^c r_c^i = v_c + \omega \times r$$

with $r = R_0^c r_c^i$ that is projected on the base.

Second, if a rigid body is pivoted at a fixed point, the torque τ due to a force f acting at a point of the body is given by

$$\tau = r \times f,$$

where r is the radial vector tailed at the pivoting point with its arrow at the force acting point.

Now, the **inertia tensor** for a rigid body with respect to an arbitrary frame a is defined in classical mechanics by

$$\Gamma_a = \begin{pmatrix} I_{xx} & -I_{xy} & -I_{xz} \\ -I_{yx} & I_{yy} & -I_{yz} \\ -I_{zx} & -I_{zy} & I_{zz} \end{pmatrix}, \tag{3.2}$$

where

$$I_{xx} = \int_M (y^2 + z^2)dm, \quad I_{yy} = \int_M (z^2 + x^2)dm, \quad I_{zz} = \int_M (x^2 + y^2)dm,$$

and

$$I_{xy} = I_{yx} = \int_M xydm, \quad I_{yz} = I_{zy} = \int_M yzdm, \quad I_{zx} = I_{xz} = \int_M zxdm.$$

All the values of x, y and z in the above integrals are the coordinates at each locating point of the differential mass dm to be integrated over the rigid body with respect to frame a. If the origin of frame a is defined at the mass center c, then we denote this special inertia tensor by Γ_c.

Furthermore, let $r = (x\ y\ z)^T$ be a radial vector with its arrow at the locating point of each dm and its tail at the origin of frame a. Then, the above 3 by 3 inertia tensor (3.2) can be rewritten in a more compact form:

$$\Gamma_a = \int_M (r^T r I_3 - rr^T)dm, \tag{3.3}$$

where I_3 is the 3 by 3 identity.

Once we define the positive-definite symmetric inertia tensor Γ_a that is referred to frame a for a rigid body, the scalar **moment of inertia** about any unit vector ξ_a that is projected on frame a can be immediately determined by the following quadratic form:

$$I_{\xi_a} = \xi_a^T \Gamma_a \xi_a. \tag{3.4}$$

In a special case, if $\xi_a = (1 \ 0 \ 0)^T$, i.e., the unit vector of the x-axis of frame a, then, by substituting the inertia tensor definition (3.2) into (3.4), we obtain $I_{\xi_a} = I_{xx}$. Similarly, $I_{\xi_a} = I_{yy}$ and I_{zz} if $\xi_a = (0 \ 1 \ 0)^T$ and $\xi_a = (0 \ 0 \ 1)^T$, respectively.

In classical mechanics, it is well-known that the above moment of inertia I_{ξ_a} can also be determined by a "parallel shift" for the moment of inertia I_{ξ_c} about the axis ξ_c that is parallel to ξ_a but passes through the mass center of the rigid body, i.e.,

$$I_{\xi_a} = I_{\xi_c} + md^2,$$

where m is the mass of the rigid body and d is the shortest distance between the two parallel axes ξ_a and ξ_c. This property is commonly referred to as the **Parallel-Axis Theorem**, or **Steiner's Theorem**.

In fact, we can extend the above parallel axis theorem from a scalar moment of inertia to a 3 by 3 inertia tensor defined by (3.2). Let $c_a = (c_x \ c_y \ c_z)^T$ be a vector that is projected on frame a with its arrow pointing to the mass center of the rigid body and its tail at the origin of frame a. Let $C_a = S(c_a) = c_a \times$ be the skew-symmetric matrix of the vector c_a. Namely,

$$C_a = \begin{pmatrix} 0 & -c_z & c_y \\ c_z & 0 & -c_x \\ -c_y & c_x & 0 \end{pmatrix}.$$

It can be shown without difficulty that the following **Tensor Parallel-Axis Theorem** is true:

$$\Gamma_a = \Gamma_c + mC_aC_a^T = \Gamma_c + mC_a^TC_a = \Gamma_c - mC_a^2. \tag{3.5}$$

This extension offers a useful procedure of the inertia tensor determination when the reference frame is parallel-shifted. For instance, if frame a is parallel-shifted to frame b by a position vector p_a^b that is projected on frame a, then the vector tailed at the origin of frame b and pointing to the mass center becomes $c_b = c_a - p_a^b$, and its skew-symmetric matrix is $C_b = S(c_b) = c_b \times$. Based on the tensor parallel axis theorem in equation (3.5),

$$\Gamma_b = \Gamma_c + mC_bC_b^T = \Gamma_a - mC_aC_a^T + mC_bC_b^T = \Gamma_a + m(C_a^2 - C_b^2). \tag{3.6}$$

In other words, $\Gamma_a + mC_a^2 = \Gamma_b + mC_b^2 = \Gamma_c$. This shows that the sum $\Gamma_a + mC_a^2$ is invariant under any parallel-shift of frame a.

If the reference frame is now rotated from frame a to frame b by a rotation matrix R_a^b but with no any shift of its origin, then

$$\Gamma_b = R_b^a \Gamma_a R_a^b = (R_a^b)^T \Gamma_a R_a^b. \tag{3.7}$$

Furthermore, if the reference frame is translated and rotated simultaneously, which can be described by a homogeneous transformation, or a dual rotation matrix from frame a to frame b:

$$H_a^b = \begin{pmatrix} R_a^b & P_a^b \\ 0_3 & 1 \end{pmatrix}, \quad \text{or} \quad \hat{R}_a^b = R_a^b + \varepsilon S_a^b,$$

then, by combining equations (3.6) and (3.7) together, we obtain

$$\Gamma_b = R_b^a[\Gamma_a + m(C_a^2 - C_b^2)]R_a^b, \tag{3.8}$$

where $C_a - C_b = P_a^b = S(p_a^b) = p_a^b \times$.

As a special case, if frame a is originated at the mass center of the rigid body, then $c_a = 0$ and $\Gamma_a = \Gamma_c$ so that equation (3.8) is back to the tensor parallel axis theorem with a rotation R_c^b, i.e.,

$$\Gamma_b = R_b^c(\Gamma_c - mC_b^2)R_c^b = R_b^c(\Gamma_c + mC_b^T C_b)R_c^b, \tag{3.9}$$

This equation shows a clear fact that *any shift of the reference frame away from the mass center will create an additional linear velocity when the rigid body is rotating.*

3.2.2 KINEMATIC PARAMETERS VS. DYNAMIC PARAMETERS

As we have studied the inertia tensor Γ_a and the coordinates of the mass center C_a with respect to frame a, all the dynamic parameters can now be augmented to form a 6 by 6 **dynamic parameter matrix** for a rigid body:

$$U_a = \begin{pmatrix} mI_3 & mC_a^T \\ mC_a & \Gamma_a \end{pmatrix}, \tag{3.10}$$

where I_3 is the 3 by 3 identity and m is the mass of the rigid body. If the origin of frame a is shifted to the mass center c, then, $C_a = 0_3$, the 3 by 3 zero matrix. In this case, the dynamic parameter matrix referred to frame c is reduced to

$$U_c = \begin{pmatrix} mI_3 & 0_3 \\ 0_3 & \Gamma_c \end{pmatrix}.$$

One can count the total number of dynamic parameters contained in U_a for a rigid body. The answer is clearly up to 10: one is the mass m, three in the skew-symmetric matrix C_a, and up to six in the symmetric inertia tensor Γ_a. If the rigid body is geometrically symmetric and the mass center is the same as its centroid, then Γ_a will have only three nonzero diagonal elements, and all the off-diagonal elements are zero so that the total number of dynamic parameters is reduced to 7. It will be further reduced to 4 if considering U_c that is referred to the mass center: one is the mass m and three are the moments of inertia I_x, I_y and I_z on the diagonal of the inertia tensor Γ_c.

However, it is quite a "cruel" fact that all the 10 dynamic parameters are uncertain, because they are not accurately measurable in the reality, called the *parameter uncertainty*. In contrast, every kinematic parameter is exactly measurable, such as the link length a and the joint off-set d as well as the twist angle α.

3.3 KINETIC ENERGY, POTENTIAL ENERGY AND LAGRANGE EQUATIONS

3.3.1 DETERMINATION OF KINETIC ENERGY

In classical mechanics, the original definition of **kinetic energy** K was initiated with a system of mass points:

Definition 3.2. A kinetic energy of a system of mass points is the sum of the kinetic energy of the points:

$$K = \frac{1}{2} \sum_{i=1}^{n} m_i \dot{r}_i^T \dot{r}_i,$$

where m_i is the mass of point i and r_i is the radial position vector of point i.

It can be easily justified that any kinetic energy increment is due to the work done by forces acting on the mass points, i.e.,

$$K(t) - K(t_0) = \int_{t_0}^{t} f \cdot dr$$

to reflect the energy conservation.

However, this original definition was based only on a single rigid body of mass points. In order to extend it to find the kinetic energy K for a multi-body system, we cannot just stay at the level of mass points. Instead, based on the Chasles theorem, the total kinetic energy K of a rigid body motion can be completely decomposed into a translational kinetic energy K_t of the mass center and a rotational kinetic energy K_r about the mass center, i.e., $K = K_t + K_r$ for a single rigid body.

Now, the translational kinetic energy can be written as $K_t = \frac{1}{2} m v_c^T v_c$ while the rotational kinetic energy is determined by $\frac{1}{2} \omega^T \Gamma_c \omega$, so that the total kinetic energy for a single rigid body is the sum of both:

$$K = K_t + K_r = \frac{1}{2} m v_c^T v_c + \frac{1}{2} \omega^T \Gamma_c \omega, \tag{3.11}$$

where v_c and ω are the linear velocity of the mass center and angular velocity of the body with respect to the fixed base, respectively, and Γ_c is the inertia tensor referred to the mass center.

By recalling the Cartesian velocity, i.e., the twist $V_c = \begin{pmatrix} v_c \\ \omega \end{pmatrix}$ and the dynamic parameter matrix U_c in the last chapter of kinematics modeling and analysis, we can now rewrite the above kinetic energy in (3.11) for a single rigid body in a more compact form:

$$K = \frac{1}{2} (v_c \ \ \omega) \begin{pmatrix} m I_3 & O_3 \\ O_3 & \Gamma_c \end{pmatrix} \begin{pmatrix} v_c \\ \omega \end{pmatrix} = \frac{1}{2} V_c^T U_c V_c,$$

where I_3 and O_3 are the 3 by 3 identity and zero matrix, respectively.

However, in robotic applications, after applying the D-H convention, a coordinate frame assigned on each link is not originated at the mass center of the link. We have to find the kinetic energy $K = K_t + K_r$ that is referred to a point a other than the mass center. Based on equation (3.1), the velocity at point a is given by

$$v_a = v_c + \omega \times r_a = v_c - \omega \times c_a,$$

where r_a is a radial vector tailed at the mass center and arrow-pointing to point a, while c_a is the vector tailed at point a with its arrow pointing to the mass center so that $c_a = -r_a$. Thus,

$$v_c = v_a + \omega \times c_a = v_a + C_a^T \omega$$

with the skew-symmetric matrix $C_a = c_a \times$. If the origin of a coordinate frame is at point a other than the mass center,

$$K = K_t + K_r = \frac{1}{2} m v_c^T v_c + \frac{1}{2} \omega^T \Gamma_c \omega$$

$$= \frac{1}{2} m (v_a + C_a^T \omega)^T (v_a + C_a^T \omega) + \frac{1}{2} \omega^T (\Gamma_a - m C_a C_a^T) \omega$$

$$= \frac{1}{2} m (v_a^T v_a + 2 v_a^T C_a^T \omega + \omega^T C_a C_a^T \omega) + \frac{1}{2} \omega^T (\Gamma_a - m C_a C_a^T) \omega$$

$$= \frac{1}{2} m v_a^T v_a + m v_a^T C_a^T \omega + \frac{1}{2} \omega^T \Gamma_a \omega.$$

Clearly, this equation can also be rewritten in a more compact form:

$$K = \frac{1}{2} (v_a \ \ \omega) \begin{pmatrix} m I_3 & m C_a^T \\ m C_a & \Gamma_a \end{pmatrix} \begin{pmatrix} v_a \\ \omega \end{pmatrix} = \frac{1}{2} V_a^T U_a V_a. \tag{3.12}$$

Therefore, for a robot arm with n links, the total kinetic energy K can be expressed as follows:

$$K = \frac{1}{2} \sum_{j=1}^{n} V_j^T U_j V_j. \tag{3.13}$$

For each link, the kinetic energy K_j is now expressed in terms of the twist of frame j that is attached on link j. In other words, the dynamic parameter matrix U_j for link j is sandwiched by the twists V_j. If each twist V_j is calculated using the Jacobian equation in differential motion: $V_j = J_j \dot{q}$, then the kinetic energy K_j for link j will be written in terms of the joint velocity vector $\dot{q}$, instead of the Cartesian velocities.

Let a **sub-Jacobian matrix** J_j for link j be defined by

$$J_j = \begin{pmatrix} p_j^0 \times r_j^0 & \cdots & p_j^{j-1} \times r_j^{j-1} & O_{3 \times (n-j)} \\ r_j^0 & \cdots & r_j^{j-1} & O_{3 \times (n-j)} \end{pmatrix} = \begin{pmatrix} s_j^0 & \cdots & s_j^{j-1} & O_{3 \times (n-j)} \\ r_j^0 & \cdots & r_j^{j-1} & O_{3 \times (n-j)} \end{pmatrix},$$
$$\tag{3.14}$$

which is 6 by n dimensional. Of course, if the i-th joint $(0 < i \leq j)$ is prismatic, then the column $\begin{pmatrix} p_j^{i-1} \times r_j^{i-1} \\ r_j^{i-1} \end{pmatrix}$ or $\begin{pmatrix} s_j^{i-1} \\ r_j^{i-1} \end{pmatrix}$ must be replaced by $\begin{pmatrix} r_j^{i-1} \\ 0_3 \end{pmatrix}$ as a default.

Then,

$$K = \frac{1}{2} \sum_{j=1}^{n} \dot{q}^T J_j^T U_j J_j \dot{q} = \frac{1}{2} \dot{q}^T \left(\sum_{j=1}^{n} J_j^T U_j J_j \right) \dot{q} = \frac{1}{2} \dot{q}^T W \dot{q}, \tag{3.15}$$

where we have defined an n by n positive-definite symmetric matrix

$$W = \sum_{j=1}^{n} J_j^T U_j J_j, \tag{3.16}$$

called an **inertial matrix** of the robot. If the total kinetic energy K is determined in terms of the joint velocity $\dot{q}$ in (3.15), it is obviously a positive-definite quadratic form of the joint tangent variables, and all the coefficients of the terms can be extracted and augmented in a certain order to form an n by n positive-definite inertial matrix for the robot [25].

Let us start using the basic equation (3.11) and gradually increase the complexity of robotic systems as examples to illustrate finding their kinetic energy K and inertial matrix W.

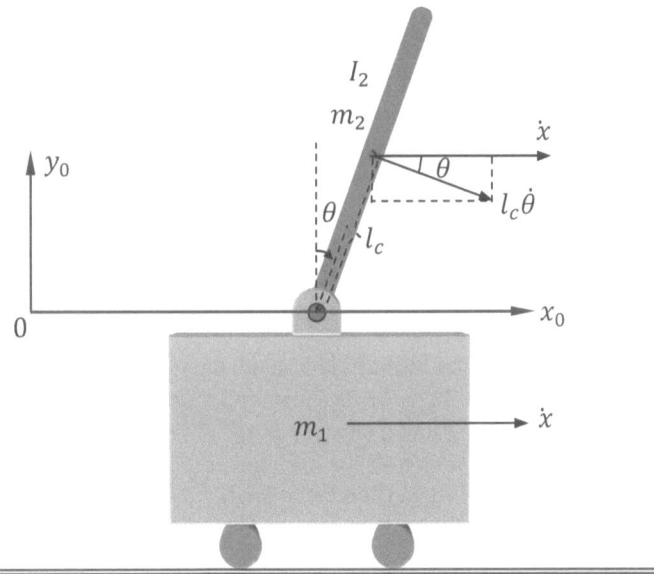

Figure 3.1   An inverted-pendulum system for finding the kinetic energy

Example 3.1. The first example is a well-known inverted pendulum system, which is also one of the most popular apparatus for demonstrations in control classes, as shown in Figure 3.1. To find its symbolical kinetic energy K, we have to determine K_1 for the linear movable cart and K_2 for the inverted pendulum. They look separated as a two-body system, but often the inverted pendulum of the second body will be imposed by the cart moving velocity of the first body due to the fact that it is a serial-chain mechanism. Since the cart is moving as simple as a linear translation without any rotation, its velocity is just $\dot{x}$ such that $K_1 = \frac{1}{2} m_1 \dot{x}^2$.

For the second body, in addition to having a linear velocity $l_c \dot{\theta}$ at its mass center due to the rotation with its angular velocity $\dot{\theta}$ according to equation (3.1), it is also imposed by $\dot{x}$ from body 1 as a parallel shift from the linear velocity of body 1. Since $l_c \dot{\theta}$ is always perpendicular to the inverted pendulum, while $\dot{x}$ is always horizontal, we have to decompose either one orthogonally in order to add these two vectors together. Let us decompose the vector $l_c \dot{\theta}$ into two orthogonal components: $l_c \dot{\theta} \cos \theta$ and $l_c \dot{\theta} \sin \theta$, as can be seen in Figure 3.1. Then, the translational kinetic energy of body 2 becomes

$$K_{t2} = \frac{1}{2} m_2 [(\dot{x} + l_c \dot{\theta} \cos \theta)^2 + (l_c \dot{\theta} \sin \theta)^2] = \frac{1}{2} m_2 (\dot{x}^2 + l_c^2 \dot{\theta}^2 + 2 l_c \dot{x} \dot{\theta} \cos \theta).$$

By adding a rotational kinetic energy $K_{r2} = \frac{1}{2} I_2 \dot{\theta}^2$, where I_2 is the moment of inertia of body 2 with respect to its mass center, the total kinetic energy turns out to be

$$K = K_1 + K_{t2} + K_{r2} = \frac{1}{2} [m_1 \dot{x}^2 + m_2 (\dot{x}^2 + l_c^2 \dot{\theta}^2 + 2 l_c \dot{x} \dot{\theta} \cos \theta) + I_2 \dot{\theta}^2].$$

This kinetic energy K is a quadratic form of the tangent variables $\dot{x}$ and $\dot{\theta}$. According to (3.15), we can extract all of its coefficients to form the following 2 by 2 positive-definite symmetric inertial matrix W of the inverted pendulum system:

$$W = \begin{pmatrix} m_1 + m_2 & m_2 l_c \cos \theta \\ m_2 l_c \cos \theta & m_2 l_c^2 + I_2 \end{pmatrix}. \tag{3.17}$$

Example 3.2. This example is to find the kinetic energy K and inertial matrix W for a 2-revolute-joint planar robot arm, as shown in Figure 3.2. Since both two links involve the translation and rotation, the example will show how link 2 is imposed by both the translation and rotation velocities from link 1. For shortening the notations, let $b_1 = a_1 - l_{c1}$ and $b_2 = a_2 - l_{c2}$. The kinetic energy $K_1 = K_{t1} + K_{r1}$ can be easily determined by

$$K_1 = \frac{1}{2} m_1 (b_1 \dot{\theta}_1)^2 + \frac{1}{2} I_1 \dot{\theta}_1^2,$$

where I_1 is the moment of inertia with respect to the mass center of link 1. Now for link 2, both its linear velocity and angular velocity are imposed by the motion of link 1. Namely, the angular velocity of link 2 is now $\dot{\theta}_1 + \dot{\theta}_2$, instead of just $\dot{\theta}_2$ so that the linear velocity of the mass center becomes $b_2 (\dot{\theta}_1 + \dot{\theta}_2)$. In addition, the imposed linear velocity from link 1 should be $a_1 \dot{\theta}_1$ through the rotation axis point of joint 2.

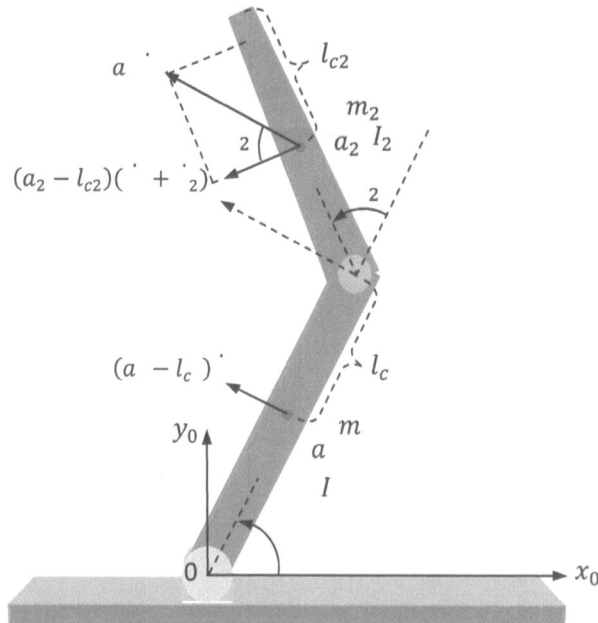

Figure 3.2 A two-link planar robot for finding the kinetic energy

In order to add two vectors together, let $a_1\dot\theta_1$ be decomposed into two orthogonal components: $a_1\dot\theta_1\cos\theta_2$ and $a_1\dot\theta_1\sin\theta_2$. Then, the kinetic energy of link 2 should be

$$K_2 = \frac{1}{2}m_2[(a_1\dot\theta_1\cos\theta_2 + b_2(\dot\theta_1 + \dot\theta_2))^2 + (a_1\dot\theta_1\sin\theta_2)^2] + \frac{1}{2}I_2(\dot\theta_1 + \dot\theta_2)^2$$

$$= \frac{1}{2}m_2[a_1^2\dot\theta_1^2 + b_2^2(\dot\theta_1 + \dot\theta_2)^2 + 2a_1b_2\dot\theta_1(\dot\theta_1 + \dot\theta_2)\cos\theta_2] + \frac{1}{2}I_2(\dot\theta_1 + \dot\theta_2)^2.$$

Thus, the total kinetic energy K should be the sum of K_1 and K_2, i.e., $K = K_1 + K_2$ for this 2-joint planar robot arm.

To determine the 2 by 2 inertial matrix W for the robot, extracting all the coefficients from the quadratic form K will yield

$$W = \begin{pmatrix} w_{11} & m_2b_2^2 + m_2a_1b_2\cos\theta_2 + I_2 \\ m_2b_2^2 + m_2a_1b_2\cos\theta_2 + I_2 & m_2b_2^2 + I_2 \end{pmatrix}, \quad (3.18)$$

where $w_{11} = m_1b_1^2 + m_2a_1^2 + m_2b_2^2 + 2m_2a_1b_2\cos\theta_2 + I_1 + I_2$, and I_2 is the moment of inertia with respect to the mass center of link 2. This result exactly matches with equation (2.21) in the last chapter.

Example 3.3. This example is to illustrate a 2-joint robot arm sitting on a cylindrical log. Including the log rolling motion on the floor, this system can be modeled as

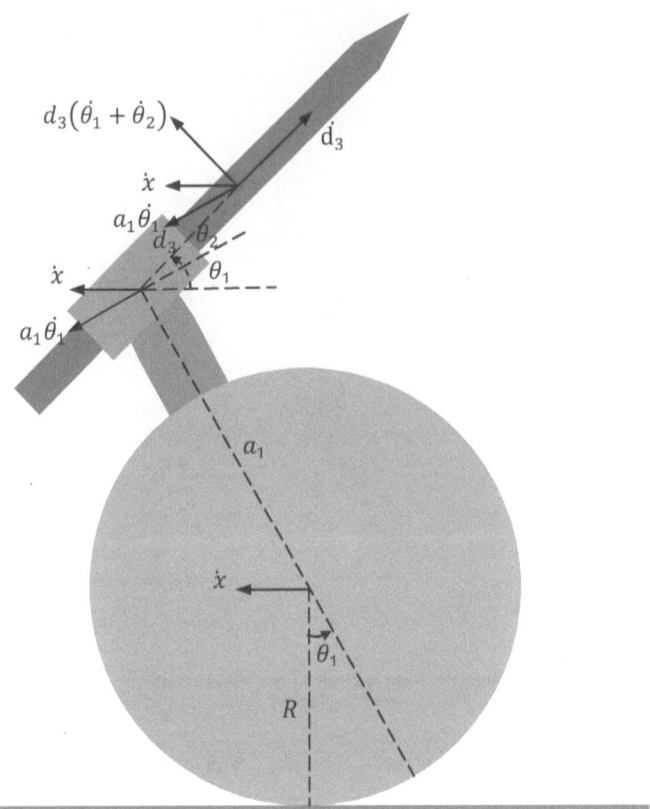

Figure 3.3 A log-arm robotic system for finding the kinetic energy

a 3-link robot. The rolling angle θ_1, however, is not independent of its translation displacement x_1. If the log has no slip with the floor, then, $x_1 = R\theta_1$, where R is the radius of the log, and θ_1 in this equation must be in radians.

If the mass center of the log is just at the circular center, the kinetic energy of the log is obviously equal to

$$K_1 = \frac{1}{2}m_1\dot{x}_1^2 + \frac{1}{2}I_1\dot{\theta}_1^2 = \frac{1}{2}(m_1R^2 + I_1)\dot{\theta}_1^2.$$

The second link is a sleeve box that is pivoted at the second joint axis to provide a tubular house for the 3rd link sliding movement. If it is pivoted just at its mass center, then, only two imposed linear velocities $a_1\dot{\theta}_1$ and $\dot{x}$ both from the log, while its angular velocity is the sum of $\dot{\theta}_2$ that is the revolute angular velocity by the joint itself and $\dot{\theta}_1$ imposed by the log. Thus, the kinetic energy should be

$$K_2 = \frac{1}{2}m_2[(\dot{x}_1 + a_1\dot{\theta}_1\cos\theta_1)^2 + (a_1\dot{\theta}_1\sin\theta_1)^2] + \frac{1}{2}I_2(\dot{\theta}_1 + \dot{\theta}_2)^2$$

$$= \frac{1}{2}m_2[(R^2+a_1^2)\dot{\theta}_1^2 + 2a_1R\dot{\theta}_1^2 c_1] + \frac{1}{2}I_2(\dot{\theta}_1+\dot{\theta}_2)^2.$$

Let d_3 be defined as a prismatic joint displacement of the mass center of link 3. Clearly, the third link has its sliding velocity $\dot{d}_3$ and a linear velocity due to the θ_2 rotation, plus two imposed velocities from the log: $a_1\dot{\theta}_1$ and $\dot{x}$. Since they are not parallel to each other, we have to project them onto an orthogonal frame for finding its kinetic energy K_3. Thus,

$$K_3 = \frac{1}{2}m_3[(\dot{d}_3 - \dot{x}c_{12} - a_1\dot{\theta}_1 c_2)^2 + (d_3(\dot{\theta}_1+\dot{\theta}_2) + \dot{x}s_{12} + a_1\dot{\theta}_1 s_2)^2] + \frac{1}{2}I_3(\dot{\theta}_1+\dot{\theta}_2)^2$$

$$= \frac{1}{2}m_3[\dot{d}_3^2 + (R^2+a_1^2)\dot{\theta}_1^2 + 2a_1R\dot{\theta}_1^2 c_1 - 2R\dot{d}_3\dot{\theta}_1 c_{12} - 2a_1\dot{d}_3\dot{\theta}_1 c_2 + d_3^2(\dot{\theta}_1+\dot{\theta}_2)^2 +$$

$$+ 2d_3R(\dot{\theta}_1^2+\dot{\theta}_1\dot{\theta}_2)s_{12} + + 2a_1d_3(\dot{\theta}_1^2+\dot{\theta}_1\dot{\theta}_2)s_2] + \frac{1}{2}I_3(\dot{\theta}_1+\dot{\theta}_2)^2.$$

Therefore, by extracting all the coefficients from the quadratic form $K = K_1 + K_2 + K_3$, the inertial matrix W for this 3-joint log-arm system can be determined as follows:

$$W = \begin{pmatrix} w_{11} & w_{12} & -m_3(Rc_{12}+a_1c_2) \\ w_{21} & m_3d_3^2+I_2+I_3 & 0 \\ -m_3(Rc_{12}+a_1c_2) & 0 & m_3 \end{pmatrix}, \qquad (3.19)$$

where

$$w_{11} = m_1R^2 + m_2(R^2+a_1^2+2a_1Rc_1) +$$

$$+ m_3(R^2+a_1^2+2a_1Rc_1+d_3^2+2d_3Rs_{12}+2a_1d_3s_2) + I_1 + I_2 + I_3,$$

and

$$w_{12} = w_{21} = m_3(d_3^2 + d_3Rs_{12} + a_1d_3s_2) + I_2 + I_3.$$

The above three examples are all the 2D planar robotic systems. One can relatively easily identify and determine each velocity vector in the 2D planar case, no matter it is produced by the link itself, or imposed by the lower links. Once the case is extended to a 3D robotic system, only relying on the geometrical inspection would be far lacking the confidence to determine each velocity precisely.

It is also observable from the above examples that the most challenging issue is how to identify and exactly determine the velocity vectors that are imposed by the lower links. As the analysis of kinetic energy is getting upper links for a robot manipulator, one could soon be confused or even lost in a 3D case, and any single mistake would blow up all of the efforts. Therefore, we demand a mathematical algorithm to help for easing the symbolical derivation or numerical computation to find the inertial matrix W of a robot. Equation (3.16) just offers a perfect algorithm to avoid any subjective mistake due to the complicated geometrical inspection, and fully rely on the mathematics towards a successful result, especially in the case of a 3D robot with a large number of joints, like a humanoid robotic system.

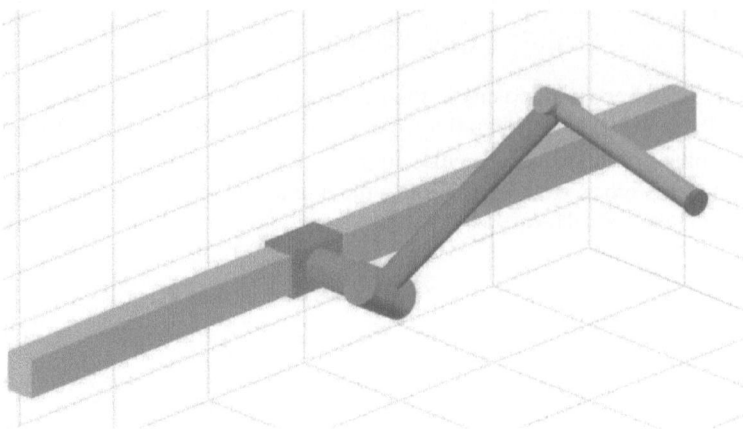

Figure 3.4 The (4+1)-joint beam-robot manipulator to find its inertial matrix W

Example 3.4. Let us now revisit the (4+1)-joint robot manipulator, as shown in Figure 3.4 that is an animation model at a time instant of testing, and can also refer to Figure 2.5 as a theoretical analysis model. The joint positions at the testing time instant are set as $d_1 = 0$, $\theta_2 = -30°$, $\theta_3 = -30°$, and $\theta_4 = 20°$. The tool disc angle still keeps $\theta_5 = 0$. Because this is a typical 3D robotic system, we now apply (3.16) to calculate its 5 by 5 inertial matrix W numerically in MATLAB^{TM}. Since equations (3.14) and (3.16) require inverting matrices for almost every step of homogeneous transformations, one has to prepare those matrices as well as the dynamic parameter matrix U_j of each link in programming. The following MATLAB^{TM} code is to illustrate how to program the algorithm based on (3.14) and (3.16), and the resultant numerical 5 by 5 inertial matrix W at the testing time instant is then printed out afterwards:

```
clear
d5=0.8; a3=1; d2=0.4;        % In meters
q=[0 -30 -30 20]'*pi/180;    % The testing joint positions

s2=sin(q(2)); c2=cos(q(2));
s3=sin(q(3)); c3=cos(q(3));
s4=sin(q(4)); c4=cos(q(4));
s34=sin(q(3)+q(4)); c34=cos(q(3)+q(4));
d1=q(1);          % For abbreviation

A1=[1 0 0 0; 0 0 1 0; 0 -1 0 d1+15; 0 0 0 1];
A2=[c2 0 -s2 0; s2 0 c2 0; 0 -1 0 d2; 0 0 0 1];
A3=[c3 -s3 0 a3*c3; s3 c3 0 a3*s3; 0 0 1 0; 0 0 0 1];
A4=[c4 0 s4 0; s4 0 -c4 0; 0 1 0 0; 0 0 0 1];
A5=[1 0 0 0; 0 1 0 0; 0 0 1 d5; 0 0 0 1];
    % One-step homogeneous transformations and also set theta_5=0

A10=[A1(1:3,1:3)' -A1(1:3,1:3)'*A1(1:3,4); 0 0 0 1];
A21=[A2(1:3,1:3)' -A2(1:3,1:3)'*A2(1:3,4); 0 0 0 1];
```

```
A32=[A3(1:3,1:3)' -A3(1:3,1:3)'*A3(1:3,4); 0 0 0 1];
A43=[A4(1:3,1:3)' -A4(1:3,1:3)'*A4(1:3,4); 0 0 0 1];
A54=[A5(1:3,1:3)' -A5(1:3,1:3)'*A5(1:3,4); 0 0 0 1];
    % To invert all the one-step homegeneous transformations

A20=A21*A10;  A30=A32*A20;  A40=A43*A30;  A50=A54*A40;
A31=A32*A21;  A41=A43*A31;  A51=A54*A41;
A42=A43*A32;  A52=A54*A42;
A53=A54*A43;

J1=[[A10(1:3,3);zeros(3,1)] zeros(6,4)]; % Notice the first joint is prismatic
J2=[[A20(1:3,3);zeros(3,1)] ...
    [cross(A21(1:3,4),A21(1:3,3)); A21(1:3,3)] zeros(6,3)];
J3=[[A30(1:3,3);zeros(3,1)] [cross(A31(1:3,4),A31(1:3,3)); A31(1:3,3)] ...
    [cross(A32(1:3,4),A32(1:3,3)); A32(1:3,3)] zeros(6,2)];
J4=[[A40(1:3,3);zeros(3,1)] [cross(A41(1:3,4),A41(1:3,3)); A41(1:3,3)] ...
    [cross(A42(1:3,4),A42(1:3,3)); A42(1:3,3)]] ...
    [cross(A43(1:3,4),A43(1:3,3)); A43(1:3,3)] zeros(6,1)];
J5=[[A50(1:3,3);zeros(3,1)] [cross(A51(1:3,4),A51(1:3,3)); A51(1:3,3)] ...
    [cross(A52(1:3,4),A52(1:3,3)); A52(1:3,3)] ...
    [cross(A53(1:3,4),A53(1:3,3)); A53(1:3,3)]] ...
    [cross(A54(1:3,4),A54(1:3,3)); A54(1:3,3)]];
    % Prepare all the sub-Jacobian matrices

C1=[0 -0.1 0;0.1 0 0;0 0 0]; C2=[0 0 0.2;0 0 0;-0.2 0 0];
C3=[0 0 0;0 0 0.5;0 -0.5 0]; C4=[0 -0.4 0;0.4 0 0;0 0 0]; C5=zeros(3);
    % Defining all the mass center coordinates in skew-symmetric matrices

U1=[1*eye(3) C1'; C1 diag([0.1 0.1 0.1])];
U2=[1*eye(3) C2'; C2 diag([0.12 0.08 0.12])];
U3=[2.5*eye(3) 2.5*C3'; 2.5*C3 diag([0.25 0.4 0.4])];
U4=[2*eye(3) 2*C4'; 2*C4 diag([0.3 0.3 0.2])];
U5=[0.4*eye(3) 0.4*C5'; 0.4*C5 diag([0.06 0.06 0.04])];
    % Dynamic parameter matrices

W=J1'*U1*J1+J2'*U2*J2+J3'*U3*J3+J4'*U4*J4+J5'*U5*J5
    % Finally compute and print out the inertial matrix W
```

After running the above code in MATLAB^{TM}, the resultant 5 by 5 inertial matrix is immediately printed out, as given below:

```
W =

    6.9000   -2.5691    1.4640    0.5515         0
   -2.5691    2.1570         0         0    0.0394
    1.4640         0    4.1821    0.9991         0
    0.5515         0    0.9991    0.6160         0
         0    0.0394         0         0    0.0400
```

Through this MATLAB^{TM} programming example, it can be evidently convinced that equation (3.16) along with equation (3.14), indeed, offer an exclusive numerical algorithmic advantage to determine the inertial matrix W especially for a large-scaled multi-joint robotic system.

3.3.2 POTENTIAL ENERGY DUE TO GRAVITY AND OTHER FORMS

In general, a vector field $f(q)$ in $\mathbb{R}^n$ can be categorized into **conservative and non-conservative fields**.

Definition 3.3. A vector field f is conservative if and only if its integral along any path depends only on the two endpoints a and b, but not on the shape of the path, i.e.,

$$\int_a^b f(q) \cdot dq = P(b) - P(a).$$

To test a vector field f to see if it is conservative or not, the best way is to check if the integral along any closed loop C is equal to zero or not, i.e.,

$$\oint_C f(q) \cdot dq = 0 ? \tag{3.20}$$

If it is zero, we can always choose two distinct points a and b on C to split the closed loop C into two segments $C_1 + C_2 = C$ such that

$$\oint_C f(q) \cdot dq = \int_{a(C_1)}^b f(q) \cdot dq + \int_{b(C_2)}^a f(q) \cdot dq = 0,$$

This implies that

$$\int_{a(C_1)}^b f(q) \cdot dq = -\int_{b(C_2)}^a f(q) \cdot dq = \int_{a(C_2)}^b f(q) \cdot dq$$

for any arbitrary $C = C_1 + C_2$ with two common endpoints a and b. Based on Definition 3.3, $f(q)$ is a conservative vector field.

If the vector field is in 3-space $\mathbb{R}^3$, equation (3.20) can have its differential form: $\nabla \times f = 0$. According to calculus, it is called a *curl* of the vector field f. Often, the curl operator can be rewritten by a skew-symmetric differential operator, i.e.,

$$\nabla \times f = \begin{pmatrix} 0 & -\frac{\partial}{\partial z} & \frac{\partial}{\partial y} \\ \frac{\partial}{\partial z} & 0 & -\frac{\partial}{\partial x} \\ -\frac{\partial}{\partial y} & \frac{\partial}{\partial x} & 0 \end{pmatrix} f.$$

For instance, for any smooth scalar field $h(q)$ with $q = (x \ y \ z)^T$, its **gradient vector** is always conservative, because

$$\nabla \times \frac{\partial h(q)}{\partial q} = \begin{pmatrix} 0 & -\frac{\partial}{\partial z} & \frac{\partial}{\partial y} \\ \frac{\partial}{\partial z} & 0 & -\frac{\partial}{\partial x} \\ -\frac{\partial}{\partial y} & \frac{\partial}{\partial x} & 0 \end{pmatrix} \begin{pmatrix} \frac{\partial h}{\partial x} \\ \frac{\partial h}{\partial y} \\ \frac{\partial h}{\partial z} \end{pmatrix} \equiv 0.$$

In contrast, the vector field $g = (y \ -x \ z)^T$ is not conservative, because

$$
\nabla \times g = \begin{pmatrix} 0 & -\frac{\partial}{\partial z} & \frac{\partial}{\partial y} \\ \frac{\partial}{\partial z} & 0 & -\frac{\partial}{\partial x} \\ -\frac{\partial}{\partial y} & \frac{\partial}{\partial x} & 0 \end{pmatrix} \begin{pmatrix} y \\ -x \\ z \end{pmatrix} = \begin{pmatrix} 0 \\ 0 \\ -2 \end{pmatrix},
$$

which is a nonzero vector.

If a conservative vector field $f(q)$ represents a force vector or a torque vector, then based on Definition 3.3, there exists a **potential energy** P that is a scalar function $P(q)$ such that

$$
f = -\frac{\partial P}{\partial q}.
$$

Thus, any conservative force or torque vector field can always be determined by the above relation in terms of a potential energy P.

For example, a single body with mass m that is placed at a level of height $h(q)$ has its *gravitational potential energy* $P = mgh(q)$, where $g = 9.806$ m/sec^2 is the gravitational acceleration on the earth. Then,

$$
f_g = -\frac{\partial P}{\partial q} = -mg\frac{\partial h(q)}{\partial q}
$$

represents a force along the vector q. In particular, if this vector q is the height vector hz_0, then $f = -mgz_0$ is the gravity of the body, where z_0 is the unity vector of the z-axis of the base coordinate system.

Another example is a robot arm with n links. If the mass center of link i has the height $h_i(q)$ that is actually the z-component of the mass center position vector p_0^{ci}, then the joint torque vector due to gravity can be determined by

$$
\tau_g = -\frac{\partial P(q)}{\partial q} = -\sum_{i=1}^{n} m_i g \frac{\partial h_i(q)}{\partial q}.
$$

When making a kinematics model for a robot arm by the D-H convention, we can only find the position vector p_0^i with its arrow pointing at the origin of frame i that is attached on link i. In order to find p_0^{ci} whose arrow is pointing at the mass center of link i, we need first to project the vector c_i to the base frame 0, where c_i is tailed at the origin of frame i and arrow-pointing to the mass center of link i so that,

$$
p_0^{ci} = p_0^i + R_0^i c_i \quad \text{for } i = 1, \cdots, n.
$$

As an alternative way, we can also find the **center of gravity (CG)** first for a large-scale robotic system, such as a humanoid robot, and then to determine the conservative joint torque vector τ. If we know the mass center position p_0^{ci} and the mass m_i for link i, then, the position vector of the CG of the entire robot with respect to the base should be

$$
p_0^{cg} = \frac{\sum_{i=1}^{n} m_i p_0^{ci}}{\sum_{i=1}^{n} m_i} = \frac{\sum_{i=1}^{n} m_i p_0^{ci}}{M}
$$

with the total mass of the robot $M = m_1 + \cdots + m_n$.

The total gravitational potential energy P for the robot can also be found as $P = Mgh_{cg}$, where h_{cg} is the z-component of the CG position vector p_0^{cg}. If this h_{cg} is a function of the joint position vector q, the joint torque due to gravity can be determined by

$$\tau_g = -\frac{\partial P}{\partial q} = -Mg\frac{\partial h_{cg}}{\partial q}.$$

In fact, finding the CG position vector p_0^{cg} is key important for a humanoid robot in its balance control process under a fall avoidance criterion.

In addition to gravity, a spring system can also be modeled in the same way. Suppose that each joint of a robot has a torsional spring with its spring constant k_i. Then, each joint is acted with a spring torque $\tau_{sp}^i = -k_i\theta_i = -k_iq_i$, and the total joint torque vector due to the torsional spring is

$$\tau_{sp} = -\begin{pmatrix} k_1 & & O \\ & \ddots & \\ O & & k_n \end{pmatrix} q = -Kq$$

with the n by n diagonal spring constant matrix K. Obviously, its potential energy should be the following quadratic form:

$$P_{sp} = \frac{1}{2}q^T Kq.$$

The conservative joint torque vector can also be determined by

$$\tau_{sp} = -\frac{\partial P_{sp}}{\partial q} = -\frac{1}{2}(K + K^T)q = -Kq,$$

because of the symmetry of K.

There are many different forms of potential energies and their corresponding conservative forces and torques. A multi-coil system after each coil gets a current in has the following potential energy stored in the system:

$$P_m = \frac{1}{2}I^T LI,$$

where the current vector $I = (i_1 \cdots i_n)^T$, and the n by n symmetric inductance matrix L contains every self-inductance on the diagonal and every mutual inductance on the off-diagonal corners. If x is a movable displacement involved in some coil, then the magnetic force can be determined by

$$f_m = -\frac{\partial P_m}{\partial x} = -\frac{1}{2}I^T\left(\frac{\partial L}{\partial x}\right)I.$$

Similarly, the potential energy P_e stored in a multi-capacitor system can be expressed by

$$P_e = \frac{1}{2}V^T CV,$$

where the voltage vector $V = (v_1 \cdots v_n)^T$, and the n by n diagonal capacitance matrix C contains every capacitance of the n capacitors on the diagonal of the matrix C without mutual static electric interaction. If x is a movable displacement, such as a distance between two parallel plates of some capacitor, then the static electric force can be found as

$$f_e = -\frac{\partial P_e}{\partial x} = -\frac{1}{2}V^T \left(\frac{\partial C}{\partial x}\right) V.$$

However, if a force or torque is non-conservative, the work done by such force or torque will no longer be returned back to add up the total energy of the system. Therefore, we will only consider conservative forces and torques in our future dynamics modeling and analysis.

3.3.3 CONSISTENCY BETWEEN THE LAGRANGE AND GEODESIC EQUATIONS

The great Italian mathematician and astronomer Joseph-Louis Lagrange (1736–1813), a later naturalized French, made multiple significant contributions to the fields of mathematical analysis, number theory, and classical and celestial mechanics. He was one of the key creators of the **Calculus of Variations** to derive and solve the well-known Euler-Lagrange equations for extrema of functionals.

The initial notion and motivation in calculus of variations were to find an extremal curve (arc length) to minimize the so-called **action** that is defined by

$$S(\gamma) = \int_a^b L(q(t), \dot{q}(t)) dt,$$

which is a function of curves γ so that $S(\gamma)$ is a functional, instead of just a function.

Suppose that an extremal curve $q(t)$ for $t \in [a,b]$ is found such that any small curve change $y = q + \varepsilon \eta$ can always make $S(q) \leq S(q + \varepsilon \eta)$, where $\eta(t)$ is an arbitrary curve with the zero endpoints, i.e., $\eta(a) = \eta(b) = 0$. Thus,

$$\frac{dS(q + \varepsilon \eta)}{d\varepsilon}\bigg|_{\varepsilon=0} = \int_a^b \frac{dL}{d\varepsilon}\bigg|_{\varepsilon=0} dt = 0.$$

However,

$$\frac{dL}{d\varepsilon} = \frac{dL}{dy}\frac{dy}{d\varepsilon} + \frac{dL}{d\dot{y}}\frac{d\dot{y}}{d\varepsilon} = \frac{dL}{dy}\eta + \frac{dL}{d\dot{y}}\dot{\eta}.$$

Then, applying the integration by parts on the second term, we obtain

$$\int_a^b \frac{dL}{d\varepsilon}\bigg|_{\varepsilon=0} dt = \int_a^b \frac{dL}{dq}\eta dt + \frac{dL}{d\dot{q}}\eta\bigg|_a^b - \int_a^b \frac{d}{dt}\left(\frac{dL}{d\dot{q}}\right)\eta dt = 0.$$

Since $\frac{dL}{d\dot{q}}\eta\big|_a^b = 0$, considering η is an arbitrary curve connecting to the two endpoints, we conclude that

$$\frac{dL}{dq} - \frac{d}{dt}\left(\frac{dL}{d\dot{q}}\right) = 0. \tag{3.21}$$

This is the well-known Euler-Lagrange equation to solve for the extremal curve $q(t)$ to minimize the action $S(\gamma)$ [1, 26, 27].

The scalar function $L(q, \dot{q})$ inside the integral of the action $S(\gamma)$ is called a **Lagrangian**. If it is defined as a Riemannian metric $L(q, \dot{q}) = \frac{1}{2} w_{ij} \dot{q}^i \dot{q}^j = \frac{1}{2} \langle \dot{q}, \dot{q} \rangle_w$ in a tensor form endowed on a smooth manifold M^n, then the first term of the Euler-Lagrange equation (3.21) is

$$\frac{dL}{dq^k} = \frac{1}{2} \frac{\partial w_{ij}}{\partial q^k} \dot{q}^i \dot{q}^j,$$

while the second term becomes

$$\frac{d}{dt}\left(w_{jk} \dot{q}^j\right) = w_{jk} \ddot{q}^j + \frac{\partial w_{jk}}{\partial q^i} \dot{q}^i \dot{q}^j.$$

Combining the two terms together yields

$$w_{jk} \ddot{q}^j + \left(\frac{\partial w_{jk}}{\partial q^i} - \frac{1}{2} \frac{\partial w_{ij}}{\partial q^k}\right) \dot{q}^i \dot{q}^j = 0.$$

Furthermore, it can be easily verified that

$$\frac{\partial w_{jk}}{\partial q^i} \dot{q}^i \dot{q}^j = \frac{1}{2}\left(\frac{\partial w_{ik}}{\partial q^j} + \frac{\partial w_{jk}}{\partial q^i}\right) \dot{q}^i \dot{q}^j.$$

By recalling the Christoffel symbol in Chapter 2,

$$\Gamma^l_{ij} = \frac{1}{2} w^{kl}\left(\frac{\partial w_{ik}}{\partial q^j} + \frac{\partial w_{jk}}{\partial q^i} - \frac{\partial w_{ij}}{\partial q^k}\right),$$

we reach to the geodesic equation (2.27), i.e.,

$$\ddot{q}^l + \Gamma^l_{ij} \dot{q}^i \dot{q}^j = 0.$$

Therefore, the above justification has shown the following theorem:

Theorem 3.2. *If the Lagrangian in the action $S(\gamma)$ is a Riemannian metric on some manifold, i.e., $L(q, \dot{q}) = w_{ij} \dot{q}^i \dot{q}^j = \langle \dot{q}, \dot{q} \rangle_w$, then the Euler-Lagrange equation for the extremal curves is equivalent to the geodesic equation related to the metric w_{ij}.*

In the above theorem and proof, with a half and without a half in the coefficient of Riemannian metric does not affect the equivalence conclusion, because both the Euler-Lagrange and geodesic equations are homogeneous. If we translate the tensor form to the matrix form, then $\frac{1}{2} \dot{q}^T W \dot{q}$ in the proof is just a kinetic energy K.

It can also be justified that if the Lagrangian $L = K - P$, the difference between a kinetic energy K and potential energy P, then the Euler-Lagrange equation (3.21) will be non-homogeneous, which is consistent to the Newton's equation of dynamics as well as the Hamilton's principle in classical mechanics. Based on equations (2.30) and (2.31), if taking all the external forces/torques into consideration, the Euler-Lagrange equation with $L = K - P$ in the matrix form turns out to be

$$W\ddot{q} + \frac{1}{2}(\dot{W} + W_d^T - W_d)\dot{q} = W\ddot{q} + \left(W_d^T - \frac{1}{2}W_d\right)\dot{q} = \tau + \tau_g, \qquad (3.22)$$

where W_d is the **derivative matrix**, as defined in equation (2.28), i.e.,

$$W_d = \begin{pmatrix} \dot{q}^T \frac{\partial W}{\partial q^1} \\ \vdots \\ \dot{q}^T \frac{\partial W}{\partial q^n} \end{pmatrix}, \qquad (3.23)$$

τ is any external force/torque, and $\tau_g = -\frac{\partial P}{\partial q}$ is the force/torque due to gravity.

3.4 DYNAMIC FORMULATIONS FOR ROBOTIC SYSTEMS

3.4.1 DETERMINATION OF CENTRIFUGAL AND CORIOLIS TERMS

According to the dynamic equation in the matrix form, as given in (3.22), finding the inertial matrix W for a robotic system to be dynamically modeled is pivotally important. The first term $W\ddot{q}$ is an inertial force/torque, while the second term

$$\frac{1}{2}(\dot{W} + W_d^T - W_d)\dot{q} = \left(W_d^T - \frac{1}{2}W_d\right)\dot{q}$$

represents all the centrifugal and Coriolis forces/torques. In order to compute them, one must determine the derivative matrix W_d first, which requires to take derivatives of the inertial matrix W with respect to each joint position q_i. Clearly, if we have a symbolical form of W, then those derivatives can be found exactly. However, if only a numerical inertial matrix W is available, we would either take advantage of the dual number computation to determine exactly the numerical derivative matrix W_d if MATLAB^{TM} could have the dual number algorithm subroutine built-in, or a lower-order approximation may have to be programmed in MATLAB^{TM}.

Let us now look back each example as illustrated in the previous section for finding the inertial matrix W. The first three examples have found the symbolical forms of W, because they are all in a 2D planar case, while the last one has to calculate its W numerically due to the 3D complication. Although using equation (3.16) with the sub-Jacobian matrices equation in (3.14) can compute an accurate inertial matrix W numerically, no matter how large scale the robot is, the determination of the derivative matrix W_d may still have to face a lower-order approximation challenge.

In the first inverted pendulum example, its 2 by 2 inertial matrix W has been symbolically found in (3.17), which is a function of the pendulum rotating angle θ only and independent of the cart displacement x. Thus, the first derivative $\frac{\partial W}{\partial x} = 0$. The second one can be readily derived as

$$\frac{\partial W}{\partial \theta} = \begin{pmatrix} 0 & -m_2 l_c \sin \theta \\ -m_2 l_c \sin \theta & 0 \end{pmatrix}.$$

Then, the derivative matrix becomes

$$W_d = \begin{pmatrix} (\dot{x} \ \dot{\theta}) \frac{\partial W}{\partial x} \\ (\dot{x} \ \dot{\theta}) \frac{\partial W}{\partial \theta} \end{pmatrix} = \begin{pmatrix} 0 & 0 \\ -m_2 l_c \sin\theta\,\dot{\theta} & -m_2 l_c \sin\theta\,\dot{x} \end{pmatrix}.$$

Substituting this W_d into the dynamic equation (3.22) yields

$$W\ddot{q} + \left(W_d^T - \frac{1}{2} W_d \right)\dot{q} = W\ddot{q} + \begin{pmatrix} 0 & -m_2 l_c \sin\theta\,\dot{\theta} \\ \frac{1}{2} m_2 l_c \sin\theta\,\dot{\theta} & -\frac{1}{2} m_2 l_c \sin\theta\,\dot{x} \end{pmatrix}\begin{pmatrix} \dot{x} \\ \dot{\theta} \end{pmatrix}$$

$$= W\ddot{q} - \begin{pmatrix} m_2 l_c \sin\theta\,\dot{\theta}^2 \\ 0 \end{pmatrix} = \tau + \tau_g.$$

As can be seen from this result, all the Coriolis terms containing $\dot{x}\dot{\theta}$ are canceled out and only a centrifugal term containing $\dot{\theta}^2$ remains in the dynamic equation. Moreover, since the height of the inverted pendulum mass center is $h = l_c \cos\theta$, the joint torque due to gravity becomes

$$\tau_g = -\frac{\partial m_2 g h}{\partial q} = \begin{pmatrix} 0 \\ m_2 g l_c \sin\theta \end{pmatrix}.$$

Finally, the complete dynamic equation for the inverted pendulum system is achieved to be

$$W\ddot{q} - \begin{pmatrix} m_2 l_c \sin\theta\,\dot{\theta}^2 \\ m_2 g l_c \sin\theta \end{pmatrix} = \tau,$$

where W is given by (3.17).

The second example is a planar robot arm with two revolute joints, and its inertial matrix W has also been found symbolically in (3.18), which is a function of the second joint angle θ_2 only. Thus,

$$\frac{\partial W}{\partial \theta_1} = 0, \quad \text{and} \quad \frac{\partial W}{\partial \theta_2} = \begin{pmatrix} -2m_2 a_1 b_2 s_2 & -m_2 a_1 b_2 s_2 \\ -m_2 a_1 b_2 s_2 & 0 \end{pmatrix}.$$

Then,

$$W_d = \begin{pmatrix} (\dot{\theta}_1 \ \dot{\theta}_2) \frac{\partial W}{\partial \theta_1} \\ (\dot{\theta}_1 \ \dot{\theta}_2) \frac{\partial W}{\partial \theta_2} \end{pmatrix} = \begin{pmatrix} 0 & 0 \\ -m_2 a_1 b_2 s_2 (2\dot{\theta}_1 + \dot{\theta}_2) & -m_2 a_1 b_2 s_2 \dot{\theta}_1 \end{pmatrix}.$$

The centrifugal and Coriolis terms can be found as

$$c(q,\dot{q}) = \left(W_d^T - \frac{1}{2} W_d \right)\dot{q} = m_2 a_1 b_2 s_2 \begin{pmatrix} 0 & -(2\dot{\theta}_1 + \dot{\theta}_2) \\ \frac{1}{2}(2\dot{\theta}_1 + \dot{\theta}_2) & -\frac{1}{2}\dot{\theta}_1 \end{pmatrix}\begin{pmatrix} \dot{\theta}_1 \\ \dot{\theta}_2 \end{pmatrix}$$

$$= m_2 a_1 b_2 s_2 \begin{pmatrix} -2\dot{\theta}_1 \dot{\theta}_2 - \dot{\theta}_2^2 \\ \dot{\theta}_1^2 \end{pmatrix}.$$

According to Figure 3.2, the gravitational potential energy can be derived as

$$P = m_1 g b_1 s_1 + m_2 g (a_1 s_1 + b_2 s_{12}),$$

and the joint torque due to gravity is found to be

$$\tau_g = -\frac{\partial P}{\partial q} = -\begin{pmatrix} m_1 g b_1 c_1 + m_2 g(a_1 c_1 + b_2 c_{12}) \\ m_2 g b_2 c_{12} \end{pmatrix}. \tag{3.24}$$

Then, the final dynamic equation for the two-link planar robot arm becomes

$$W\ddot{q} + c(q,\dot{q}) - \tau_g = \tau.$$

The third example is the 2-joint robot arm sitting on the cylindrical log to form a 3-joint planar robotic system. Its inertial matrix W is symbolically derived and given in equation (3.19). Then,

$$\frac{\partial W}{\partial \theta_1} = \begin{pmatrix} -2(m_2 + m_3)a_1 R s_1 + 2m_3 d_3 R c_{12} & m_3 d_3 R c_{12} & m_3 R s_{12} \\ m_3 d_3 R c_{12} & 0 & 0 \\ m_3 R s_{12} & 0 & 0 \end{pmatrix},$$

$$\frac{\partial W}{\partial \theta_2} = \begin{pmatrix} 2m_3 d_3 (R c_{12} + a_1 c_2) & m_3 d_3 (R c_{12} + a_1 c_2) & m_3 (R s_{12} + a_1 s_2) \\ m_3 d_3 (R c_{12} + a_1 c_2) & 0 & 0 \\ m_3 (R s_{12} + a_1 s_2) & 0 & 0 \end{pmatrix},$$

and

$$\frac{\partial W}{\partial d_3} = \begin{pmatrix} 2m_3 (d_3 + R s_{12} + a_1 s_2) & m_3 (2d_3 + R s_{12} + a_1 s_2) & 0 \\ m_3 (2d_3 + R s_{12} + a_1 s_2) & 2m_3 d_3 & 0 \\ 0 & 0 & 0 \end{pmatrix}.$$

In addition, by a direct geometrical inspection on Figure 3.3, the gravitational potential energy can be found as

$$P = (m_2 + m_3)ga_1 c_1 + m_3 g d_3 s_{12}.$$

Then, the joint torque due to gravity is determined by

$$\tau_g = -\frac{\partial P}{\partial q} = -\begin{pmatrix} -(m_2 + m_3)ga_1 s_1 + m_3 g d_3 c_{12} \\ m_3 g d_3 c_{12} \\ m_3 g s_{12} \end{pmatrix}.$$

After all the derivatives are calculated by hand, no more derivatives need to be taken and everything is now just asking MATLAB^{TM} to finish the remaining numerical computations, such as to calculate W_d and $c(q,\dot{q})$.

In order to verify the accuracy of the approximation in finding the derivative matrix W_d, let us work on this log-arm system by employing the following *two-point symmetric difference quotient* numerical computation for each derivative $\frac{\partial W}{\partial q_i}$ for $i = 1, \cdots, n$:

$$\frac{\partial W}{\partial q_i} \approx \frac{W(q + h_i) - W(q - h_i)}{2h},$$

where h is the time step size, and h_i means that h is added or subtracted only on the i-th element of q. Since the symbolical form of W_d is available, we can compare the accuracy between the symbolical result and the two-point approximation outcome.

Let the step size be $h = 0.01$, and let the joint positions be $q = (0.2\ 0.3\ 0.2)^T$, where the first two are the angles in radians and the third one is d_3 in meters. Also, set the joint velocity $\dot{q} = (-1\ 1.5\ 0.2)^T$ at a time instant. Then, the accurate results W and W_d by using the above symbolical equations are calculated from MATLAB^{TM} as follows:

```
W =

    35.3554      1.4157     -4.6212
     1.4157      1.0467          0
    -4.6212           0     2.0000

Wd =

     1.8043     -0.3510     -0.9589
    -0.0930     -0.9242     -1.8454
    -0.5227     -1.4454          0
```

Whereas the two-point approximation result for W_d is printed out as

```
Wd_2 =

     1.8042     -0.3510     -0.9588
    -0.0930     -0.9242     -1.8454
    -0.5227     -1.4454          0
```

After comparing the two W_d results, it appears to be acceptable for such a two-point approximation at the medium joint velocity $\dot{q}$.

We now have more confidence to adopt the same approximation algorithm to find the numerical derivative matrix W_d for the (4+1)-joint robot manipulator from MATLAB^{TM}. Under the same joint positions, and set the joint velocity to be $\dot{q} = (0.1\ 0.7\ -0.8\ -0.8\ 0)^T$, where the first element is the sliding speed $\dot{d}_1$ along the beam in m/sec, and the rest are all the joint angular speeds in rad/sec, but set both $\dot{\theta}_5 = 0$ and $\dot{\theta}_5 = 0$. The following printed results show both W and W_d values from MATLAB^{TM}:

```
W =

     6.9000     -2.5691      1.4640      0.5515           0
    -2.5691      2.1570           0           0      0.0394
     1.4640           0      4.1821      0.9991           0
     0.5515           0      0.9991      0.6160           0
          0      0.0394           0           0      0.0400

Wd_2 =

          0           0           0           0           0
     1.7544     -0.1483     -0.2536     -0.0955           0
    -0.6662      2.4032     -0.1483      0.0097      0.0049
    -0.8242      1.1517     -2.5161     -0.8322      0.0049
          0           0           0           0           0
```

Obviously, this W is exactly same as the previous result due to the same joint positions. Also, the first row of W_d is zero because the sliding position d_1 has no contribution to the inertial matrix W so that $\frac{\partial W}{\partial d_1} = 0$.

One can also employ the following five-point approximation to find the derivative matrix W_d:

$$\frac{\partial W}{\partial q_i} \approx \frac{W(q-2h) - 8W(q-h) + 8W(q+h) - W(q+2h)}{12h},$$

if the accuracy becomes more critical. However, for the regular purpose of finding W_d, the two-point method is good enough.

3.4.2 DYNAMICS MODELING FOR A VARIETY OF ROBOTIC SYSTEMS

Since a complete kinematics model for a robotic system is the infrastructure for its dynamics modeling, for an open serial-chain robot, its kinematics model can always be developed by following the D-H convention, and its dynamics modeling can then be achieved as a follow-up effort. Obviously, the major procedure and formulation given in equation (3.22) cannot be fulfilled without a completion of kinematics modeling. Now, is the systematic D-H convention only applicable to the robotic systems with open serial-chain mechanism? In this section, we will provide three typical examples to illustrate and convince that the D-H convention can also be a powerful tool to model the kinematics for the closed serial/parallel hybrid-chain robotic systems. Let us start looking at a planar 3-bar closed mechanical system, as depicted in Figure 3.5.

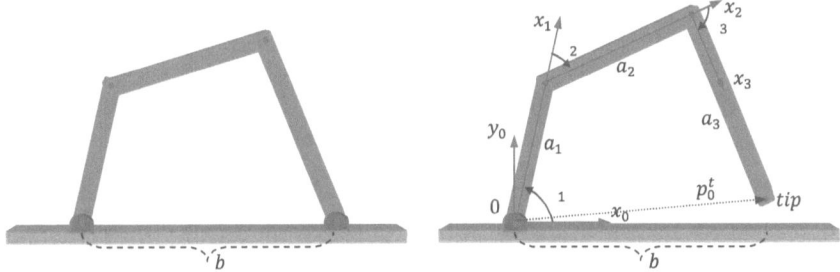

Figure 3.5   A three-bar closed serial-chain planar system

Example 3.5. In mechanical engineering, there was a well-known equation, called the Grubler's formula, to determine the net d.o.f. or the degree of mobility r for any type of chained multi-body systems with n joints and l links except the fixed base link [6, 25, 31, 32]. This formula can be expressed as follows:

$$r = D(l-n) + \sum_{i=1}^{n} f_i,$$

where each f_i is the number of axes for joint i, and $D = 3$ in the 2D planar cases and $D = 6$ in the 3D cases.

The above Grubler's formula for most of the open serial-chain robotic systems is quite trivial, because their number of joints n is always equal to the number of links so that the net d.o.f. r is just equal to the total number of axes. Now, let us try to find r for this closed 3-bar mechanism. Since $n = 4$, $l = 3$ and $D = 3$ in this particular planar case, the first term of the Grubler's formula is -3. Because each joint only offers a single axis of movement, the second term is 4, and thus, $r = 1$. This indicates that the 3-bar closed-chain planar system can have only a single d.o.f. of motion.

In fact, one can intuitively observe from Figure 3.5 that if the right-side bottom joint on the floor is removed to make the tip point of link 3 floating, then it becomes a typical 3-joint or 3-link planar open serial-chain robot arm. Its 2-dimensional tip position vector with respect to the base is p_0^t, as depicted in the right picture of Figure 3.5. By following the D-H convention, we can find an explicit form of p_0^t below:

$$p_0^t = \begin{pmatrix} a_1 c_1 + a_2 c_{12} + a_3 c_{123} \\ a_1 s_1 + a_2 s_{12} + a_3 s_{123} \end{pmatrix}. \tag{3.25}$$

Its Jacobian matrix can also be derived by

$$J_0 = \frac{\partial p_0^t}{\partial q} = \begin{pmatrix} -a_1 s_1 - a_2 s_{12} - a_3 s_{123} & -a_2 s_{12} - a_3 s_{123} & -a_3 s_{123} \\ a_1 c_1 + a_2 c_{12} + a_3 c_{123} & a_2 c_{12} + a_3 c_{123} & a_3 c_{123} \end{pmatrix},$$

which is 2 by 3 dimensional. Based on the analysis of redundant robotic systems in Chapter 2, the 3-bar system after the tip point is floated possesses $(r = 1)$ degree of redundancy, or the null space N^r is $(r = 1)$-dimensional, which is exactly consistent to what was determined from the Grubler's formula.

Actually, the tip point of the 3-link open serial-chain robot is connected to the right-side bottom joint. This fact implies that the tip position $p_0^t = \begin{pmatrix} b \\ 0 \end{pmatrix}$ is locked as a constant vector. In other words, the last link (link 3) of the 3-joint open serial-chain robot can only be rotated about its tip point at the axis of the bottom joint, but cannot change its position. Thus, the 2D linear velocity of the tip point $v = \dot{p}_0^t = 0$. Under this situation, the Jacobian equation becomes homogeneous:

$$J_0 \dot{q} = \dot{p}_0^t = 0 \quad \text{such that} \quad \dot{q} = (I - J_0^+ J_0)\eta.$$

Namely, the homogeneous Jacobian equation has only a null solution without the rank solution.

Since the dimension of the null space is $r = 1$, this 3-bar mechanism has only a single degree of mobility. We may install a motor to drive any one of the four joints to control one d.o.f. motion output $y = h(q) \in \mathbb{R}^1$. Suppose that let θ_1 be controlled so that both $s_1 = \sin \theta_1$ and $c_1 = \cos \theta_1$ are given. Also, let the motion output $y = h(q)$, as a control target, be the orientation of the top bar, i.e., the angle of the x_2-axis with respect to the horizontal direction x_0-axis, which should be $y = \theta_1 + \theta_2$. Then, based on equation (3.25) and noticing that $p_0^t = \begin{pmatrix} b \\ 0 \end{pmatrix}$,

$$\begin{cases} a_2 c_{12} + a_3 c_{123} = b - a_1 c_1 = p_1 \\ a_2 s_{12} + a_3 s_{123} = -a_1 s_1 = p_2. \end{cases}$$

This actually leads to a problem of inverse kinematics (IK) to solve for $y = \theta_1 + \theta_2$. Let both sides of each equation be squared before adding them together. We obtain

$$a_2^2 + a_3^2 + 2a_2 a_3 c_3 = (b - a_1 c_1)^2 + (-a_1 s_1)^2 = p_1^2 + p_2^2,$$

and thus,

$$c_3 = \cos \theta_3 = \frac{p_1^2 + p_2^2 - a_2^2 - a_3^2}{2a_2 a_3}.$$

This system has a multi-configuration issue due to its singularity. If we keep the closed 3-bar mechanism being always a convex shape, as displayed in Figure 3.5, the angle θ_3 from x_2 to x_3 should be negative, but it is in the 3rd or 4th quadrant, i.e., $-180° < \theta_3 < 0°$ so that

$$s_3 = \sin \theta_3 = -\sqrt{1 - \cos^2 \theta_3}.$$

Once we determine both s_3 and c_3, substituting the trigonometric identities $c_{123} \equiv c_{12} c_3 - s_{12} s_3$ and $s_{123} \equiv s_{12} c_3 + c_{12} s_3$ into the above simultaneous equation yields

$$\begin{cases} (a_2 + a_3 c_3) c_{12} - a_3 s_3 s_{12} = p_1 \\ (a_2 + a_3 c_3) s_{12} + a_3 s_3 c_{12} = p_2, \end{cases}$$

or

$$\begin{pmatrix} a_2 + a_3 c_3 & -a_3 s_3 \\ a_3 s_3 & a_2 + a_3 c_3 \end{pmatrix} \begin{pmatrix} c_{12} \\ s_{12} \end{pmatrix} = \begin{pmatrix} p_1 \\ p_2 \end{pmatrix}.$$

Thus, both s_{12} and c_{12} can be solved uniquely except at the singular point, and

$$y = h(q) = \theta_1 + \theta_2 = \text{atan2}(s_{12}, c_{12}),$$

where $\text{atan2}(\cdot, \cdot)$ is the four-quadrant arc tangent function in MATLAB^{TM}.

The above approach is the IK based Cartesian motion planning to find the output $y = \theta_1 + \theta_2$ in terms of the input $u = \theta_1$. However, an alternative way is the null-solution based differential motion planning to find the update of each angle, i.e.,

$$\dot{q} = (I - J_0^+ J_0) \eta,$$

where η is the gradient vector of a potential function $p(q)$, which can be defined as a distance square between a desired θ_1^d and the actual θ_1, i.e., $p(q) = (\theta_1^d - \theta_1)^2$ to be minimized. Then,

$$\eta = \frac{\partial p(q)}{\partial q} = 2 \begin{pmatrix} \theta_1 - \theta_1^d \\ 0 \\ 0 \end{pmatrix}.$$

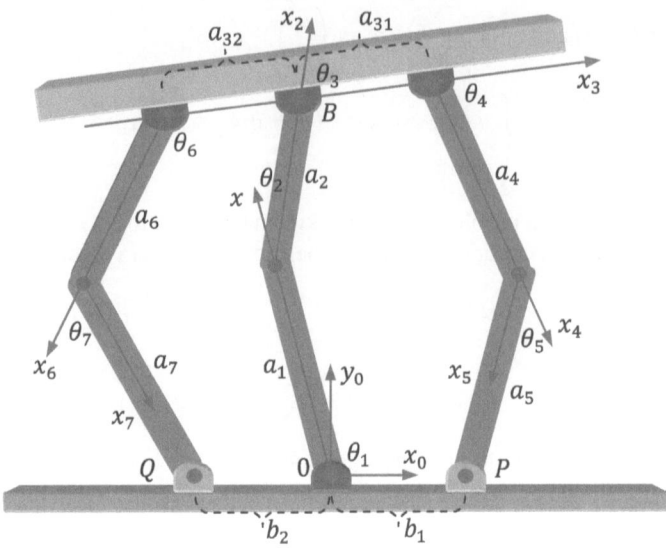

Figure 3.6　A 2D planar three-leg closed serial/parallel hybrid-chain system

Example 3.6. The second example is a 3-leg closed planar mechanical system, as shown in Figure 3.6. The number of joints is $n = 9$ and the number of links except the base link is $l = 7$. Since each joint offers a single axis and the entire system is moving in 2D space, based on the Grubler formula, the net d.o.f., or the degree of mobility $r = 3(7 - 9) + 9 = 3$.

Under the similar idea to the first example, we can imagine this mechanism as an open serial/parallel hybrid-chain planar system. Namely, among the three joints that touch down on the floor, let the middle one be assigned as a fixed base with frame 0 attached, while the two side joints P and Q are floating by taking off their joint connectors from the floor. Then, this system becomes an open hybrid-chain mechanism with P and Q as two feet of the left and right legs. Similar to the kinematics modeling procedure for a humanoid robot in Chapter 2, where we assigned 6 joints to represent the entire humanoid body translations and rotations before reaching to the H-triangle at the hip area, and then to split the frame assignment into 4 paths: the left leg, right leg, left arm and right arm, as illustrated in Figure 2.7. In this planar 3-leg system, after the central leg reaches its third link that is a common top bar for all the three legs, the assignment is divided to two opposite directions: one is going to the left leg through a_{31}, a_4, and a_5 to reach the left foot P, and the other one is going to the right leg via a_{32}, a_6, and a_7 to arrive at the right foot Q. This virtually open parallel/serial hybrid-chain structure can be modeled by the D-H convention. Similar to Table 2.8, its the D-H table is given in Table 3.1.

Joint Angle θ_i	Joint Offset d_i	Twist Angle α_i	Link Length a_i
θ_1	0	0	a_1
θ_2	0	0	a_2

θ_i	d_i	α_i	a_i	θ_i	d_i	α_i	a_i
θ_3	0	0	$-a_{32}$	θ_3	0	0	a_{31}
θ_6	0	0	a_6	θ_4	0	0	a_4
θ_7	0	0	a_7	θ_5	0	0	a_5

Table 3.1

The D-H table for the 3-leg closed-chain planar system

Since it is a planar system, every z-axis is normal to the paper, and each one-step homogeneous transformation has a common matrix form for $i = 1, \cdots, 7$,

$$A_{i-1}^i = \begin{pmatrix} c_i & -s_i & 0 & a_i c_i \\ s_i & c_i & 0 & a_i s_i \\ 0 & 0 & 1 & 0 \\ 0 & 0 & 0 & 1 \end{pmatrix}.$$

However, the homogeneous transformation from link 2 to link 3 is split to the left leg and right leg, and their matrices should be

$$A_2^{3l} = \begin{pmatrix} c_3 & -s_3 & 0 & a_{31}c_3 \\ s_3 & c_3 & 0 & a_{31}s_3 \\ 0 & 0 & 1 & 0 \\ 0 & 0 & 0 & 1 \end{pmatrix}, \quad \text{and} \quad A_2^{3r} = \begin{pmatrix} c_3 & -s_3 & 0 & -a_{32}c_3 \\ s_3 & c_3 & 0 & -a_{32}s_3 \\ 0 & 0 & 1 & 0 \\ 0 & 0 & 0 & 1 \end{pmatrix}.$$

Similar to the first example, the position vector of the tip point P with respect to the base can be derived as

$$p_0^P = \begin{pmatrix} a_1 c_1 + a_2 c_{12} + a_{31} c_{123} + a_4 c_{1234} + a_5 c_{12345} \\ a_1 s_1 + a_2 s_{12} + a_{31} s_{123} + a_4 s_{1234} + a_5 s_{12345} \end{pmatrix}.$$

Also, the position vector of the tip point Q with respect to the base can be derived by

$$p_0^Q = \begin{pmatrix} a_1 c_1 + a_2 c_{12} - a_{32} c_{123} + a_6 c_{1236} + a_7 c_{12367} \\ a_1 s_1 + a_2 s_{12} - a_{32} s_{123} + a_6 s_{1236} + a_7 s_{12367} \end{pmatrix}.$$

Their Jacobian matrices J_0^P and J_0^Q can be directly determined by taking their derivatives with respect to q. Then, the two Jacobian matrices are partitioned into four Jacobian blocks: J_0^{3l} for the first three links ended at the origin of frame 3, J_0^{3r} for the first three links ended at the origin of frame 6, J_0^{45} for link 4 and link 5, and J_0^{67} for link 6 and link 7.

After the above preparation, the overall Jacobian matrix for this virtual 3-leg open hybrid-chain planar robot can be augmented by

$$J_0 = \begin{pmatrix} J_0^{3l} & J_0^{45} & O \\ J_0^{3r} & O & J_0^{67} \end{pmatrix},$$

where O is the 2 by 2 zero matrix. Obviously, this overall Jacobian is a 4 by 7 matrix, which will offer a 3-dimensional null space N^3. Since the linear velocities $v = 0$ at both the virtual floating points P and Q, the Jacobian equation $J_0 \dot{q} = 0$ is homogeneous so that its rank solution vanishes, while the null solution is given by:

$$\dot{q} = (I - J_0^+ J_0)\eta$$

with the 7 by 7 identity I. At this end, the degree of redundancy $r = 3$ is exactly consistent to the degree of mobility by the Grubler's formula.

This conclusion also implies that such a 3-leg planar mechanism can be controlled by three joint drives as a 3-dimensional input u to meet a 3-dimensional desired output y. For instance, we may pick the 2D position vector p_0^B and a scalar orientation of the top bar (link 3) as the output y, while the bottom three joints P, the base origin 0 and Q are actuated to drive and control the system to match the desired y. In this case, since the scalar orientation of the top bar should be $\beta_{13} = \theta_1 + \theta_2 + \theta_3$ based on the kinematic model,

$$y = h(q) = \begin{pmatrix} a_1 c_1 + a_2 c_{12} \\ a_1 s_1 + a_2 s_{12} \\ \theta_1 + \theta_2 + \theta_3 \end{pmatrix}. \tag{3.26}$$

Also, according to the kinematic model, the angle at point P from x_0 to x_5 axis is $q_{05} = \theta_1 + \theta_2 + \theta_3 + \theta_4 + \theta_5$, while the angle at point Q from x_0 to x_7 axis is $q_{07} = \theta_1 + \theta_2 + \theta_3 + \theta_6 + \theta_7$. We may thus define a potential function to be minimized as follows:

$$p(q) = \frac{1}{2}\left[(\theta_1^d - \theta_1)^2 + (q_{05}^d - q_{05})^2 + (q_{07}^d - q_{07})^2 \right],$$

such that the gradient vector becomes

$$\eta = \frac{\partial p(q)}{\partial q} = \begin{pmatrix} 1 & 1 & 1 \\ 0 & 1 & 1 \\ 0 & 1 & 1 \\ 0 & 1 & 0 \\ 0 & 1 & 0 \\ 0 & 0 & 1 \\ 0 & 0 & 1 \end{pmatrix} \begin{pmatrix} \theta_1 - \theta_1^d \\ q_{05} - q_{05}^d \\ q_{07} - q_{07}^d \end{pmatrix}.$$

Alternatively, we can also define a potential function to minimize the distance between the real output y and the desired output y^d, i.e.,

$$p(q) = \frac{1}{2}(y^d - y)^T(y^d - y).$$

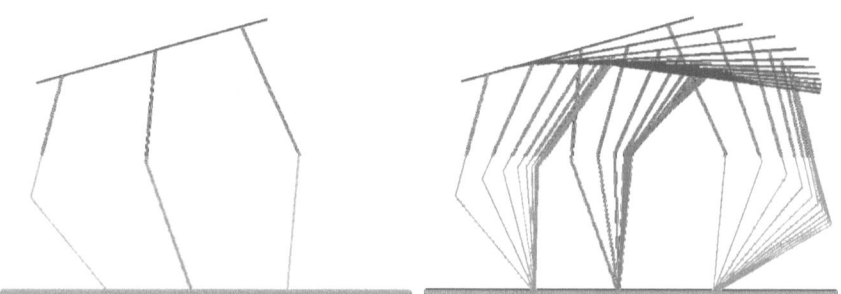

Figure 3.7 An animation study for the planar three-leg closed hybrid-chain system

Its gradient vector becomes:

$$\eta = \frac{\partial p(q)}{\partial q} = \left(\frac{\partial y}{\partial q}\right)^T (y - y^d),$$

where the output y is given by equation (3.26).

An animation study has been performed in MATLAB^{TM}. The plots are depicted in Figure 3.7, where the left picture is the initial posture, and the right one is an animation recording as the null solution controls both the position and orientation of the top bar. At the initial posture,

$$y(0) = \begin{pmatrix} p_0^B(0) \\ \beta_{13}(0) \end{pmatrix} = \begin{pmatrix} -0.2723 \\ 1.7366 \\ 0.2618 \end{pmatrix},$$

while the desired output was specified as

$$y^d = \begin{pmatrix} 0.4 \\ 1.2 \\ -0.4 \end{pmatrix},$$

where all the lengths are in meters and the angle β_{13} is in radians.

Example 3.7. This new system has a 3D closed parallel/serial hybrid-chain mechanism, as shown in Figure 3.8. The bottom three joints are all universal joints, each of which offers three axes. Each of the three legs has a prismatic joint of variable length plus a revolute joint to connect it to the top platform. The three revolute joint axes form an equilateral triangle on the top platform. Actually, they can also form different desired shapes, such as a rectangle or a right triangle, etc. [25].

Based on the Grubler's formula, the total number of joints for this particular 3D system is $n = 9$ (each leg has 3 joints), the total number of links is $l = 7$, including the top platform but excluding the bottom base panel. Because each universal joint

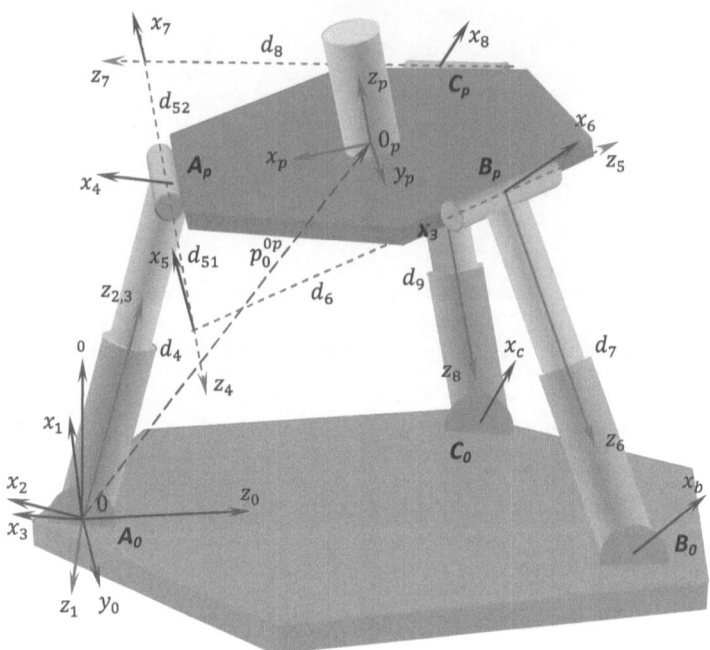

Figure 3.8   A 3D three-leg closed serial/parallel hybrid-chain system

offers 3 axes plus one prismatic axis and one revolute axis for each leg, the total number of axes is $3 \times 5 = 15$. Therefore,

$$r = D(l - n) + \sum_{i=1}^{n} f_i = -12 + 15 = 3,$$

where $D = 6$ due to the motion in 3D space.

We now adopt the same way to virtually float the bottom two joints B_0 and C_0, and leave the universal joint A_0 intact as the origin of the base coordinate frame. Then, this virtual model can be imagined as an open hybrid-chain system. Namely, starting from the base frame, along the serial-chain stem with total 5 axes: 3 at joint A_0, one is the prismatic piston and the last one is the revolute axis to reach the top platform. Then, the stem is split into two branches on the top platform: one is going to the virtually floating point B_0 via the revolute axis at B_p and the prismatic piston along the z_6 axis, and the other one is going to the virtually floating point C_0 via the revolute axis at C_p and the prismatic piston along the z_8 axis, as can be seen in Figure 3.8. For such an open serial/parallel hybrid-chain model, by applying the D-H convention, we can have its D-H table in Table 3.2.

Joint Angle θ_i	Joint Offset d_i	Twist Angle α_i	Link Length a_i
θ_1	0	$-90°$	0
θ_2	0	$90°$	0
θ_3	0	0	0
$\theta_4 = 0$	d_4	$-90°$	0

θ_i	d_i	α_i	a_i	θ_i	d_i	α_i	a_i
θ_5	d_{51}	$120°$	0	θ_5	$-d_{52}$	$-120°$	0
θ_6	d_6	$-90°$	0	θ_8	$-d_8$	$-90°$	0
$\theta_7 = 0$	d_7	0	0	$\theta_9 = 0$	d_9	0	0

Table 3.2

The D-H table for the 3D closed serial/parallel hybrid-chain system

Once the D-H parameter table is ready, every one-step homogeneous transformation matrix can be found below:

$$
A_0^1 = \begin{pmatrix} c_1 & 0 & -s_1 & 0 \\ s_1 & 0 & c_1 & 0 \\ 0 & -1 & 0 & 0 \\ 0 & 0 & 0 & 1 \end{pmatrix}, \quad
A_1^2 = \begin{pmatrix} c_2 & 0 & s_2 & 0 \\ s_2 & 0 & -c_2 & 0 \\ 0 & 1 & 0 & 0 \\ 0 & 0 & 0 & 1 \end{pmatrix},
$$

$$
A_2^3 = \begin{pmatrix} c_3 & -s_3 & 0 & 0 \\ s_3 & c_3 & 0 & 0 \\ 0 & 0 & 1 & 0 \\ 0 & 0 & 0 & 1 \end{pmatrix}, \quad
A_3^4 = \begin{pmatrix} 1 & 0 & 0 & 0 \\ 0 & 0 & 1 & 0 \\ 0 & -1 & 0 & d_4 \\ 0 & 0 & 0 & 1 \end{pmatrix}.
$$

Starting link 5, i.e., the top platform, the main stem is split into two paths:

$$
A_4^{51} = \begin{pmatrix} c_5 & \frac{1}{2}s_5 & \frac{\sqrt{3}}{2}s_5 & 0 \\ s_5 & -\frac{1}{2}c_5 & -\frac{\sqrt{3}}{2}c_5 & 0 \\ 0 & \frac{\sqrt{3}}{2} & -\frac{1}{2} & d_{51} \\ 0 & 0 & 0 & 1 \end{pmatrix}, \quad
A_4^{52} = \begin{pmatrix} c_5 & \frac{1}{2}s_5 & -\frac{\sqrt{3}}{2}s_5 & 0 \\ s_5 & -\frac{1}{2}c_5 & \frac{\sqrt{3}}{2}c_5 & 0 \\ 0 & -\frac{\sqrt{3}}{2} & -\frac{1}{2} & -d_{52} \\ 0 & 0 & 0 & 1 \end{pmatrix}.
$$

Then,

$$
A_5^6 = \begin{pmatrix} c_6 & 0 & -s_6 & 0 \\ s_6 & 0 & c_6 & 0 \\ 0 & -1 & 0 & d_6 \\ 0 & 0 & 0 & 1 \end{pmatrix}, \quad
A_6^7 = \begin{pmatrix} 1 & 0 & 0 & 0 \\ 0 & 1 & 0 & 0 \\ 0 & 0 & 1 & d_7 \\ 0 & 0 & 0 & 1 \end{pmatrix},
$$

$$
A_7^8 = \begin{pmatrix} c_8 & 0 & -s_8 & 0 \\ s_8 & 0 & c_8 & 0 \\ 0 & -1 & 0 & -d_8 \\ 0 & 0 & 0 & 1 \end{pmatrix}, \quad
A_8^9 = \begin{pmatrix} 1 & 0 & 0 & 0 \\ 0 & 1 & 0 & 0 \\ 0 & 0 & 1 & d_9 \\ 0 & 0 & 0 & 1 \end{pmatrix}.
$$

With the same procedure as that in the last example, we try to determine the Jacobian matrices: J_0^{51}, J_0^{52}, J_0^{67}, and J_0^{89}. Then, the overall Jacobian matrix for this virtual 3D open hybrid-chain robotic model can be augmented as

$$J_0 = \begin{pmatrix} J_0^{51} & J_0^{67} & O \\ J_0^{52} & O & J_0^{89} \end{pmatrix}. \tag{3.27}$$

Since each position vector of the floating tip points B_0 and C_0 is 3-dimensional, and the total number of axes for the virtual open hybrid-chain system is $n = 9$, obviously, this overall Jacobian is a 6 by 9 matrix, which offers an $(r = 3)$-dimensional null space N^3. This result is once again shows the consistency with the answer from the Grubler's formula.

In summery, for a closed-chain system, if n is the total number of joints, n_g is the total number of joints fixed on the ground, and D_0 is the dimension of motion space, then the net d.o.f., or the degree of mobility r of the system can be determined by

$$r = \left(\sum_{i=1}^{n-n_g+1} f_i \right) - D_0(n_g - 1). \tag{3.28}$$

The first term should be the count of all the axes of every joint except the virtually floated ones from the floor. For the current example, $n = 9$, $n_g = 3$, and $D_0 = 3$. After taking away the virtually floated joints from the floor, the total number of axes is counted as follows: 3 at the universal joint A_0, 3 for the three prismatic pistons, and 3 for the top revolute joints. Thus,

$$\sum_{i=1}^{n-n_g+1} f_i = 9, \quad \text{and} \quad r = 9 - 3(3 - 1) = 3.$$

Because $r = 3$ for this 3D closed-chain system, we can actuate any 3 out of all the axes to meet the 3-dimensional output requirement. Thus, both the input u and output $y = h(q)$ are 3-dimensional. In contrast, the joint position vector is 9-dimensional, i.e.,

$$q = (\theta_1 \ \theta_2 \ \theta_3 \ d_4 \ \theta_5 \ \theta_6 \ d_7 \ \theta_8 \ d_9)^T.$$

Let us now choose the three prismatic pistons d_4, d_7 and d_9 as the input drives. Likewise, the Jacobian equation should be homogeneous $J_0 \dot{q} = 0$ so that its null solution $\dot{q} = (I - J_0^+ J_0)\eta$ is the only non-trivial joint velocity vector to update every joint position at each sampling point. We can also define a 3D position vector p_0^{op} that is arrow-pointing to the origin of frame p, the center of the top platform, as the output $y = p_0^{op}$.

If one wishes to find the orientation of the top platform R_0^p with respect to frame 0, one can just rotate frame 4 by the angle θ_5 about the z_4 axis before transforming it to the pre-defined frame p. Namely,

$$R_0^p = R_0^4 \begin{pmatrix} c_5 & -s_5 & 0 \\ s_5 & c_5 & 0 \\ 0 & 0 & 1 \end{pmatrix} \begin{pmatrix} 0 & 0 & 1 \\ 1 & 0 & 0 \\ 0 & 1 & 0 \end{pmatrix}.$$

Once the output $y = h(q)$ is defined, we can now use the null solution to control the entire system towards a desired output target y^d. To do so, a potential function is defined by

$$p(q) = \frac{1}{2}(y^d - y)^T(y^d - y) = \frac{1}{2}(y^d - p_0^{op})^T(y^d - p_0^{op})$$

that should be minimized. Then, the gradient vector η can be readily calculated by

$$\eta = \frac{\partial p(q)}{\partial q} = \left(\frac{\partial p_0^{op}}{\partial q}\right)^T(p_0^{op} - y^d).$$

Once again, a 3D animation study has also been performed in MATLAB^{TM} to verify the effectiveness of the open hybrid-chain virtual model and null-space control scheme. Figure 3.9 depicts a so-called $3+3$ robot model system that consists of the 3D 3-leg closed parallel-chain platform and a regular 3-joint open serial-chain robot arm on the top. Obviously, the bottom 3-leg platform offers a 3-dimensional position while the top 3-joint arm is maneuvered to meet the orientation requirement.

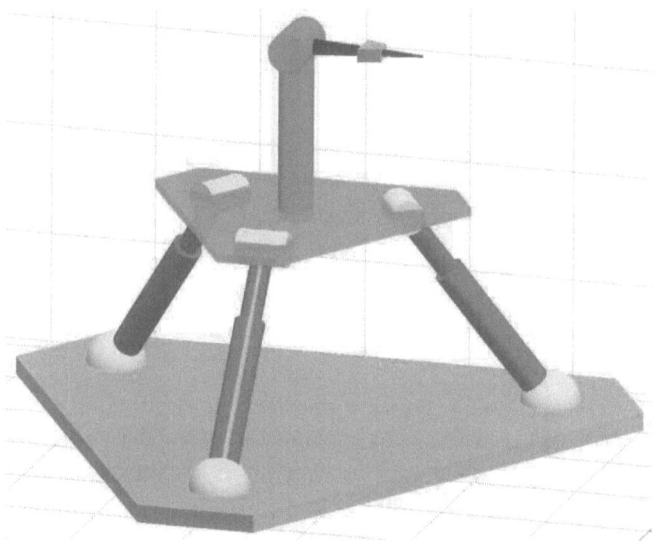

Figure 3.9 A 3+3 hybrid-chain robotic system is ready for animation study

With the 6 d.o.f. motion control, this $3+3$ hybrid-chain robotic system can operate any tasks for industrial applications. The animation study just demonstrates the robot drawing a panda on the board, as shown in Figure 3.10.

The above three examples exhibit a solid evidence that *most closed-chain mechanisms can be modeled as an open hybrid-chain robotic system*. Therefore, their

Figure 3.10 The 3+3 hybrid-chain robot is drawing a panda on the board in the animation study

dynamics modeling can follow the same procedure as discussed in the previous sections to find each inertial matrix W by equation (3.16) and the corresponding derivative matrix W_d by (3.23) through the kinematics modeling process to determine every sub-Jacobian matrix using equation (3.14).

For this particular $3 + 3$ hybrid-chain robotic system, in order to find each sub-Jacobian matrix, we need to follow a kinematic tree flowchart, as depicted in Figure 3.11. The entire robotic system conceivably consists of a tree stem and 3 branches. The tree stem covers 5 links: the first two are concentrated in the universal joint at A_0, and the 3rd one is the lower sleeve part of the piston with frame 3 (x_3-z_3) as its attached coordinate system. Then, frame 4 (x_4-z_4) is attached on link 4. The last link is the top platform with frame 5 attached. We can easily convert frame 5 to frame p at the platform center, see Figure 3.8.

Both branch 1 and branch 2 have only two links for each, and they are attached with their own coordinate frames: frame 6 (x_6-z_6) and frame b (x_b-z_b) for the virtu-

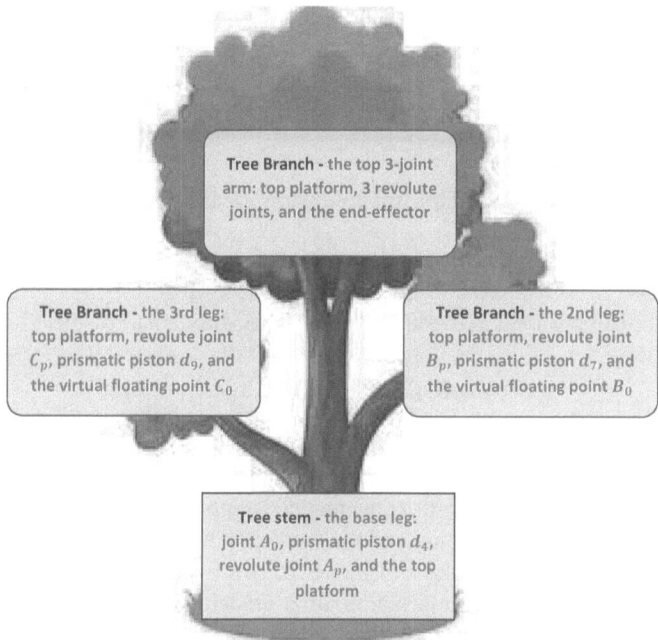

Figure 3.11 A kinematics and dynamics modeling tree for the 3+3 hybrid-chain robot

ally floated foot B_0, and frame 8 (x_8-z_8) and frame c (x_c-z_c) for the virtually floated foot C_0. Branch 3 is the top 3-joint arm that has three links before reaching its end-effector, as seen in Figure 3.9.

Once the attached coordinate frame is identified for each link, calculating every sub-Jacobian matrix is just to follow equation (3.14). For instance, to find the 6 by 5 nonzero portion of the sub-Jacobian matrix J_p for the top platform with frame p attached, based on (3.14) and noticing that axis 4 is a prismatic piston,

$$J_p = \begin{pmatrix} p_p^0 \times r_p^0 & p_p^1 \times r_p^1 & p_p^2 \times r_p^2 & r_p^3 & p_p^4 \times r_p^4 \\ r_p^0 & r_p^1 & r_p^2 & 0_3 & r_p^4 \end{pmatrix},$$

where the i-th p_p^i and r_p^i are the 4th and 3rd columns of the homogeneous transformation matrix A_p^i, respectively, for $i = 0, \cdots, 4$.

Often, most such closed hybrid-chain mechanisms belong to an **under-actuated systems** category, which means that only a few out of all the axes are actuated, and the rest remains passive without any actuators or drives. In other words, the dimension of the input is less than the total number of axes. In this particular 3+3 hybrid-chain robotic system, the total number of axes is $n = 12$, while only 6 of them are actuated, which are the three prismatic pistons and the three revolute joints of the top arm. Using the above sub-Jacobian matrices J_i for $i = 1, \cdots, 12$, the inertial matrix W that is constructed by equation (3.16) should be a 12 by 12 large positive-definite symmetric matrix. After calculating the 12 by 12 derivative matrix W_d by equation (3.23), the entire dynamic equation becomes

$$W\ddot{q} + \left(W_d^T - \frac{1}{2}W_d\right)\dot{q} - \tau_g = \tau,$$

where the axis position vector q is the following 12-dimensional column vector:

$$q = (\theta_1 \ \theta_2 \ \theta_3 \ d_4 \ \theta_5 \ \theta_6 \ d_7 \ \theta_8 \ d_9 \ \theta_{10} \ \theta_{11} \ \theta_{12})^T,$$

while the 12-dimensional axis-torque vector τ has only a half of the 12 elements to be nonzero:

$$\tau = (0 \ 0 \ 0 \ f_4 \ 0 \ 0 \ f_7 \ 0 \ f_9 \ \tau_{10} \ \tau_{11} \ \tau_{12})^T.$$

All the zero elements in the torque vector will play a role in internal dynamic constraints, called an **internal dynamics**, which is very common for such an under-actuated system if the virtual floating model is applied. The interesting topic on under-actuated systems modeling and the corresponding state-feedback control will be further studied and discussed in Chapter 6. However, if a dual modeling procedure is adopted other than the virtual floating model, the passive orientation of the top platform for the 3-leg closed parallel-chain system must be determined as a required IK algorithm. Therefore, the 3+3 hybrid-chain robotic system will be revisited again in Chapter 5 to develop an alternative way to model both its kinematics and dynamics under **the principle of duality**.

REFERENCES

1. Arnold, V., (1978) Mathematical Methods of Classical Mechanics. Springer-Verlag, New York.
2. Abraham, R. and Marsden, J., (1978) Foundations of Mechanics. The Benjamin/Cummings Publishing Company.
3. Marsden, J. and Ratiu, T., (1994) Introduction to Mechanics and Symmetry. Springer-Verlag, New York.
4. Asada, H. and Slotine, J., (1986) Robot Analysis and Control. John Wiley & Sons, New York.
5. Spong, M. and Vidyasagar, M., (1989) Robot Dynamics and Control. John Wiley & Sons, New York.
6. Murray, R., Li, Z. and Sastry, S., (1994) A Mathematical Introduction to Robotic Manipulation. CRC Press, Boca Raton, London, New York.
7. Luh, J., Walker, M. and Paul, R., (1980) On-Line Computational Scheme for Mechanical Manipulators. *ASME Journal of Dynamic Systems, Measurement and Control*. Vol.102, June, pp. 69-76.
8. Brockett, R., (1984) Robotic Manipulators and the Product of Exponentials Formula. In: Fuhrman P (ed) Mathematical Theory of Networks and Systems. Springer-Verlag, New York.
9. Brockett, R., (1990) Some Mathematical Aspects of Robotics. *Robotics, AMS Short Course Lecture Notes*, Vol.41, Providence, American Mathematical Society.

10. Brockett, R., Stokes, A. and Park, F., (1993) A Geometrical Formulation of the Dynamic Equations Describing Kinematic Chains. *Proc. IEEE Int. Conf. Robotics and Automation*, Atlanta, GA., pp. 637-641.

11. Koditschek, D., (1985) Robot Kinematics and Coordinate Transformations. *Proc. 1985 IEEE Conf. on Decision and Control*, Florida, Dec., pp. 1-4.

12. Gu, Edward Y.L. and Loh, N., (1988) Dynamic Modeling and Control by Utilizing an Imaginary Robot Model. *IEEE Journal of Robotics and Automation*, Vol.4, No.5, Oct., pp. 532-540.

13. Gu, Edward Y.L., (1991) Modeling and Simplification for Dynamic Systems with Testing Procedures and Metric Decomposition. *Proc. 1991 IEEE International Conference on Systems, Man, and Cybernetics*, Charlottesville, VA, Oct., pp. 487-492.

14. Featherstone, R., (1987) Robot Dynamics Algorithms. Kluwer Publisher, Boston, MA.

15. Rodriguez, G., Jain, A. and Kreutz-Delgado, K., (1991) A Spatial Operator Algebra for Manipulator Modeling and Control. *International Journal on Robotics Research*, Vol.10, pp. 371-381.

16. Rodriguez, G. and Kreutz-Delgado, K., (1992) Spatial Operator Factorization and Inversion of the Manipulator Mass Matrix. *IEEE Transactions on Robotics and Automation*, Vol.8, pp. 65-76.

17. Angeles, J., (1992) The Design of Isotropic Manipulator Architectures in the Presence of Redundancies. *International Journal of Robotics Research*, Vol.11, No.3, June, pp. 196-201.

18. Angeles, J. and Lopez-Cajun, C., (1992) Kinematic Isotropy and Conditioning Index of Serial Robotic Manipulators. *International Journal of Robotics Research*, Vol.11, No.6, Dec., pp. 560-571.

19. Spong, M., (1992) Remarks on Robot Dynamics: Canonical Transformations and Riemannian Geometry. *Proc. IEEE Int. Conf. Robotics and Automation*, Nice, France.

20. Park, F., (1994) Computational Aspects of the Product-of-Exponentials Formula for Robot Kinematics. *IEEE Transactions on Automatic Control*, Vol.39, No.3, March, pp. 643-647.

21. Park, F., Bobrow, J. and Ploen, S., (1995) A Lie Group Formulation of Robot Dynamics. *International Journal of Robotics Research*, Vol.14, No.6, Dec., pp. 609-618.

22. McCarthy, J., (1990) Introduction to Theoretical Kinematics. MIT Press.

23. Gu, Edward Y.L., (2000) Configuration Manifolds and Their Applications to Robot Dynamic Modeling and Control. *IEEE Transactions on Robotics and Automation*, Vol.16, No.5, October, pp. 517-527.

24. Gu, Edward Y.L., (2000) A Configuration Manifold Embedding Model for Dynamic Control of Redundant Robots. *International Journal of Robotics Research*, Vol.19, No.3, March, pp. 289-304.

25. Gu, Edward Y.L., (2013) A Journey from Robot to Digital Human. Springer, Heidelberg, New York.

26. Landau, L. and Lifshitz, E., (1960) Mechanics. Addison-Wesley.

27. Dubrovin, B., Fomenko, A. and Novikov, S., (1992) Modern Geometry – Methods and Applications, Part I - The Geometry of Surfaces, Transformation Groups, and Fields. Springer-Verlag, New York.

28. Dubrovin, B., Fomenko, A. and Novikov, S., (1992) Modern Geometry – Methods and Applications, Part II - The Geometry and Topology of Manifolds. Springer-Verlag, New York.

29. Bredon, G., (1993) Topology and Geometry. Springer-Verlag, New York.

30. Berger, M. and Gostiaux, B., (1988) Differential Geometry: Manifolds, Curves, and Surfaces. Springer-Verlag, New York.
31. Khalil, W., Ibrahim, O., (2007) General Solution for the Dynamics Modeling of Parallel Robots. *Journal of Intelligent and Robotic Systems*, Vol.49, pp. 19-37.
32. Merlet, J., (2006) Parallel Robots, 2nd Edition. Springer, Dordrecht, The Netherlands.

4 Advanced Dynamics Modeling

4.1 THE CONFIGURATION MANIFOLD AND ISOMETRIC EMBEDDING

The concept of configuration manifold (C-manifold) and isometric embedding has been introduced from the mathematical perspective in Chapter 2. We now give a formal definition of the C-manifold:

Definition 4.1. A **configuration manifold** M^n for an n-order dynamic system is a Riemannian manifold that satisfies the following assertions:

1. An n-tuple local coordinate system $q = \{q^1, \cdots, q^n\}$ is defined in an open neighborhood $U \subset M^n$ at each point on M^n;
2. The basis of the topology of M^n is a collection of open subsets of $B(3) \times SO(3)$, where $B(3) \subset \mathbb{R}^3$ is a compact subspace;
3. The kinetic energy $K = \frac{1}{2} w_{ij} \dot{q}^i \dot{q}^j$ of the dynamic system is endowed on M^n as a Riemannian metric.

Based on the above definition, the kinetic energy K for a dynamic system is the Riemannian metric endowed on the C-manifold. According to Theorem 3.2, any shortest arc length represented by the local coordinate trajectory $q(t)$ obeys the geodesic equation related to the same metric as well as the Euler-Lagrange equation. Since a C-manifold M^n is confined by a bounded and closed subspace $B(3) \times SO(3)$, we desire to embed such a compact C-manifold into a Euclidean m-space with a minimum m but being spacious enough to keep the Riemannian metric preserved [1–3].

If such an isometric embedding $z = \zeta(q)$ can be found for a given n-order dynamic system, such as an n-joint robot manipulator, then,

$$\dot{z} = \frac{\partial \zeta}{\partial q} \dot{q} = J \dot{q}$$

with a Jacobian matrix J of the C-manifold M^n. Based on differential geometry [4–7],

$$K = \frac{1}{2} w_{ij} \dot{q}^i \dot{q}^j = \frac{1}{2} \langle \dot{q}, \dot{q} \rangle_w = \frac{1}{2} \langle \dot{z}, \dot{z} \rangle_\delta = \frac{1}{2} \delta_{ij} \dot{z}^i \dot{z}^j,$$

where δ_{ij} is the Kronecker symbol that is defined by

$$\delta_{ij} = \begin{cases} 1 & \text{if } i = j \\ 0 & \text{if } i \neq j. \end{cases}$$

This implies that

$$w_{ij} = \delta_{kl} \frac{\partial \zeta^k}{\partial q^i} \frac{\partial \zeta^l}{\partial q^j},$$

which is also given in equation (2.13) in a tensor form. The matrix form for the above metric relation has been given explicitly by equation (2.14) in Chapter 2. Namely,

$$W = J^T J,$$

where J is the genuine mathematical Jacobian matrix for the embedding $z = \zeta(q)$, instead of the traditional "Jacobian matrix" in kinematic differential motion models. The above metric factorization indicates that the kinetic energy

$$K = \frac{1}{2} \dot{q}^T W \dot{q} = \frac{1}{2} \dot{q}^T J^T J \dot{q} = \frac{1}{2} \dot{z}^T \dot{z}.$$

Furthermore, since $\dot{z} = \frac{\partial \zeta}{\partial q} \dot{q} = J \dot{q}$,

$$\frac{\partial \zeta}{\partial q} = \frac{d}{dt} \frac{\partial \zeta}{\partial q} = J. \tag{4.1}$$

This equation becomes a key important differential relation that any non-mathematical Jacobian matrix is disqualified to be held.

Let us now substitute $K = \frac{1}{2} \dot{q}^T J^T J \dot{q} = \frac{1}{2} \dot{z}^T \dot{z}$ into the Euler-Lagrange equation as a part of Lagrangian $L(q, \dot{q})$. First, by noticing that $W = J^T J$ is a symmetric matrix,

$$\frac{\partial K}{\partial \dot{q}} = \frac{1}{2} [J^T J + (J^T J)^T] \dot{q} = J^T J \dot{q}.$$

Then,

$$\frac{d}{dt} \frac{\partial K}{\partial \dot{q}} = J^T J \ddot{q} + \dot{J}^T J \dot{q} + J^T \dot{J} \dot{q}.$$

For the second term of the Euler-Lagrange equation,

$$\frac{\partial K}{\partial q} = \left(\frac{\partial \dot{\zeta}}{\partial q} \right)^T \dot{\zeta} = \dot{J}^T J \dot{q}.$$

Thus,

$$\frac{d}{dt} \frac{\partial K}{\partial \dot{q}} - \frac{\partial K}{\partial q} = J^T J \ddot{q} + J^T \dot{J} \dot{q} = J^T \ddot{z}.$$

By filling every related form of potential energy into the Lagrangian $L = K - P$, we obtain a non-homogeneous Euler-Lagrange equation that is seen in a Euclidean m-space:

$$J^T \ddot{z} = \tau + \tau_g. \tag{4.2}$$

Therefore, we have sufficiently justified the following theorem:

Theorem 4.1. *If an isometric embedding $z = \zeta(q)$ can be found to embed the C-manifold M^n of an n-order dynamic system into an ambient Euclidean m-space $\mathbb{R}^m$ with $m \geq n$ to preserve its Riemannian metric, i.e., the kinetic energy K of the system, then, its Euler-Lagrange equation (3.22) is equivalent to*

$$J^T \ddot{z} = \tau + \tau_g,$$

where J is the Jacobian matrix of the embedding $J = \frac{\partial \zeta}{\partial q}$.

Equation (4.2) looks like a Newton's second law either $f = m\ddot{x}$ in a translational case or $\tau = I\ddot{\theta}$ in a rotational single body case, or resemble to a robot statics equation given by (2.66). It should be not surprised that every complicated nonlinear and highly coupled equation seen in non-Euclidean space can be reduced as a basic form if it is seen in Euclidean space. This result also shows that *such a compact appearance of dynamic equation is attributed to the isometric embedding*. Therefore, to find an isometric embedding $z = \zeta(q)$ with the minimum dimension m is essential and also pivotal for dynamics model reduction.

The above justification for Theorem 4.1 reveals a clear fact that the Euler-Lagrange dynamic equation could not be simplified without the property given in equation (4.1). In other words, *J must be a genuine mathematical Jacobian matrix in order to possess such a privilege property so that the dynamics model reduction can be succeeded*. This is also the same reason why we cannot find any 6-dimensional unified vector to uniquely and one-to-one represent both position and orientation. However, if we insist to realize such a unification, the dimension m of an ambient Euclidean space has to be greater than the C-manifold dimension n, i.e., $m > n$. This is the cost one has to pay for seeking a successful isometric embedding in higher dimension [3].

At the energy conservation standpoint, we can also justify the compact dynamic equation (4.2). In fact, since the kinetic energy $K = \frac{1}{2}\dot{z}^T\dot{z}$, the kinetic power should be $\dot{K} = \dot{z}^T\ddot{z}$. Because $\dot{z} = J\dot{q}$, $\dot{K} = \dot{q}^T J^T \ddot{z}$. On the other hand, $\dot{K} = \dot{q}^T\tau$ in the joint space. Therefore, $\tau = J^T\ddot{z}$ so that the compact dynamic equation can also be achieved.

Let us now revisit the 2-link planar robot arm, as shown in Figure 2.3 that has been investigated in Chapter 2. A successful isometric embedding for its C-manifold M^2 has been found in equation (2.19), and let us rewrite it as follows:

$$\begin{cases} z^1 = b_1 c_1 + b_2 c_{12} \\ z^2 = b_1 s_1 + b_2 s_{12} \\ z^3 = h_1(\theta_1 + \theta_2) \\ z^4 = h_2\theta_1. \end{cases}$$

Its Jacobian matrix is 4 by 2 dimensional:

$$J = \frac{\partial \zeta}{\partial q} = \begin{pmatrix} -b_1 s_1 - b_2 s_{12} & -b_2 s_{12} \\ b_1 c_1 + b_2 c_{12} & b_2 c_{12} \\ h_1 & h_1 \\ h_2 & 0 \end{pmatrix},$$

With this explicit form of the above isometric embedding $z = \zeta(q)$, we can easily determine its acceleration vector:

$$
\ddot{z} = \begin{pmatrix} -b_1 c_1 \dot{\theta}_1^2 - b_1 s_1 \ddot{\theta}_1 - b_2 c_{12} (\dot{\theta}_1 + \dot{\theta}_2)^2 - b_2 s_{12} (\ddot{\theta}_1 + \ddot{\theta}_2) \\ -b_1 s_1 \dot{\theta}_1^2 + b_1 c_1 \ddot{\theta}_1 - b_2 s_{12} (\dot{\theta}_1 + \dot{\theta}_2)^2 + b_2 c_{12} (\ddot{\theta}_1 + \ddot{\theta}_2) \\ h_1 (\ddot{\theta}_1 + \ddot{\theta}_2) \\ h_2 \ddot{\theta}_1 \end{pmatrix}.
$$

Then, multiplying the above two results yields $J^T \ddot{z} = \tau + \tau_g$ that is the complete dynamic equation of the 2-link planar robot arm, and the torque due to gravity can be found in equation (3.24). All the dynamic parameters involved in the embedding: b_1, b_2, h_1 and h_2 have also been well-determined by a direct comparison between the Riemannian metric $W = J^T J$ of the C-manifold M^2 and the 2-link robot inertial matrix W, as given in equation (2.22).

In summary, a qualified isometric embedding $z = \zeta(q)$ should satisfy the following conditions:

1. It must meet the embedding definition 2.3, i.e., the mapping $\zeta : M^n \to \mathbb{R}^m$ must be one-to-one;
2. The inner product $\frac{1}{2} \dot{z}^T \dot{z}$ should be exactly equal to the kinetic energy $K = \frac{1}{2} \dot{q}^T W \dot{q}$ of the system;
3. Every dynamic parameter involved in $z = \zeta(q)$ must be constant, and no one can be time-varying.

4.2   HOW TO FIND AN ISOMETRIC EMBEDDING

As aforementioned, we have to seek a qualified isometric embedding by expanding the ambient Euclidean space to be higher dimensional due to the "disappointed" but irresistible fact that finding a unified 6-dimensional vector to uniquely represent both position and orientation has almost no hope. If it could be possible, then this 6-dimensional vector would have already been a qualified isometric embedding and the 6-dimensional C-manifold M^6 for a 6-joint robotic system could be directly isometrically embedded into $\mathbb{R}^6$ without any complication. However, the reality is "cruel", and the complication of rotation is that root cause. Let us now step by step explore and develop a general form of the qualified isometric embedding $z = \zeta(q)$ to realize the compact dynamic equation in (4.2).

Lemma 4.1. *If a single rigid body is confined in 4 d.o.f. motion with its C-manifold $M^4 \subset B(3) \times SO(2)$, i.e., the unity axis k of its angular velocity $\omega = \dot{\phi} k$ is fixed, then, M^4 can be isometrically embedded into a Euclidean 4-space $\mathbb{R}^4$ by*

$$
z = \zeta(q) = \begin{pmatrix} \sqrt{m} p_0^c \\ \sqrt{I_k^c} \phi \end{pmatrix}, \tag{4.3}
$$

provided that the rank of its Jacobian matrix $rank(J) = rank(\frac{\partial \zeta}{\partial q}) = 4$, where p_0^c is the 3-dimensional mass center position of the body with respect to the fixed base, and I_k^c

is the moment of inertia of the body about the fixed unity axis k passing through the mass center.

Proof. Since the single rigid body has confined motion within a compact region $B(3) \times SO(2)$, its C-manifold is also compact, i.e., bounded and closed. Its total kinetic energy should be

$$K = \frac{1}{2}mv_c^T v_c + \frac{1}{2}\omega^T \Gamma_c \omega,$$

as given by equation (3.11). Because the linear velocity of the mass center is $v_c = \dot{p}_0^c$, and the angular velocity is $\omega = \dot{\phi}k$ based on equation (2.51),

$$\dot{z} = \begin{pmatrix} \sqrt{m}\dot{p}_0^c \\ \sqrt{I_k}\dot{\phi} \end{pmatrix} \quad \text{such that} \quad \dot{z}^T\dot{z} = m\dot{p}_0^{cT}\dot{p}_0^c + I_k\dot{\phi}^2.$$

According to equation (3.4), $I_k = k^T\Gamma_c k$ so that $I_k\dot{\phi}^2 = \dot{\phi}k^T\Gamma_c\dot{\phi}k = \omega^T\Gamma_c\omega$. Therefore,

$$\frac{1}{2}\dot{z}^T\dot{z} = \frac{1}{2}m\dot{p}_0^{cT}\dot{p}_0^c + \frac{1}{2}\omega^T\Gamma_c\omega = K.$$

Furthermore, since the unity axis k for the angular velocity ω is fixed on the rigid body, the moment of inertia $I_k = k^T\Gamma_c k$ is a constant parameter. Due to the full rank of the Jacobian matrix J, the mapping $\zeta : M^4 \to \mathbb{R}^4$ is one-to-one and qualified to be an embedding. □

In fact, for a 6 d.o.f. motion in the special Euclidean space $E(3) = \mathbb{R}^3 \times SO(3)$, the angular velocity $\omega = \dot{\phi}k$ cannot always rotate about a fixed axis k in general, unless the rotation is confined on a 2D plane, where every motion can have only one d.o.f. of rotation in addition to the 3 d.o.f. of translation so that the motion is actually confined in a compact region $B(3) \times SO(2)$. Obviously, the Lie group $SO(2)$ is only 1-dimensional, and thus, this lemma is basically applied for a rigid body in a 2D planar motion case. Therefore, the 3-dimensional isometric embedding in (4.3) is well-suitable for every planar revolute-joint robot arm.

This is also the major reason why the last link, as a single rigid body, can be isometrically embedded by the first three equations of (2.19) in the 2-link robot case. The top two equations are actually the position of the mass center p_0^c times a dynamic parameter, while the third one $z^3 = h_1(\theta_1 + \theta_2)$ is the same as (4.3) in the above lemma with the angle $\phi = \theta_1 + \theta_2$ due to the last link imposed by the first link rotation θ_1 in addition to its own rotation θ_2. However, the last equation $h_2\theta_1$ in (2.19) is added as part of the contribution from link 1, which can also be interpreted to compensate for the one-to-one requirement due to the multi-configuration issue, as discussed in Chapter 2.

We now take a close look at a revolute-prismatic-revolute (RPR) type 3-joint planar robot arm, as shown in Figure 4.1. This time, we adopt equation (3.16) along with the sub-Jacobian formula (3.14) to determine the inertial matrix W for this RPR-type planar arm. First, its D-H table can be found by:

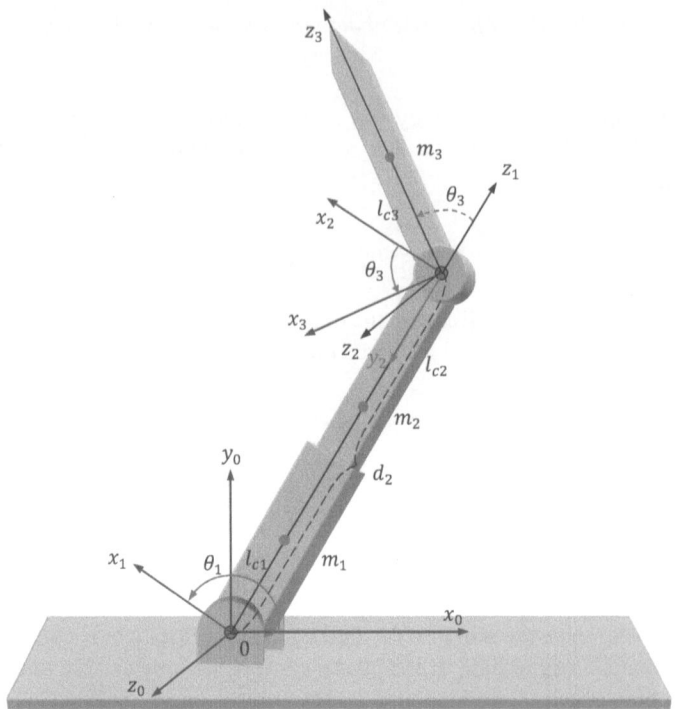

Figure 4.1 An RPR 3-joint planar robot arm

Joint Angle θ_i	Joint Offset d_i	Twist Angle α_i	Link Length a_i
θ_1	0	$+90°$	0
$\theta_2 = 0$	d_2	$-90°$	0
θ_3	0	$+90°$	0

Once the D-H parameter table is ready, every one-step homogeneous transformation matrix can be derived as follows:

$$A_0^1 = \begin{pmatrix} c_1 & 0 & s_1 & 0 \\ s_1 & 0 & -c_1 & 0 \\ 0 & 1 & 0 & 0 \\ 0 & 0 & 0 & 1 \end{pmatrix}, A_1^2 = \begin{pmatrix} 1 & 0 & 0 & 0 \\ 0 & 0 & 1 & 0 \\ 0 & -1 & 0 & d_2 \\ 0 & 0 & 0 & 1 \end{pmatrix}, A_2^3 = \begin{pmatrix} c_3 & 0 & s_3 & 0 \\ s_3 & 0 & -c_3 & 0 \\ 0 & 1 & 0 & 0 \\ 0 & 0 & 0 & 1 \end{pmatrix}.$$

According to equation (3.14), we need to prepare all the necessary homogeneous transformations in order to find its three sub-Jacobian matrices J_1, J_2 and J_3. Clearly, the first subrobot only contains link 1, the second subrobot consists of both link 1 and link 2, and the third one covers the entire three-joint planar robot. Thus, with a few symbolic computations and noticing that $A_j^i = (A_i^j)^{-1}$ has to be derived by the inverse equation (2.49) in Chapter 2, we obtain

$$A_1^0 = \begin{pmatrix} c_1 & s_1 & 0 & 0 \\ 0 & 0 & 1 & 0 \\ s_1 & -c_1 & 0 & 0 \\ 0 & 0 & 0 & 1 \end{pmatrix}, \quad A_2^0 = \begin{pmatrix} c_1 & s_1 & 0 & 0 \\ -s_1 & c_1 & 0 & d_2 \\ 0 & 0 & 1 & 0 \\ 0 & 0 & 0 & 1 \end{pmatrix},$$

$$A_2^1 = \begin{pmatrix} 1 & 0 & 0 & 0 \\ 0 & 0 & -1 & d_2 \\ 0 & 1 & 0 & 0 \\ 0 & 0 & 0 & 1 \end{pmatrix}, \quad A_3^0 = \begin{pmatrix} c_{13} & s_{13} & 0 & d_2 s_3 \\ 0 & 0 & 1 & 0 \\ s_{13} & -c_{13} & 0 & -d_2 c_3 \\ 0 & 0 & 0 & 1 \end{pmatrix},$$

$$A_3^1 = \begin{pmatrix} c_3 & 0 & -s_3 & d_2 s_3 \\ 0 & 1 & 0 & 0 \\ s_3 & 0 & c_3 & -d_2 c_3 \\ 0 & 0 & 0 & 1 \end{pmatrix}, \quad A_3^2 = \begin{pmatrix} c_3 & s_3 & 0 & 0 \\ 0 & 0 & 1 & 0 \\ s_3 & -c_3 & 0 & 0 \\ 0 & 0 & 0 & 1 \end{pmatrix}.$$

where $s_{13} = \sin(\theta_1 + \theta_3)$ and $c_{13} = \cos(\theta_1 + \theta_3)$ are the short notations.

Now, based on equation (3.14) and noticing that the second joint of the robot is prismatic, the three sub-Jacobian matrices can be determined by

$$J_1 = \begin{pmatrix} 0 & 0 & 0 \\ 0 & 0 & 0 \\ 0 & 0 & 0 \\ 0 & 0 & 0 \\ 1 & 0 & 0 \\ 0 & 0 & 0 \end{pmatrix}, \quad J_2 = \begin{pmatrix} d_2 & 0 & 0 \\ 0 & -1 & 0 \\ 0 & 0 & 0 \\ 0 & 0 & 0 \\ 0 & 0 & 0 \\ 1 & 0 & 0 \end{pmatrix}, \quad J_3 = \begin{pmatrix} d_2 c_3 & -s_3 & 0 \\ 0 & 0 & 0 \\ d_2 s_3 & c_3 & 0 \\ 0 & 0 & 0 \\ 1 & 0 & 1 \\ 0 & 0 & 0 \end{pmatrix}.$$

After all the three symbolical sub-Jacobian matrices are ready, the preparation task remains to define the dynamic parameter matrix U_j for each of the three links. It should be emphasized that every parameter in the matrices U_j's is referred to each individual link frame that has been assigned by the D-H convention. Thus, according to Figure 4.1, the mass center coordinates vector of each link can be determined by $c_1 = (0\ 0\ l_{c1})^T$, $c_2 = (0\ l_{c2}\ 0)^T$ and $c_3 = (0\ 0\ l_{c3})^T$. After converting them into their skew-symmetric matrices $C_j = S(c_j)$ for $j = 1, 2, 3$, by following equation (3.10), we have

$$U_1 = \begin{pmatrix} m_1 & 0 & 0 & 0 & m_1 l_{c1} & 0 \\ 0 & m_1 & 0 & -m_1 l_{c1} & 0 & 0 \\ 0 & 0 & m_1 & 0 & 0 & 0 \\ 0 & -m_1 l_{c1} & 0 & I_{x1} & 0 & 0 \\ m_1 l_{c1} & 0 & 0 & 0 & I_{y1} & 0 \\ 0 & 0 & 0 & 0 & 0 & I_{z1} \end{pmatrix},$$

$$U_2 = \begin{pmatrix} m_2 & 0 & 0 & 0 & 0 & -m_2 l_{c2} \\ 0 & m_2 & 0 & 0 & 0 & 0 \\ 0 & 0 & m_2 & m_2 l_{c2} & 0 & 0 \\ 0 & 0 & m_2 l_{c2} & I_{x2} & 0 & 0 \\ 0 & 0 & 0 & 0 & I_{y2} & 0 \\ -m_2 l_{c2} & 0 & 0 & 0 & 0 & I_{z2} \end{pmatrix},$$

$$U_3 = \begin{pmatrix} m_3 & 0 & 0 & 0 & m_3 l_{c3} & 0 \\ 0 & m_3 & 0 & -m_3 l_{c3} & 0 & 0 \\ 0 & 0 & m_3 & 0 & 0 & 0 \\ 0 & -m_3 l_{c3} & 0 & I_{x3} & 0 & 0 \\ m_3 l_{c3} & 0 & 0 & 0 & I_{y3} & 0 \\ 0 & 0 & 0 & 0 & 0 & I_{z3} \end{pmatrix},$$

where each inertia tensor Γ_j is diagonal under the assumption that every link is geo-metrically symmetric and mass-density uniformly distributed.

Now, sandwiching each U_j by J_j^T and J_j yields

$$W_1 = J_1^T U_1 J_1 = \begin{pmatrix} I_{y1} & 0 & 0 \\ 0 & 0 & 0 \\ 0 & 0 & 0 \end{pmatrix},$$

$$W_2 = J_2^T U_2 J_2 = \begin{pmatrix} m_2 d_2^2 - 2 m_2 l_{c2} d_2 + I_{z2} & 0 & 0 \\ 0 & m_2 & 0 \\ 0 & 0 & 0 \end{pmatrix},$$

and $W_3 = J_3^T U_3 J_3 =$

$$\begin{pmatrix} m_3 d_2^2 + 2 m_3 l_{c3} d_2 c_3 + I_{y3} & -m_3 l_{c3} s_3 & m_3 l_{c3} d_2 c_3 + I_{y3} \\ -m_3 l_{c3} s_3 & & \\ m_3 l_{c3} d_2 c_3 + I_{y3} & m_3 & -m_3 l_{c3} s_3 \\ -m_3 l_{c3} s_3 & I_{y3} \end{pmatrix}.$$

By adding all the above three W_j's together, we achieve the complete inertial matrix $W = W_1 + W_2 + W_3$ for this RPR-type 3-joint planar robot arm. Note that the moments of inertia I_{y1}, I_{z2} and I_{y3} inside each W_j here are about the y_1-axis, z_2-axis and y_3-axis, respectively. According to the parallel-axis theorem, $I_{y1} = I_1 + m_1 l_{c1}^2$, $I_{z2} = I_2 + m_2 l_{c2}^2$ and $I_{y3} = I_3 + m_3 l_{c3}^2$, each of which can be converted to referring to the respective mass center.

We now turn to seek an overall isometric embedding $z = \zeta(q)$ that can cover all the three links of the robot. Based on Lemma 4.3, the first step is to construct a following 3-dimensional mapping for link 3:

$$z_3 = \zeta_3(q) = \begin{pmatrix} \sqrt{m_3} p_0^{c3} \\ \sqrt{I_3} (\theta_1 + \theta_3) \end{pmatrix}.$$

Based on the D-H table and homogeneous transformations, we can readily derive without difficulty that

$$p_0^{c3} = \begin{pmatrix} d_2 s_1 + l_{c3} s_{13} \\ -d_2 c_1 - l_{c3} c_{13} \end{pmatrix}.$$

According to (4.3) again, the isometric embedding for link 2 can be found as

$$z_2 = \zeta_2(q) = \begin{pmatrix} \sqrt{m_2} p_0^{c2} \\ \sqrt{I_2} \theta_1 \end{pmatrix} = \begin{pmatrix} \sqrt{m_2}(d_2 - l_{c2}) s_1 \\ -\sqrt{m_2}(d_2 - l_{c2}) c_1 \\ \sqrt{I_2} \theta_1 \end{pmatrix},$$

where the rotation of link 2 is imposed by θ_1. Since the first link has only a rotation about the origin of frame 1, the isometric embedding can be simplified by

$$z_1 = \zeta_1(q) = \sqrt{I_{z1}}\,\theta_1 = \sqrt{I_1 + m_1 l_{c1}^2}\,\theta_1,$$

which can be directly merged into $z_2 = \zeta_2(q)$ of link 2. Thus, the overall isometric embedding of the entire 3-joint planar robot can be augmented and combined as follows:

$$z = \zeta(q) = \begin{pmatrix} \sqrt{m_3}(d_2 s_1 + l_{c3} s_{13}) \\ -\sqrt{m_3}(d_2 c_1 + l_{c3} c_{13}) \\ \sqrt{I_3}(\theta_1 + \theta_3) \\ \sqrt{m_2}(d_2 - l_{c2}) s_1 \\ -\sqrt{m_2}(d_2 - l_{c2}) c_1 \\ \sqrt{I_1 + I_2 + m_1 l_{c1}^2}\,\theta_1 \end{pmatrix},$$

which is 6-dimensional. Its 6 by 3 Jacobian matrix becomes

$$J = \frac{\partial \zeta}{\partial q} = \begin{pmatrix} \sqrt{m_3}(d_2 c_1 + l_{c3} c_{13}) & \sqrt{m_3} s_1 & \sqrt{m_3} l_{c3} c_{13} \\ \sqrt{m_3}(d_2 s_1 + l_{c3} s_{13}) & -\sqrt{m_3} c_1 & \sqrt{m_3} l_{c3} s_{13} \\ \sqrt{I_3} & 0 & \sqrt{I_3} \\ \sqrt{m_2}(d_2 - l_{c2}) c_1 & \sqrt{m_2} s_1 & 0 \\ \sqrt{m_2}(d_2 - l_{c2}) s_1 & -\sqrt{m_2} c_1 & 0 \\ \sqrt{I_1 + I_2 + m_1 l_{c1}^2} & 0 & 0 \end{pmatrix}.$$

Thus, the Riemannian metric turns out to be

$$W = J^T J = \begin{pmatrix} w_{11} & -m_3 l_{c3} s_3 & m_3 l_{c3} d_2 c_3 + m_3 l_{c3}^2 + I_3 \\ -m_3 l_{c3} s_3 & m_3 + m_2 & -m_3 l_{c3} s_3 \\ m_3 l_{c3} d_2 c_3 + m_3 l_{c3}^2 + I_3 & -m_3 l_{c3} s_3 & m_3 l_{c3}^2 + I_3 \end{pmatrix},$$

where $w_{11} = m_3 d_2^2 + m_3 l_{c3}^2 + I_3 + 2 m_3 l_{c3} d_2 c_3 + m_2 (d_2 - l_{c2})^2 + I_1 + I_2 + m_1 l_{c1}^2$. This Riemannian metric exactly agrees with the inertial matrix $W = W_1 + W_2 + W_3$ of this RPR-type 3-joint planar robot.

It is conceivable that the last link of any serial-chain robot is always the busiest rigid body over all the links, the kinematics of which is the function of every joint position. We call the kinematically busiest body for a multi-body dynamic system the **dominant body**. The above example typically shows an important fact that if one follows equation (4.3) for the last link of the robot, it may have already been a qualified embedding, but has not been isometric yet until every motion tribute from the lower links are all augmented and combined. Specifically for the relatively simple 3-joint planar robot case in this example, we can successfully find an overall isometric embedding and enjoy the reduced compact dynamic model (4.2). However, as the case is extended to a 3D robot, to find an overall isometric embedding becomes much more challenging.

Let us now study a new manipulator that is also a 3-joint and RPR-type robot but is in 3D motion, as depicted in Figure 4.2. Its D-H table can be easily determined as follows:

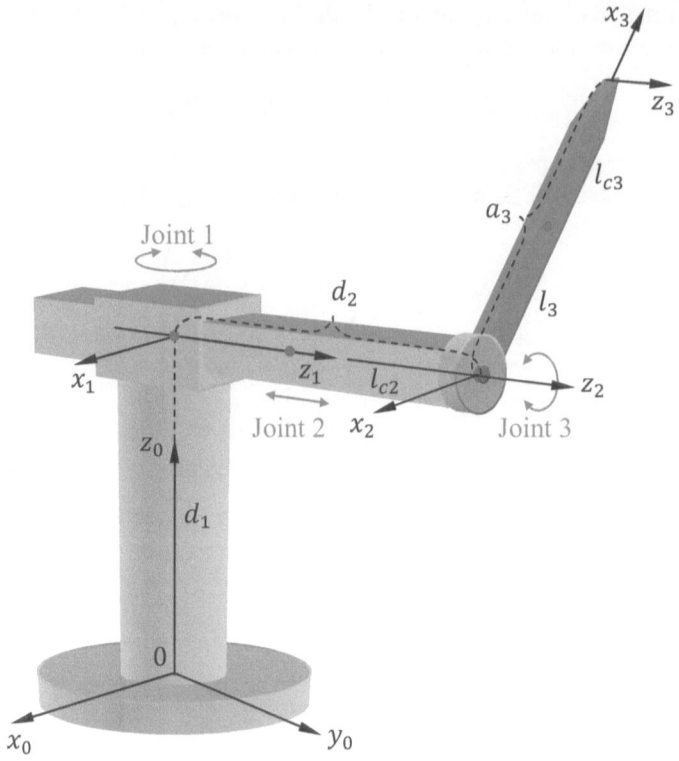

Figure 4.2 An RPR 3-joint 3D robot manipulator

Joint Angle θ_i	Joint Offset d_i	Twist Angle α_i	Link Length a_i
θ_1	d_1	$-90°$	0
$\theta_2 = 0$	d_2	0	0
θ_3	0	0	a_3

Based on the D-H table, all the three one-step homogeneous transformations can be formulated:

$$A_0^1 = \begin{pmatrix} c_1 & 0 & -s_1 & 0 \\ s_1 & 0 & c_1 & 0 \\ 0 & -1 & 0 & d_1 \\ 0 & 0 & 0 & 1 \end{pmatrix}, \quad A_1^2 = \begin{pmatrix} 1 & 0 & 0 & 0 \\ 0 & 1 & 0 & 0 \\ 0 & 0 & 1 & d_2 \\ 0 & 0 & 0 & 1 \end{pmatrix},$$

$$\text{and} \quad A_2^3 = \begin{pmatrix} c_3 & -s_3 & 0 & a_3c_3 \\ s_3 & c_3 & 0 & a_3s_3 \\ 0 & 0 & 1 & 0 \\ 0 & 0 & 0 & 1 \end{pmatrix}.$$

Then,

$$A_0^3 = A_0^1 A_1^2 A_2^3 = \begin{pmatrix} c_1 c_3 & -c_1 s_3 & -s_1 & a_3 c_1 c_3 - d_2 s_1 \\ s_1 c_3 & -s_1 s_3 & c_1 & a_3 s_1 c_3 + d_2 c_1 \\ -s_3 & -c_3 & 0 & d_1 - a_3 s_3 \\ 0 & 0 & 0 & 1 \end{pmatrix}.$$

It can be seen that the last column of the above A_0^3 is the robot tip position p_0^3, i.e., the origin of frame 3 with respect to the base frame 0, while the upper left 3 by 3 block R_0^3 is the orientation of frame 3 referred to the base. Similar to the last example, in order to derive the symbolical inertial matrix W for this 3D robot, we adopt equation (3.16) again along with the sub-Jacobian matrix formation (3.14). As required, every inverse of the homogeneous transformations is needed to be derived,

$$A_1^0 = \begin{pmatrix} c_1 & s_1 & 0 & 0 \\ 0 & 0 & -1 & d_1 \\ -s_1 & c_1 & 0 & 0 \\ 0 & 0 & 0 & 1 \end{pmatrix}, \quad A_2^0 = \begin{pmatrix} c_1 & s_1 & 0 & 0 \\ 0 & 0 & -1 & d_1 \\ -s_1 & c_1 & 0 & -d_2 \\ 0 & 0 & 0 & 1 \end{pmatrix},$$

$$A_2^1 = \begin{pmatrix} 1 & 0 & 0 & 0 \\ 0 & 1 & 0 & 0 \\ 0 & 0 & 1 & -d_2 \\ 0 & 0 & 0 & 1 \end{pmatrix}, \quad A_3^0 = \begin{pmatrix} c_1 c_3 & s_1 c_3 & -s_3 & d_1 s_3 - a_3 \\ -c_1 s_3 & -s_1 s_3 & -c_3 & d_1 c_3 \\ -s_1 & c_1 & 0 & -d_2 \\ 0 & 0 & 0 & 1 \end{pmatrix},$$

$$A_3^1 = \begin{pmatrix} c_3 & s_3 & 0 & -a_3 \\ -s_3 & c_3 & 0 & 0 \\ 0 & 0 & 1 & -d_2 \\ 0 & 0 & 0 & 1 \end{pmatrix}, \quad A_3^2 = \begin{pmatrix} c_3 & s_3 & 0 & -a_3 \\ -s_3 & c_3 & 0 & 0 \\ 0 & 0 & 1 & 0 \\ 0 & 0 & 0 & 1 \end{pmatrix}.$$

Once all the homogeneous transformations are prepared, the three sub-Jacobian matrices can be found based on (3.14) with a notice that the second joint is prismatic, and they are

$$J_1 = \begin{pmatrix} 0 & 0 & 0 \\ 0 & 0 & 0 \\ 0 & 0 & 0 \\ 0 & 0 & 0 \\ -1 & 0 & 0 \\ 0 & 0 & 0 \end{pmatrix}, \quad J_2 = \begin{pmatrix} -d_2 & 0 & 0 \\ 0 & 0 & 0 \\ 0 & 1 & 0 \\ 0 & 0 & 0 \\ -1 & 0 & 0 \\ 0 & 0 & 0 \end{pmatrix}, \quad J_3 = \begin{pmatrix} -d_2 c_3 & 0 & 0 \\ d_2 s_3 & 0 & a_3 \\ a_3 c_3 & 1 & 0 \\ -s_3 & 0 & 0 \\ -c_3 & 0 & 0 \\ 0 & 0 & 1 \end{pmatrix}.$$

The next step is to define the following three dynamic parameter matrices:

$$U_1 = \begin{pmatrix} m_1 & 0 & 0 & 0 & 0 & 0 \\ 0 & m_1 & 0 & 0 & 0 & 0 \\ 0 & 0 & m_1 & 0 & 0 & 0 \\ 0 & 0 & 0 & I_{x1} & 0 & 0 \\ 0 & 0 & 0 & 0 & I_{y1} & 0 \\ 0 & 0 & 0 & 0 & 0 & I_{z1} \end{pmatrix},$$

$$U_2 = \begin{pmatrix} m_2 & 0 & 0 & 0 & -m_2 l_{c2} & 0 \\ 0 & m_2 & 0 & m_2 l_{c2} & 0 & 0 \\ 0 & 0 & m_2 & 0 & 0 & 0 \\ 0 & m_2 l_{c2} & 0 & I_{x2} & 0 & 0 \\ -m_2 l_{c2} & 0 & 0 & 0 & I_{y2} & 0 \\ 0 & 0 & 0 & 0 & 0 & I_{z2} \end{pmatrix},$$

$$U_3 = \begin{pmatrix} m_3 & 0 & 0 & 0 & 0 & 0 \\ 0 & m_3 & 0 & 0 & 0 & -m_3 l_{c3} \\ 0 & 0 & m_3 & 0 & m_3 l_{c3} & 0 \\ 0 & 0 & 0 & I_{x3} & 0 & 0 \\ 0 & 0 & m_3 l_{c3} & 0 & I_{y3} & 0 \\ 0 & -m_3 l_{c3} & 0 & 0 & 0 & I_{z3} \end{pmatrix},$$

and once again all the inertia tensors Γ_j's are considered to be diagonal under the assumption that each link is geometrically symmetric and mass-density uniformly distributed.

It is time now to calculate the inertial matrix W_j for each link.

$$W_1 = J_1^T U_1 J_1 = \begin{pmatrix} I_{y1} & 0 & 0 \\ 0 & 0 & 0 \\ 0 & 0 & 0 \end{pmatrix},$$

$$W_2 = J_2^T U_2 J_2 = \begin{pmatrix} m_2 d_2^2 - 2m_2 d_2 l_{c2} + I_{y2} & 0 & 0 \\ 0 & m_2 & 0 \\ 0 & 0 & 0 \end{pmatrix},$$

and $W_3 = J_3^T U_3 J_3 =$

$$\begin{pmatrix} w_{11} & m_3(a_3 - l_{c3})c_3 & m_3 d_2(a_3 - l_{c3})s_3 \\ m_3(a_3 - l_{c3})c_3 & m_3 & 0 \\ m_3 d_2(a_3 - l_{c3})s_3 & 0 & m_3 a_3^2 - 2m_3 a_3 l_{c3} + I_{z3} \end{pmatrix},$$

where $w_{11} = m_3 d_2^2 + m_3(a_3 - 2l_{c3})a_3 c_3^2 + I_{x3} s_3^2 + I_{y3} c_3^2$.

Since the moments of inertia I_{y2}, I_{y3} and I_{z3} are currently referred to their individual link frames, instead of their mass centers, we may employ the parallel-axis theorem to convert them to refer to the axes passing through the mass centers. Namely,

$$I_{y2} = I_{y2}^c + m_2 l_{c2}^2, \quad I_{y3} = I_{y3}^c + m_3 l_{c3}^2, \quad \text{and} \quad I_{z3} = I_{z3}^c + m_3 l_{c3}^2.$$

The other two: I_{y1} and I_{x3} will have no change because both have already been referred to the axes passing through their mass centers. In addition, the last link length $a_3 = l_{c3} + l_3$, as indicated in Figure 4.2, such that $a_3 - l_{c3} = l_3$. Then, W_1 keeps the same, while the W_2 and W_3 are updated to be

$$W_2 = J_2^T U_2 J_2 = \begin{pmatrix} m_2(d_2 - l_{c2})^2 + I_{y2}^c & 0 & 0 \\ 0 & m_2 & 0 \\ 0 & 0 & 0 \end{pmatrix},$$

and

$$W_3 = J_3^T U_3 J_3 = \begin{pmatrix} m_3 d_2^2 + m_3 l_3^2 c_3^2 + I_{x3} s_3^2 + I_{y3}^c c_3^2 & m_3 l_3 c_3 & m_3 d_2 l_3 s_3 \\ m_3 l_3 c_3 & m_3 & 0 \\ m_3 d_2 l_3 s_3 & 0 & m_3 l_3^2 + I_{z3}^c \end{pmatrix}.$$

Finally, the total inertial matrix for the 3-joint 3D robot should be $W = W_1 + W_2 + W_3$.

Because we have already had the homogeneous transformation A_0^3, which consists of

$$p_0^3 = \begin{pmatrix} a_3 c_1 c_3 - d_2 s_1 \\ a_3 s_1 c_3 + d_2 c_1 \\ d_1 - a_3 s_3 \end{pmatrix}, \quad \text{and} \quad R_0^3 = \begin{pmatrix} c_1 c_3 & -c_1 s_3 & -s_1 \\ s_1 c_3 & -s_1 s_3 & c_1 \\ -s_3 & -c_3 & 0 \end{pmatrix}.$$

Since this robot has three joints, it can only guarantee a 3D translational motion as the output. Its kinematic Jacobian matrix can be derived by

$$J_0 = \frac{\partial p_0^3}{\partial q} = \begin{pmatrix} -a_3 s_1 c_3 - d_2 c_1 & -s_1 & -a_3 c_1 s_3 \\ a_3 c_1 c_3 - d_2 s_1 & c_1 & -a_3 s_1 s_3 \\ 0 & 0 & -a_3 c_3 \end{pmatrix}.$$

Due to the determinant $\det(J_0) = a_3 d_2 c_3$, the robot has singular points at $\theta_3 = \pm 90°$ as the last link is vertically up or down. Based on Theorem 2.9, it has a multi-configuration case, and we have to compensate it for finding an overall qualified embedding.

Let us now try the mass center position vector p_0^{c3} and a linear combination of the three columns r_x, r_y and r_z of the rotation matrix R_0^3 to form an isometric embedding for link 3, i.e.,

$$z_3 = \zeta_3(q) = \begin{pmatrix} \sqrt{m_3} \, p_0^{c3} \\ b_1 r_x \\ b_2 r_y \\ b_3 r_z \end{pmatrix}.$$

Its Jacobian is an 11 by 3 matrix (the last element of r_z is zero):

$$J_3 = \begin{pmatrix} -\sqrt{m_3}(l_3 s_1 c_3 + d_2 c_1) & -\sqrt{m_3} s_1 & -\sqrt{m_3} l_3 c_1 s_3 \\ \sqrt{m_3}(l_3 c_1 c_3 - d_2 s_1) & \sqrt{m_3} c_1 & -\sqrt{m_3} l_3 s_1 s_3 \\ 0 & 0 & -\sqrt{m_3} l_3 c_3 \\ -b_1 s_1 c_3 & 0 & -b_1 c_1 s_3 \\ b_1 c_1 c_3 & 0 & -b_1 s_1 s_3 \\ 0 & 0 & -b_1 c_3 \\ b_2 s_1 s_3 & 0 & -b_2 c_1 c_3 \\ -b_2 c_1 s_3 & 0 & -b_2 s_1 c_3 \\ 0 & 0 & b_2 s_3 \\ -b_3 c_1 & 0 & 0 \\ -b_3 s_1 & 0 & 0 \end{pmatrix}.$$

Thus, the Riemannian metric for link 3 becomes

$$W_3 = J_3^T J_3 = \begin{pmatrix} m_3 l_3^2 c_3^2 + m_3 d_2^2 + b_1^2 c_3^2 + b_2^2 s_3^2 + b_3^2 & m_3 l_3 c_3 & m_3 d_2 l_3 s_3 \\ m_3 l_3 c_3 & m_3 & 0 \\ m_3 d_2 l_3 s_3 & 0 & m_3 l_3^2 + b_1^2 + b_2^2 \end{pmatrix}.$$

Let us augment z_3 of link 3 by z_2 of the prismatic link with the mass center position p_0^{c2} together to form a 13-dimensional final version of the overall isometric embedding:

$$z = \zeta(q) = \begin{pmatrix} \sqrt{m_3} p_0^{c3} \\ b_1 r_x \\ b_2 r_y \\ b_3 r_z \\ -\sqrt{m_2}(d_2 - l_{c2})s_1 \\ \sqrt{m_2}(d_2 - l_{c2})c_1 \end{pmatrix}.$$

After its 13 by 3 Jacobian matrix J and the 3 by 3 Riemanian metric $W = J^T J$ are derived, all the dynamic parameters can be matched with those in the inertial matrix $W = W_1 + W_2 + W_3$ of the 3D robot, and under the condition $I_{z3}^c = I_{y3}^c$ for link 3, they are

$$b_1 = \sqrt{I_{z3}^c - I_{x3}/2}, \quad b_2 = \sqrt{I_{x3}/2}, \quad b_3 = \sqrt{b_2^2 + I_{y2}^c + I_{y1}}.$$

If the condition is $I_{z3}^c = I_{y3}^c + I_{x3}$ for link 3, then

$$b_1 = \sqrt{I_{y3}^c}, \quad b_2 = \sqrt{I_{x3}}, \quad b_3 = \sqrt{I_{y2}^c + I_{y1}}.$$

In summary, through the above two typical examples and their detailed kinematics and dynamics modeling, analyses and discussions, let us now state and prove the following theorem and then provide the theorem with mathematical and physical explanations and interpretations:

Theorem 4.2. *For a single rigid body in up to 6 d.o.f. motion within a compact region of $B(3) \times SO(3)$, its up to $(n = 6)$-dimensional configuration manifold (C-manifold) M^n with the local coordinates q can be isometrically embedded into a Euclidean 9-space $\mathbb{R}^9$ by*

$$z = \zeta(q) = \begin{pmatrix} \sqrt{m} p_0^c(q) \\ r_{c(0)}^1(q) \\ r_{c(0)}^2(q) \end{pmatrix}, \tag{4.4}$$

provided that the Jacobian matrix of $\zeta(q)$ is full-ranked, where $r_{c(0)}^1$ and $r_{c(0)}^2$ are two 3-dimensional radial vectors both tailed at the mass center and their two arrow points are fixed on the body but not lined up with the mass center of the body, and they are all projected on the base, while p_0^c is the 3 by 1 mass center position vector with respect to the base.

Proof. First, the C-manifold M^n is compact and non-singular due to its full-ranked Jacobian matrix. Then,

$$\dot{z} = J\dot{q} = \begin{pmatrix} \sqrt{m} \dot{p}_0^c \\ \dot{r}_{c(0)}^1 \\ \dot{r}_{c(0)}^2 \end{pmatrix}.$$

Since each $r^i_{c(0)} = R^c_0 r^i_c$ and r^i_c is fixed on the rigid body, $\dot{r}^i_{c(0)} = \dot{R}^c_0 r^i_c + R^c_0 \dot{r}^i_c$, and the second term should vanish. However, the first term $\dot{R}^c_0 r^i_c = \dot{R}^c_0 R^c_c R^c_0 r^i_c = \Omega_{(0)} r^i_{c(0)} = \omega_{(0)} \times r^i_{c(0)} = R^c_0 (\omega \times r^i_c)$, so that

$$\dot{z}^T \dot{z} = m \dot{p}^{cT}_0 \dot{p}^c_0 + (\omega \times r^1_c)^T (\omega \times r^1_c) + (\omega \times r^2_c)^T (\omega \times r^2_c).$$

After the inner product, the projection matrix R^c_0 has simply been canceled out due to $R^c_0 (R^c_0)^T = I$. Thus, each r^i_c in the above inner product can be projected onto any frame, and, of course, the best projection frame is its own link frame.

Let a skew-symmetric matrix $S(r^i_c) = r^i_c \times$ for $i = 1, 2$. Then

$$\frac{1}{2} \dot{z}^T \dot{z} = \frac{1}{2} m \dot{p}^{cT}_0 \dot{p}^c_0 + \frac{1}{2} \omega^T [S(r^1_c)^T S(r^1_c) + S(r^2_c)^T S(r^2_c)] \omega.$$

Because for each i, $r^i_c = (r^i_x \ r^i_y \ r^i_z)^T$, and

$$S(r^i_c)^T S(r^i_c) = \begin{pmatrix} 0 & r^i_z & -r^i_y \\ -r^i_z & 0 & r^i_x \\ r^i_y & -r^i_x & 0 \end{pmatrix} \begin{pmatrix} 0 & -r^i_z & r^i_y \\ r^i_z & 0 & -r^i_x \\ -r^i_y & r^i_x & 0 \end{pmatrix}$$

$$= \begin{pmatrix} r^{i2}_y + r^{i2}_z & -r^i_x r^i_y & -r^i_z r^i_x \\ -r^i_x r^i_y & r^{i2}_z + r^{i2}_x & -r^i_y r^i_z \\ -r^i_z r^i_x & -r^i_y r^i_z & r^{i2}_x + r^{i2}_y \end{pmatrix} = r^{iT}_c r^i_c I_3 - r^i_c r^{iT}_c,$$

where I_3 is the 3 by 3 identity. This implies that

$$\frac{1}{2} \dot{z}^T \dot{z} = \frac{1}{2} m \dot{p}^{cT}_0 \dot{p}^c_0 + \frac{1}{2} \omega^T \sum_{i=1}^{2} (r^{iT}_c r^i_c I_3 - r^i_c r^{iT}_c) \omega.$$

If we can adjust the two arrow points for both r^1_c and r^2_c such that

$$\sum_{i=1}^{2} (r^{iT}_c r^i_c I_3 - r^i_c r^{iT}_c) = \Gamma_c = \int_M (r^T_c r_c I_3 - r_c r^T_c) dm, \qquad (4.5)$$

the inertia tensor of the rigid body based on equation (3.3), then

$$\frac{1}{2} \dot{z}^T \dot{z} = \frac{1}{2} m \dot{p}^{cT}_0 \dot{p}^c_0 + \frac{1}{2} \omega^T \Gamma_c \omega = K.$$

This proves that (4.4) is an isometric embedding to send M^6 to $\mathbb{R}^9$. □

This theorem actually suggests a **3-point model**: one is the mass center, and two are the arrow points of the radial vector r^1_c and r^2_c. They are not lined up so that the three points can uniquely determine the orientation of the rigid body. If a coordinate frame is defined with the origin at the mass center c and fixed on the rigid body, the orientation of the body can be described by a rotation matrix R^c_0 that is seen at the base. The three coordinate axes, as seen at the body frame, can be represented by the three column vectors x_c, y_c and z_c of the identity I_3. Thus, each radial vector r_c^i seen at the body frame is formed by a linear combination of the three axes of the body frame, i.e.,

$$r_c^i = R_c^0 r_{c(0)}^i = b_1^i x_c + b_2^i y_c + b_3^i z_c = \begin{pmatrix} b_1^i \\ b_2^i \\ b_3^i \end{pmatrix}, \quad \text{for } i = 1, 2.$$

Therefore, the adjustment of the two arrow points to meet equation (4.5) is actually to solve for the six coefficients b_1^i, b_2^i and b_3^i for $i = 1, 2$ in terms of the six moments of inertia I_{xx}, I_{yy}, and I_{zz}, as well as I_{xy}, I_{yz} and I_{zx} in the inertia tensor Γ_c.

However, a more straightforward way is simply to augment all the three columns r_x, r_y and r_z of R_0^c to form a 9 by 1 vector with each carrying an individual coefficient. Then, the new isometric embedding is formed as a following 12 by 1 column vector:

$$z = \zeta(q) = \begin{pmatrix} \sqrt{m} p_0^c(q) \\ b_1 r_x(q) \\ b_2 r_y(q) \\ b_3 r_z(q) \end{pmatrix}. \tag{4.6}$$

With such an augmentation, though adding three more dimensions, the arrow point adjustment can be significantly simplified as a special case, i.e.,

$$r_c^1 = b_1 x_c = \begin{pmatrix} b_1 \\ 0 \\ 0 \end{pmatrix}, \quad r_c^2 = b_2 y_c = \begin{pmatrix} 0 \\ b_2 \\ 0 \end{pmatrix}, \quad r_c^3 = b_3 z_c = \begin{pmatrix} 0 \\ 0 \\ b_3 \end{pmatrix}.$$

Then,

$$\sum_{i=1}^3 (r_c^{iT} r_c^i I_3 - r_c^i r_c^{iT}) = \begin{pmatrix} b_2^2 + b_3^2 & 0 & 0 \\ 0 & b_3^2 + b_1^2 & 0 \\ 0 & 0 & b_1^2 + b_2^2 \end{pmatrix}$$

becomes a diagonal matrix. If one defines $b_2^2 + b_3^2 = I_{xx}$, $b_3^2 + b_1^2 = I_{yy}$, and $b_1^2 + b_2^2 = I_{zz}$, the inertia tensor Γ_c is a diagonal matrix that is well-suitable to modeling many geometrically symmetric in shape and mass density uniformly distributed rigid bodies. In this case, all the coefficients can be solved uniquely as

$$\begin{cases} b_1^2 = 0.5(I_{yy} + I_{zz} - I_{xx}) \\ b_2^2 = 0.5(I_{zz} + I_{xx} - I_{yy}) \\ b_3^2 = 0.5(I_{xx} + I_{yy} - I_{zz}). \end{cases} \tag{4.7}$$

4.3 APPLICATIONS TO ROBOT DYNAMICS MODELING

We have thus far developed an isometric embedding formula for a planar robot in equation (4.3), and also more general forms of isometric embedding in equations (4.4) and (4.6) for 3D robotic systems. Each form of isometric embedding has not been too far away from the targeting inertial matrix W_j of link j for an n-joint robotic system. In other words, if we wish to find the inertial matrix W_j of link j for any $j = 1, \cdots, n$, we need first to derive a symbolical form of the homogeneous transformation A_0^j, and then to construct the isometric embedding $z_j = \zeta_j(q)$ by following (4.4) or (4.6). After that, taking derivative to find its Jacobian matrix $J_j = \frac{\partial \zeta_j}{\partial q}$, the inertial matrix $W_j = J_j^T J_j$ for link j can be immediately achieved.

Obviously, the nonzero block in the n by n inertial matrix W_j of link j is only j by j, and thus, the first link has only a single nonzero scalar in W_1. This means that as j is counting down from n to $j = 1$, W_j is getting simpler. The total inertial matrix of the n-joint robot should be the sum of every W_j, i.e.,

$$W = \sum_{j=1}^{n} W_j = \sum_{j=1}^{n} J_j^T J_j = \sum_{j=1}^{n} \left(\frac{\partial \zeta_j}{\partial q} \right)^T \frac{\partial \zeta_j}{\partial q}. \tag{4.8}$$

The above equation provides us with a new alternative approach to determining the inertial matrix W for a robot. The entire procedure completely relies on the homogeneous transformation A_0^j for $j = 1, \cdots, n$, and all the four columns of each A_0^j are then augmented together to form an isometric embedding $z_j = \zeta_j(q)$ for link j by equation (4.6) towards finding the corresponding $W_j = J_j^T J_j$. This procedure evidently shows that *the kinematic model of a robotic system is key essential for developing its dynamic model*. Once again, at a mathematical perspective, *kinematics determines a topological structure of the robot C-manifold, while dynamics is just to linearly combine all the kinematic entities (positions and orientations) with dynamic parameters as a geometrical "decoration"* [3].

Under the same reason, the compact dynamic equation (4.2) can also be decomposed as follows:

$$\tau + \tau_g = J^T \ddot{z} = \sum_{j=1}^{n} J_j^T \ddot{z}_j.$$

Since for each link, $J_j^T \ddot{z}_j = J_j^T J_j \ddot{q} + J_j^T \dot{J}_j \dot{q} = W_j \ddot{q} + c_j(q, \dot{q})$, clearly, the centrifugal and Coriolis term

$$c(q, \dot{q}) = \left(W_d^T - \frac{1}{2} W_d \right) \dot{q} = \sum_{j=1}^{n} (J_j^T \dot{J}_j) \dot{q}.$$

Now, let each joint position q_i be replaced by the dual number $q_i + \varepsilon \dot{q}_i$ for $i = 1, \cdots, n$. Then, based on equation (2.37),

$$J^T \hat{J}(q + \varepsilon \dot{q}) = J^T (J + \varepsilon \dot{J}) = W + \varepsilon J^T \dot{J},$$

and its real part is just the inertial matrix, while the dual part after post-multiplied by $\dot{q}$ will become the centrifugal and Coriolis term $c(q, \dot{q}) = J^T \dot{J} \dot{q}$. Furthermore, if it is post-multiplied by $\dot{q} + \varepsilon \ddot{q}$, then,

$$J^T (J + \varepsilon \dot{J})(\dot{q} + \varepsilon \ddot{q}) = W \dot{q} + \varepsilon (W \ddot{q} + J^T \dot{J} \dot{q}) = p + \varepsilon (\tau + \tau_g),$$

The Approaches	The Strength	The Shortcoming
Directly finding the kinetic energy K	Effective for many simple planar robot cases	It will be soon confused as no. of joints increasing
Sub-Jacobian (3.14) with equation (3.16)	Excellent for numerical computations, number of joints can be very high	Require to invert a lot of homogeneous transformations
Isometric embedding equation (4.3), (4.4) or (4.6)	Excellent for both symbolical and numerical derivations with easier finding centrifugal and Coriolis $c(q, \dot{q})$	Number of joints is medium but cannot be too high

Table 4.1

A comparison among three approaches to finding the inertial matrix W

where the real part $p = W\dot{q}$ is the momentum of the robotic system and the dual part is the entire Euler-Lagrange equation. Therefore, if all the dual number operations would be built-in MATLAB^{TM}, the isometric embedding based dynamic model of robotic systems could be programmed and numerically calculated in computer much more effectively.

Table 4.1 gives a comparison among the three approaches to determining the inertial matrix W for a robotic system.

By summing up what we have investigated thus far, the method of isometric embedding to deal with the dynamics modeling for an open serial-chain robotic system has the following features and possible new applications:

1. The last link is always the "busiest" rigid body, the kinematics of which depends on every joint position and velocity. If we can find an isometric embedding $z_n = \zeta_n(q)$ for the last link n, as a **dominant body** of the robotic system, it will not only be the key factor to derive the inertial matrix W_n and its dynamic equation $J_n^T \ddot{z}_n$ of link n, but will also have a chance to cover the overall inertial matrix W of the robot by adjusting the necessary dynamic parameters. If the robot has a multi-configuration issue, the last link isometric embedding $z_n = \zeta_n(q)$ may have to be augmented by more isometric embeddings $z_j = \zeta_j(q)$ from the lower links. Then, the dimension of the overall isometric embedding $z = \zeta(q)$ has to be expanded.

2. Regardless the dimension expansion, to find an isometric embedding $z_j = \zeta_j(q)$ for each link of an n-joint open serial-chain robotic system is actually an easy job. Namely, derive the homogeneous transformation A_0^j for link j first once the D-H table and all the one-step homogeneous transformations are prepared. Then, by following either equation (4.3) for a planar robot case, or (4.4) or (4.6) for 3D robotic systems, the isometric embedding $z_j = \zeta_j(q)$ can be constructed by augmenting every column of A_0^j with certain dynamic parameters combined. However, it is important that these columns must be projected on the base frame.

3. As pointed above, the isometric embedding $z_n = \zeta_n(q)$ for the last link n may not always be able to cover the entire robot inertial matrix W. However, if a robot manipulator is carrying a heavy tool or load, such as a big gripper for material handling tasks or a heavy welding gun for car body spot-welding applications, we want to know how much additional joint torque should be imposed for the robot to handle this heavy-load operation? In this case, we definitely cannot use the statics equation given in (2.66) to answer this question, because the heavy tool or load is now moving in 3D space, instead of just being acted by a static wrench. With the isometric embedding $z_n = \zeta_n(q)$ for link n as a dominant body, we can thus determine the imposed additional joint torque vector due to the heavy tool/load dynamic motion. Namely,

$$\tau_{tool} = J_n^T \ddot{z}_n - \tau_g,$$

where τ_g is the joint torque vector due to gravity of the tool.

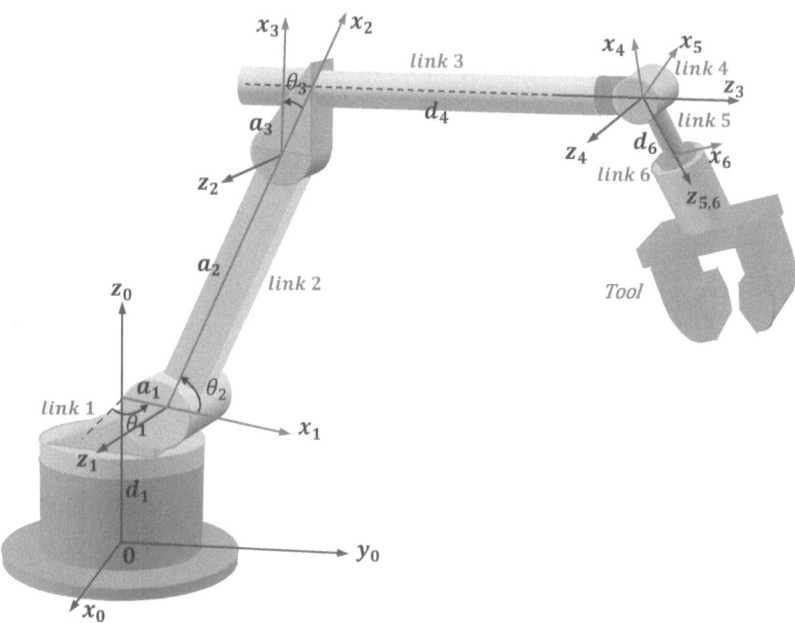

Figure 4.3   A 6-revolute-joint industrial robot manipulator

Let us take a look at a real industrial robot with six revolute joints, as shown in Figure 4.3. We are aiming to find the isometric embedding $z_6 = \zeta_6(q)$ for link 6, the last kinematic "busiest" link, which is often mounted with a heavy tool, such as a welding gun or a big gripper, or carrying a heavy load. Once all the z and x axes are assigned based on the D-H convention, its D-H parameter table can be immediately determined:

Joint Angle θ_i	Joint Offset d_i	Twist Angle α_i	Link Length a_i
θ_1	d_1	90°	a_1
θ_2	0	0	a_2
θ_3	0	90°	a_3
θ_4	d_4	−90°	0
θ_5	0	90°	0
θ_6	d_6	0	0

By following the D-H table, every one-step homogeneous transformation can be readily formulated,

$$
A_0^1 = \begin{pmatrix} c_1 & 0 & s_1 & a_1 c_1 \\ s_1 & 0 & -c_1 & a_1 s_1 \\ 0 & 1 & 0 & d_1 \\ 0 & 0 & 0 & 1 \end{pmatrix}, \quad
A_1^2 = \begin{pmatrix} c_2 & -s_2 & 0 & a_2 c_2 \\ s_2 & c_2 & 0 & a_2 s_2 \\ 0 & 0 & 1 & 0 \\ 0 & 0 & 0 & 1 \end{pmatrix},
$$

$$
A_2^3 = \begin{pmatrix} c_3 & 0 & s_3 & a_3 c_3 \\ s_3 & 0 & -c_3 & a_3 s_3 \\ 0 & 1 & 0 & 0 \\ 0 & 0 & 0 & 1 \end{pmatrix}, \quad
A_3^4 = \begin{pmatrix} c_4 & 0 & -s_4 & 0 \\ s_4 & 0 & c_4 & 0 \\ 0 & -1 & 0 & d_4 \\ 0 & 0 & 0 & 1 \end{pmatrix},
$$

$$
A_4^5 = \begin{pmatrix} c_5 & 0 & s_5 & 0 \\ s_5 & 0 & -c_5 & 0 \\ 0 & 1 & 0 & 0 \\ 0 & 0 & 0 & 1 \end{pmatrix}, \quad
A_5^6 = \begin{pmatrix} c_6 & -s_6 & 0 & 0 \\ s_6 & c_6 & 0 & 0 \\ 0 & 0 & 1 & d_6 \\ 0 & 0 & 0 & 1 \end{pmatrix}.
$$

Then, through a careful symbolical derivation, we obtain

$$
A_0^6 = \begin{pmatrix} R_0^6 & p_0^6 \\ 0_3 & 1 \end{pmatrix} = \begin{pmatrix} r_x^6 & r_y^6 & r_z^6 & p_0^6 \\ 0 & 0 & 0 & 1 \end{pmatrix}
$$

$$
= \begin{pmatrix} A_0^5(:,1)c_6 + A_0^5(:,2)s_6 & -A_0^5(:,1)s_6 + A_0^5(:,2)c_6 & A_0^5(:,3) & p_0^5 + d_6 A_0^5(:,3) \\ 0 & 0 & 0 & 1 \end{pmatrix},
$$

where $A_0^5(:,j)$ is the j-th column of A_0^5 without the last constant 0 or 1, and they are given in detail as follows:

$$
A_0^5(:,1) = \begin{pmatrix} (c_1 c_{23} c_4 + s_1 s_4)c_5 - c_1 s_{23} s_5 \\ (s_1 c_{23} c_4 - c_1 s_4)c_5 - s_1 s_{23} s_5 \\ s_{23} c_4 c_5 + c_{23} s_5 \end{pmatrix}, \quad
A_0^5(:,2) = \begin{pmatrix} -c_1 c_{23} s_4 + s_1 c_4 \\ -s_1 c_{23} s_4 - c_1 c_4 \\ -s_{23} s_4 \end{pmatrix},
$$

$$
A_0^5(:,3) = \begin{pmatrix} (c_1 c_{23} c_4 + s_1 s_4)s_5 + c_1 s_{23} c_5 \\ (s_1 c_{23} c_4 - c_1 s_4)s_5 + s_1 s_{23} c_5 \\ s_{23} c_4 s_5 - c_{23} c_5 \end{pmatrix},
$$

while

$$
p_0^5 = A_0^5(:,4) = \begin{pmatrix} a_1 c_1 + a_2 c_1 c_2 + a_3 c_1 c_{23} + d_4 c_1 s_{23} \\ a_1 s_1 + a_2 s_1 c_2 + a_3 s_1 c_{23} + d_4 s_1 s_{23} \\ d_1 + a_2 s_2 + a_3 s_{23} - d_4 c_{23} \end{pmatrix}.
$$

Once A_0^6 is symbolically derived, a 12-dimensional isometric embedding $z_6 = \zeta_6(q)$ for link 6, including the tool mounted on link 6, can be constructed by augmenting all the four columns of A_0^6 together along with certain dynamic parameters $\sqrt{m_6}, b_1, b_2$ and b_3 to be carried in, i.e.,

$$z_6 = \zeta_6(q) = \begin{pmatrix} \sqrt{m_6}\, p_0^{c6} \\ b_1 r_x^6 \\ b_2 r_y^6 \\ b_3 r_z^6 \end{pmatrix} = \begin{pmatrix} \sqrt{m_6}\, p_0^{c6} \\ b_1 A_0^6(:,1) \\ b_2 A_0^6(:,2) \\ b_3 A_0^6(:,3) \end{pmatrix} = \begin{pmatrix} \sqrt{m_6}\,(p_0^5 + d_{c6} A_0^5(:,3)) \\ b_1 (A_0^5(:,1)c_6 + A_0^5(:,2)s_6) \\ b_2 (-A_0^5(:,1)s_6 + A_0^5(:,2)c_6) \\ b_3 A_0^5(:,3) \end{pmatrix},$$

where each b_i can be solved by (4.7), and d_6 is now replaced by d_{c6} as the new joint offset parameter along the z_5-axis from the x_5-axis to the mass center of the mounted tool, as required by p_0^{c6} in the embedding. Finally, the 12 by 6 Jacobian matrix J_6 for the last link plus the heavy tool or load can be derived by taking derivative of $\zeta_6(q)$ with respect to q.

After the Jacobian matrix J_6 is completed, we no longer need more symbolical derivations. At this time, the compact dynamic equation (4.2) can be directly employed to perform a tool/load **trajectory-tracking control in z-space** by the following control input

$$u = \tau_{tool} = J_6^T [\ddot{z}_6^d + K_2(\dot{z}_6^d - \dot{z}_6) + K_1(z_6^d - z_6)] - \tau_g, \tag{4.9}$$

where z_6^d and its time derivatives $\dot{z}_6^d$ and $\ddot{z}_6^d$ are the desired dynamic motion trajectory of the tool/load, and $K_1 > 0$ and $K_2 > 0$ are the PD control gain constants, while the joint torque due to gravity $\tau_g = -\frac{\partial P_{tool}}{\partial q}$, and the potential energy $P_{tool} = m_6 g p_0^{c6}(3)$ requires the z-component of the position vector p_0^{c6} as well as the total mass m_6 of the tool/load. According to (4.9), the trajectory-tracking control scheme performed in the ambient Euclidean z-space after the isometric embedding is determined can be illustrated by a block diagram in Figure 4.4.

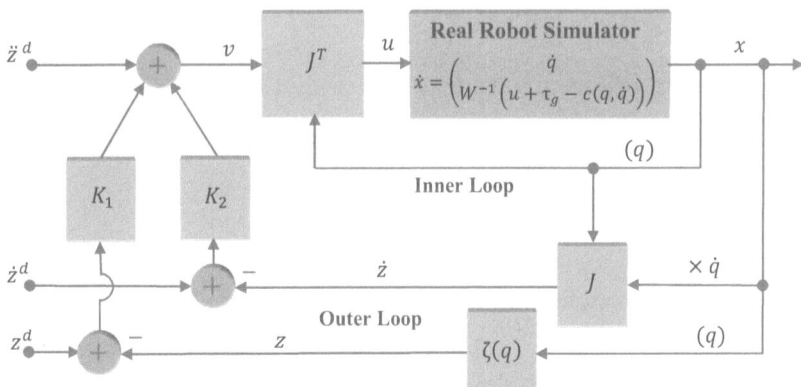

Figure 4.4   A block-diagram for the tool trajectory-tracking control in z-space

Note that the proposed trajectory-tracking control (4.9) in z-space includes both the tool translational and rotational trajectories, because the general form of an isometric embedding $z = \zeta(q)$ can always take care of the motion within the compact region $B(3) \times SO(3)$ so that it should cover both position and orientation in nature. This typical industrial robot manipulator that carries a heavy tool is depicted in Figure 4.3, which has been studied and verified by a 3D graphical animation in MATLAB^{TM}, as shown in Figure 4.5.

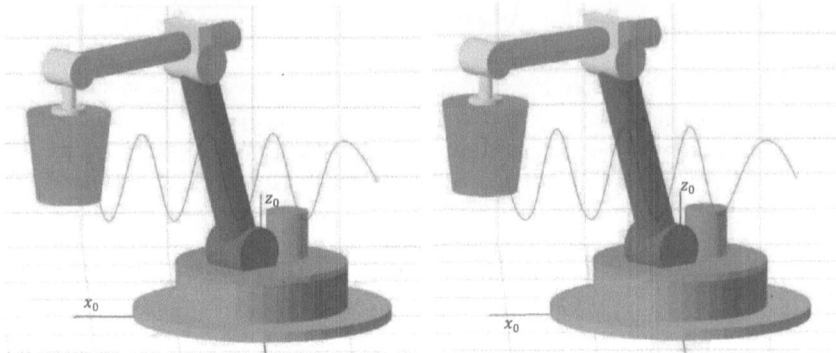

Figure 4.5 The 6-joint industrial robot carrying a heavy tool in a 3D animation study

In the simulation study, we have first tested the effectiveness of a position control of the robot while keeping the tool orientation invariant. The 12-dimensional desired embedding vector z_6^d and its time-derivatives required by equation (4.9) and depicted in Figure 4.4 were specified to allow the mass center of the heavy tool to track a desired sine wave in 3D robot work space without change of its initial orientation. At the home position, the initial orientation of the tool mass center frame with respect to the base is

$$R_0^{c6} = \begin{pmatrix} 1 & 0 & 0 \\ 0 & -1 & 0 \\ 0 & 0 & -1 \end{pmatrix}.$$

Also, the desired sine wave to be tracked is defined by the following parametric equation:

$$\begin{cases} x(t) = 0.3t - 0.5 \\ y(t) = 1.0 \\ z(t) = 0.25\sin(5t) + 1.0. \end{cases}$$

Thus, the desired embedding vector can be specified by

$$z_6^d = \begin{pmatrix} \sqrt{m_6}\,p_d^{c6} \\ b_1 r_1 \\ b_2 r_2 \\ b_3 r_3 \end{pmatrix},$$

where

$$p_d^{c6} = \begin{pmatrix} 0.3t - 0.5 \\ 1.0 \\ 0.25\sin(5t) + 1.0 \end{pmatrix}, \quad r_1 = \begin{pmatrix} 1 \\ 0 \\ 0 \end{pmatrix}, \quad r_2 = \begin{pmatrix} 0 \\ -1 \\ 0 \end{pmatrix}, \quad r_3 = \begin{pmatrix} 0 \\ 0 \\ -1 \end{pmatrix}.$$

Then, $\dot{z}_6^d$ and $\ddot{z}_6^d$ can be found by taking the first and second time-derivatives for z_6^d, respectively, and both have the last 9 zero elements due to the constant initial orientation.

The real plant in this 3D graphical animation study is to solve the Euler-Lagrange equation given by equation (3.22), and the inertial matrix W is numerically calculated in MATLAB^{TM} using the algorithm developed based on equation (3.16) along with the sub-Jacobian equation (3.14). The derivative matrix W_d in equation (3.23) is computed in a subroutine using the two-point symmetric difference quotient approximation formula. In the trajectory-tracking control equation (4.9), the two gain constants are chosen as $K_1 = 49$ and $K_2 = 14$ under the critical damping condition.

As a result, the left picture of Figure 4.5 evidently demonstrates that the mass center of the heavy tool carried by the industrial robot can follow the desired sine wave trajectory in a virtually perfect way until reaching its destination without any initial orientation change. However, this virtually perfect result was performed under a "trivial robot" condition. This means that all the link masses except the last link with the tool are shut off as zero in order to verify if the joint torque determined based only on the isometric embedding $z_6 = \zeta_6(q)$ of the last link along with the heavy tool in (4.9) can exactly control the heavy tool in 6 d.o.f. dynamic motion. Since we purposely set a large initial deviation between the desired starting point and the actual position of the tool mass center, the actual trajectory was not only quickly coming back to catch up the desired trajectory, but was also followed along the 3D desired path nearly perfectly in all the three x, y and z directions.

In addition to testing and verifying the additional robot dynamic joint torque that is imposed by a heavy tool or load, we also tried to use the control law (4.9) to drive and maneuver the entire non-trivial robot (every link mass was restored) that carried the heavy tool, the tracking result was almost the same as that under the trivial robot condition, as clearly shown in the right picture of Figure 4.5.

The second trial has been performed to track the same 3D sine wave trajectory for the tool mass center but to simultaneously control the heavy big tool to follow a desired orientation change path. It was specified by adopting the $k - \phi$ procedure, as deduced in equation (2.2) as well as its inverse solution in equation (2.50). Suppose that the tool frame is desired to rotate about a fixed unity axis defined by $k = (0 \; -\frac{\sqrt{3}}{2} \; -\frac{1}{2})^T$ that is referred to the tool frame by an angle $\phi = \omega t$ with a constant $\omega = 1$ rad/sec, then the initial orientation of the tool frame R_0^{c6} referred to the base should be post-multiplied by

$$R_r = I + \sin(\omega t)K + (1 - \cos(\omega t))K^2,$$

where $K = S(k)$ is the skew-symmetric matrix of the unity vector k. In other words, the rotation matrix $R = R_0^{c6}R_r$ is referred to the base, and each column of this R

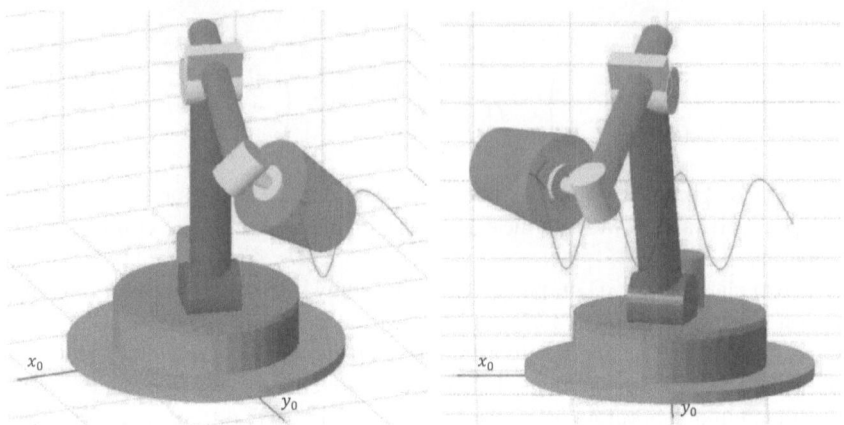

Figure 4.6 The 6-joint industrial robot carrying a heavy tool for control of both position and orientation

should form the last 9 elements of the desired isometric embedding z_6^d in addition to the first 3 elements of the same desired sine wave tracking vector p_d^{c6}. The last 9 elements of the time-derivatives of z_6^d can also be determined by

$$\dot{R} = \cos\phi\,\dot{\phi}K + \sin\phi\,\dot{\phi}K^2, \quad \text{and}$$

$$\ddot{R} = (-\sin\phi\,\dot{\phi}^2 + \cos\phi\,\ddot{\phi})K + (\cos\phi\,\dot{\phi}^2 + \sin\phi\,\ddot{\phi})K^2.$$

Fortunately, in our current case, we define $\omega = \dot{\phi}$ as a constant such that $\ddot{\phi} = 0$. Then, by splitting the three columns of each of the above $\dot{R}$ and $\ddot{R}$, the last 9 elements of the desired embedding derivatives $\dot{z}_6^d$ and $\ddot{z}_6^d$ can be filled and well-determined, respectively.

The animation study result has also showcased a great success as it can be seen from Figure 4.6, where the left picture is in the early motion phase, while the right one is nearly reaching to the destination. We have purposely set, once again, a large initial deviation between the desired and actual starting points to see how the trajectory-tracking control law can effectively maneuver the robot with a heavy big tool to strongly pull the robot back on the right track in both the trivial and non-trivial robot cases.

In summary, we conclude that the new isometric embedding approach to dealing with the dynamics modeling and control of robotic systems will become an alternative way to resolve and ease the challenges for a robotic system with higher number of joints to perform both translational and rotational control task operations. While an isometric embedding for one of the robot links may theoretically only be able to model and control this particular link in 3D dynamic motion, the simulation study shows an encourageable sign of using the isometric embedding $z_n = \zeta_n(q)$ for the last kinematically busiest link, as the dominant body, to control the entire robotic system. We will further study and discuss the effectiveness of such a control scheme again in Chapter 6.

REFERENCES

1. Gu, Edward Y.L., (2000) Configuration Manifolds and Their Applications to Robot Dynamic Modeling and Control. *IEEE Transactions on Robotics and Automation*, Vol.16, No.5, October, pp. 517-527.
2. Gu, Edward Y.L., (2000) A Configuration Manifold Embedding Model for Dynamic Control of Redundant Robots. *International Journal of Robotics Research*, Vol.19, No.3, March, pp. 289-304.
3. Gu, Edward Y.L., (2013) A Journey from Robot to Digital Human. Springer, Heidelberg, New York.
4. Dubrovin, B., Fomenko, A. and Novikov, S., (1992) Modern Geometry – Methods and Applications, Part I – The Geometry of Surfaces, Transformation Groups, and Fields. Springer-Verlag, New York.
5. Dubrovin, B., Fomenko, A., Novikov, S., (1992) Modern Geometry – Methods and Applications, Part II - The Geometry and Topology of Manifolds. Springer-Verlag, New York.
6. Bredon, G., (1993) Topology and Geometry. Springer-Verlag, New York.
7. Berger, M., and Gostiaux, B., (1988) Differential Geometry: Manifolds, Curves, and Surfaces. Springer-Verlag, New York.

5 The Principle of Duality in Kinematics and Dynamics

5.1 KINEMATIC STRUCTURES FOR STEWART PLATFORM

The closed parallel-chain robotic systems feature high speed, high rigidity and high payload, which have made themselves win wide industrial applications. The only drawback in comparison to their counterpart: the open serial-chain robotic systems, is their limited work envelopes. The Stewart platform is one of the most typical closed parallel-chain mechanisms, as shown in Figure 5.1 [1, 2].

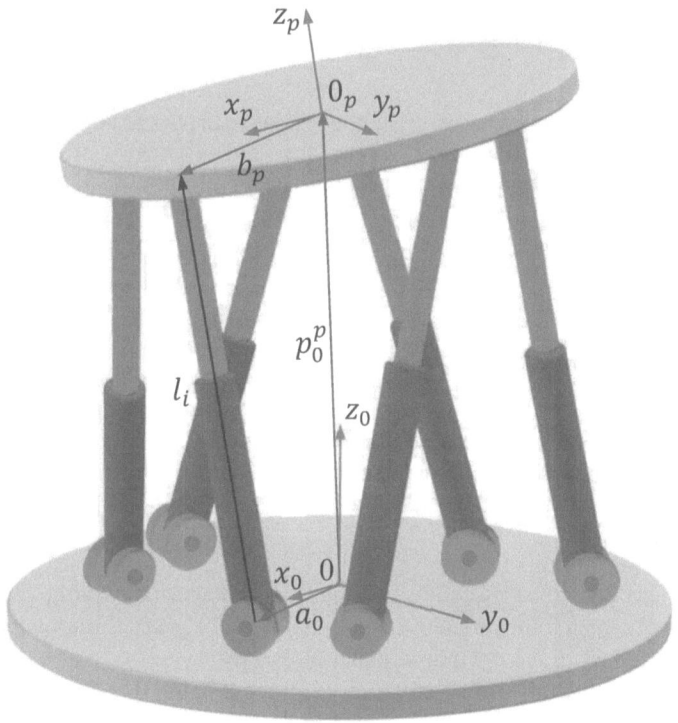

Figure 5.1 A Stewart platform as a typical parallel-chain robotic system

As we have discussed in a fairly detail in the previous chapter about the open serial-chain robot kinematics modeling and formulation, the inverse kinematics (IK) for an open serial-chain manipulator is often more challenging than its forward kinematics (FK) in general. In contrast, for a closed parallel-chain manipulator, its IK is extremely straightforward, while the FK is just opposite, not only extremely difficult, but is not even possible to have an explicit form of their FK solutions. Let us now take the Stewart platform as example to study its kinematics.

By inspecting Figure 5.1, it is observable that the total number of joints can be counted as follows: 6 for the upper joints of the six legs underneath the top panel, 6 for the prismatic pistons and $n_g = 6$ for the number of joints connected to the bottom base panel so that the amount of joints is $n = 18$. Suppose that each joint connected to the base is of universal type, while each upper joint connected to the top panel is of spherical type. Thus, the 6 upper joints offer $2 \times 6 = 12$ axes. Let one of the six joints connected to the base keep intact to offer 3 axes, while let the rest five bottom joints be virtually floated without touching on the base. Then, with each prismatic piston offers one axis, the total number of axes after $n_g - 1 = 5$ grounded joint are virtually floated will become $12 + 3 + 6 = 21$. Based on equation (3.28), the net d.o.f. for the Stewart platform is

$$r = \left(\sum_{i=1}^{n-n_g+1} f_i \right) - D_0(n_g - 1) = 21 - 3 \times 5 = 6.$$

Therefore, we plan to use the six prismatic pistons to drive and control the top panel in 6 d.o.f. motion.

Furthermore, from Figure 5.1, each of the six legs plus the given required position vector p_0^p for the origin of frame p on the top panel with respect to the base can form a quadrilateral. Thus, it is obvious to have the following vector relation:

$$l_i = p_0^p + R_0^p b_p^i - a_0^i,$$

where R_0^p is a given required orientation of the top panel with respect to the base, and both a_0^i and b_p^i are the constant parameter vectors referred to the base and frame p, respectively. If the unit vector of the i-th piston prismatic joint is l_i^o, then $l_i = d_i l_i^o$ with d_i as a prismatic joint variable so that the above vector relation becomes

$$d_i l_i^o = p_0^p + R_0^p b_p^i - a_0^i, \quad \text{for } i = 1, \cdots, 6. \tag{5.1}$$

This set of six equations is actually the inverse kinematics (IK) for the Stewart platform. Namely, each joint variable d_i is a function of the given Cartesian position p_0^p and orientation R_0^p of the top mobile panel.

If we further take time-derivatives on (5.1) by noticing that both b_p^i and a_0^i are constant vectors, then

$$\dot{d}_i l_i^o + d_i \dot{l}_i^o = \dot{p}_0^p + \dot{R}_0^p b_p^i.$$

Because for any unit vector v, $v^T v = 1$ so that $\dot{v}^T v + v^T \dot{v} = 0$. This means that any unit vector v and its time-derivative $\dot{v}$ are always orthogonal to each other, i.e., $v^T \dot{v} = 0$. Now, let us pre-multiply both sides of the above equation by $(l_i^o)^T$, we obtain

$$\dot{d}_i = (l_i^o)^T \dot{p}_0^p + (l_i^o)^T \dot{R}_0^p b_p^i, \quad \text{for } i = 1, \cdots, 6. \tag{5.2}$$

This new equation is also an IK relation but in a differential motion version for the closed parallel-chain Stewart platform.

Obviously, an IK problem for driving the Stewart platform system can be stated that given a desired position p_0^p and orientation R_0^p for frame p of the top mobile panel with respect to the base, what is each prismatic piston displacement d_i? Now, using the IK equation (5.1) in the Cartesian motion version, each d_i can be directly determined in terms of the given required p_0^p and R_0^p. However, using the IK equation (5.2) in the differential motion version, we need, in addition to p_0^p and R_0^p, both the linear velocity v_0^p and angular velocity ω_0^p of the top panel referred to the base. By inspecting equation (5.2), $v_0^p = \dot{p}_0^p$, while the skew-symmetric matrix of ω_0^p: $\Omega_0^p = \omega_0^p \times$ can be determined by $\Omega_0^p = \dot{R}_0^p R_0^{pT}$ according to equation (2.52). Thus,

$$\dot{d}_i = (l_i^o)^T v_0^p + (l_i^o)^T \Omega_0^p R_0^p b_p^i = (l_i^o)^T v_0^p + (l_i^o)^T \Omega_0^p b_0^i.$$

Because $\Omega_0^p b_0^i = \omega_0^p \times b_0^i = -b_0^i \times \omega_0^p$, the above equation for each i can be augmented to convert them to the following matrix relation:

$$\dot{q} = \begin{pmatrix} \dot{d}_1 \\ \vdots \\ \dot{d}_6 \end{pmatrix} = \begin{pmatrix} (l_1^o)^T & -(l_1^o)^T b_0^1 \times \\ \vdots & \vdots \\ (l_6^o)^T & -(l_6^o)^T b_0^6 \times \end{pmatrix} \begin{pmatrix} v_0^p \\ \omega_0^p \end{pmatrix},$$

where $\begin{pmatrix} v_0^p \\ \omega_0^p \end{pmatrix} = V_0^p$ is just the twist of the top mobile panel with respect to the base.

Let a new 6 by 6 Jacobian matrix be defined as

$$J_c \triangleq \begin{pmatrix} l_1^o & \cdots & l_6^o \\ b_0^1 \times l_1^o & \cdots & b_0^6 \times l_6^o \end{pmatrix}. \tag{5.3}$$

Then,

$$\dot{q} = \begin{pmatrix} \dot{d}_1 \\ \vdots \\ \dot{d}_6 \end{pmatrix} = J_c^T V_0^p = J_c^T \begin{pmatrix} v_0^p \\ \omega_0^p \end{pmatrix}. \tag{5.4}$$

Comparing this new Jacobian equation for the closed parallel-chain Stewart platform system with the open serial-chain robotic system $V_0 = J_0 \dot{q}$, clearly, they obey **the Principle of Duality**.

As have aforementioned, a closed parallel-chain mechanism has almost no hope to solve its forward kinematics (FK) problem in Cartesian motion. Now, at the differential motion standpoint, the following inverse of the Jacobian equation (5.4) can provide an explicit form of the FK solution in differential motion:

$$V_0^p = \begin{pmatrix} v_0^p \\ \omega_0^p \end{pmatrix} = (J_c^T)^{-1} \dot{q},$$

as long as the Jacobian matrix in (5.3) is non-singular [3–8].

Let us now in turn represent the kinematics for the Stewart platform using an isometric embedding approach, as given by equation (4.6). Then, the Jacobian matrix will be defined in a different form. Suppose that a position p_0^p and an orientation R_0^p of the top panel are given. According to (4.6), an isometric embedding vector can be constructed as follows:

$$z = \begin{pmatrix} \sqrt{m}p_0^p \\ \beta_1 r_x \\ \beta_2 r_y \\ \beta_3 r_z \end{pmatrix}, \tag{5.5}$$

where m is the mass of the top panel, r_x, r_y and r_z are the three columns of the given rotation matrix $R_0^p = (r_x \; r_y \; r_z)$, and each β_i is the constant coefficient related to the top panel inertia tensor, as described in Chapter 4. Obviously,

$$\dot{z} = \begin{pmatrix} \sqrt{m}\dot{p}_0^p \\ \beta_1 \dot{r}_x \\ \beta_2 \dot{r}_y \\ \beta_3 \dot{r}_z \end{pmatrix}. \tag{5.6}$$

Since each $R_0^p b_p^i$ in the differential motion IK (5.2) is actually the linear combination of the three column vectors of R_0^p, i.e., $R_0^p b_p^i = b_{px}^i r_x + b_{py}^i r_y + b_{pz}^i r_z$, due to the fact that each $b_p^i = (b_{px}^i \; b_{py}^i \; b_{pz}^i)^T$ seen in frame p is a constant vector, by comparing (5.2) with (5.6), we obtain

$$\dot{q} = \begin{pmatrix} \dot{d}_1 \\ \vdots \\ \dot{d}_6 \end{pmatrix} = \begin{pmatrix} l_1^{oT} \dot{p}_0^p + l_1^{oT}(b_{px}^1 \dot{r}_x + b_{py}^1 \dot{r}_y + b_{pz}^1 \dot{r}_z) \\ \vdots \\ l_6^{oT} \dot{p}_0^p + l_6^{oT}(b_{px}^6 \dot{r}_x + b_{py}^6 \dot{r}_y + b_{pz}^6 \dot{r}_z) \end{pmatrix}$$

$$= \begin{pmatrix} \frac{1}{\sqrt{m}}l_1^{oT} & \frac{b_{px}^1}{\beta_1}l_1^{oT} & \frac{b_{py}^1}{\beta_2}l_1^{oT} & \frac{b_{pz}^1}{\beta_3}l_1^{oT} \\ \vdots & \vdots & \vdots & \vdots \\ \frac{1}{\sqrt{m}}l_6^{oT} & \frac{b_{px}^6}{\beta_1}l_6^{oT} & \frac{b_{py}^6}{\beta_2}l_6^{oT} & \frac{b_{pz}^6}{\beta_3}l_6^{oT} \end{pmatrix} \dot{z}.$$

We now define another new 12 by 6 Jacobian matrix by

$$J_z = \begin{pmatrix} (l_1^o \; l_2^o \; l_3^o \; l_4^o \; l_5^o \; l_6^o)A \\ (l_1^o \; l_2^o \; l_3^o \; l_4^o \; l_5^o \; l_6^o)B_1 \\ (l_1^o \; l_2^o \; l_3^o \; l_4^o \; l_5^o \; l_6^o)B_2 \\ (l_1^o \; l_2^o \; l_3^o \; l_4^o \; l_5^o \; l_6^o)B_3 \end{pmatrix}, \tag{5.7}$$

where all the 6 by 6 diagonal constant coefficient matrices A, B_1, B_2 and B_3 are given by

$$A = \begin{pmatrix} \frac{1}{\sqrt{m}} & & 0 \\ & \ddots & \\ 0 & & \frac{1}{\sqrt{m}} \end{pmatrix}, \quad B_1 = \begin{pmatrix} \frac{b_{px}^1}{\beta_1} & & 0 \\ & \ddots & \\ 0 & & \frac{b_{px}^6}{\beta_1} \end{pmatrix},$$

$$B_2 = \begin{pmatrix} \frac{b_{py}^1}{\beta_2} & & 0 \\ & \ddots & \\ 0 & & \frac{b_{py}^6}{\beta_2} \end{pmatrix}, \quad B_3 = \begin{pmatrix} \frac{b_{pz}^1}{\beta_3} & & 0 \\ & \ddots & \\ 0 & & \frac{b_{pz}^6}{\beta_3} \end{pmatrix}.$$

Then, a new Jacobian equation in terms of the isometric embedding is finally deduced as

$$\dot{q} = \begin{pmatrix} \dot{d}_1 \\ \vdots \\ \dot{d}_6 \end{pmatrix} = J_z^T \dot{z}. \tag{5.8}$$

In comparison with the Jacobian equation for an open serial-chain robot $\dot{z} = J\dot{q}$, equation (5.8) manifests once again the Principle of Duality. However, there are a few issues behind the duality that should be addressed here:

1. The Jacobian matrix J_z defined in (5.7) seems to be constructed manually. Actually, it is a genuine mathematical Jacobian matrix, i.e., $J_z = \frac{\partial z}{\partial q}$. In fact, the isometric embedding definition in (5.5) can be directly in terms of every $d_i l_i^o$ based on the IK equation (5.1), plus the joint position vector is $q = (d_1 \cdots d_6)^T$ for this closed parallel-chain Stewart platform. Each column of the Jacobian matrix J_z should be $\frac{\partial z}{\partial d_i}$. Because each $\frac{\partial d_i l_i^o}{\partial d_i} = l_i^o$, this is the main reason why each column of J_z in (5.7) contains the unit vector l_i^o as a directional vector of the i-th piston joint.

2. It is also a quite interesting phenomenon that if the Jacobian matrix J_z in (5.7) is post-multiplied by a joint torque/force vector, i.e., $\tau = (f_1 \cdots f_6)^T$ in the case of all prismatic joints, then, each column becomes $f_i l_i^o$ as a force along the i-th piston leg. Based on the Newton's second law, it should be an acceleration so that

$$J_z \tau = \ddot{z}. \tag{5.9}$$

In fact, since every isometric embedding must obey $\frac{1}{2}\dot{z}^T\dot{z} = K$, the kinetic energy of the system, the *kinetic power* should be $\dot{K} = \dot{z}^T\ddot{z}$. If we substitute (5.9) into the kinetic power, $\dot{K} = \dot{z}^T J_z \tau = \dot{q}^T \tau$ due to equation (5.8), which exactly matches with the kinetic power $\dot{K} = \tau^T \dot{q} = \dot{q}^T \tau$ in joint space.

3. However, the isometric embedding z defined in equation (5.5) is not a full function of the joint position $q = (d_1 \cdots d_6)^T$. Instead, this 12-dimensional vector z is defined in terms of the Cartesian position p_o^p and orientation R_o^p for the top panel of the Stewart platform. While it is directly related to each piston $d_i l_i^o$, but each unit directional vector l_i^o may also be indirectly related to $q = (d_1 \cdots d_6)^T$ plus the other joint angles. It is difficult, maybe even impossible, to find such an explicit full function form $z = \zeta(q)$. The main reason is back to the forward kinematics (FK) issue. Therefore, an explicit form of the isometric embedding $z = \zeta(q)$ for the Stewart platform would be completely determined only if, like a serial-chain robotic system, the FK solution in Cartesian motion for the parallel-chain mechanism could become available.

4. The current joint position vector q covers only six prismatic piston displacements d_i's, which are just a small part of all the axis variables. In fact, after 5 out of all the 6 grounded joints have been floated as a virtual floating model, the total number of axes has been counted to be 21 for this 6-leg Stewart platform. Now, only 6 out of the 21 axes are included in the joint position vector q, and the rest $21 - 6 = 15$ axes keep passive without actuation. Therefore, the

isometric embedding model can determine a dynamic model only for the **dominant body**, such as the top mobile panel of the Stewart platform. If we wish to model the dynamics that can cover every body, the virtual model of floating grounded joints may be the only approach towards the goal.

While a closed parallel-chain mechanism has the above revealed intrinsic shortcomings in comparison to its counterpart: the open serial-chain robotic system, all the IK formulas in equations (5.1) and (5.4) as well as (5.8) can still effectively control the Stewart platform in 6 d.o.f. motion. Figure 5.2 shows a 3-3 type Stewart platform ready for animation, where 3-3 means that for the 6 parallel legs, each pair of the adjacent legs are merged to share a single universal joint on the base and share a single spherical joint underneath the top panel. Thus, there are only 3 universal joints on the base and 3 spherical joints to connect the top panel. Figure 5.3 demonstrates a realistic 6 d.o.f. animation recording for the 6 legs and top panel.

Figure 5.2   A 3-3 Stewart platform is standing at an initial posture in the animation study

5.2   KINEMATIC ANALYSIS OF DELTA CLOSED HYBRID-CHAIN ROBOTS

A Delta robot is a typical closed parallel/serial hybrid-chain manipulator that consists of three arms, each of which has a single revolute joint shoulder, a single revolute joint elbow and a single revolute joint wrist. However, each of the three forearms

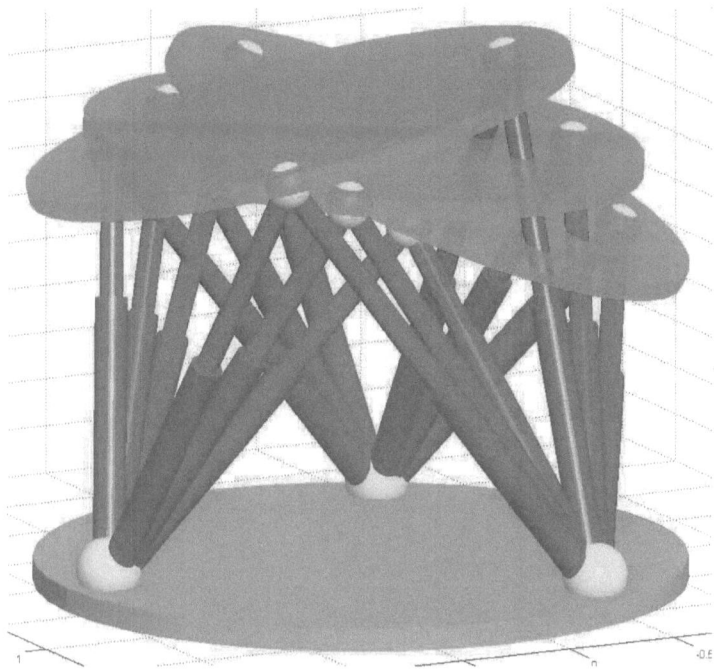

Figure 5.3 The animation recording for the 3-3 Stewart platform in 6 d.o.f. realistic motion

is using a parallelogram mechanism to ensure the lower tool plate always keeping horizontal [9]. The entire structure is very similar to the previously studied 3D 3-leg platform system, as shown in Figure 3.8, where each "knee" is a prismatic piston joint, in contrast to the revolute joint for the Delta robot. A 4-bar parallelogram mechanism of the forearm can be modeled as only one net d.o.f. Due to its high speed and high versatility, the Delta robot has won high popularity in industrial applications, such as to packaging, picking-place, and high-speed dexterous assembly, etc. While it doesn't bear much payload, its arms are very light and flexible. A Delta robot product made by Fanuc America, Inc. is shown in Figure 5.4.

For the purpose of kinematics modeling, a schematic drawing is displayed in Figure 5.5. We now use equation (3.28) to predict the net d.o.f., or the degree of mobility for this Delta robot that can offer. First, the number of grounded joints is 3 on the top fixed base panel, i.e., $n_g = 3$. This robot is in 3D motion so that $D_0 = 3$. Thus, the second term of (3.28), $D_0(n_g - 1) = 6$. The key step is to find the first term as the total number of axes after the $(n_g - 1)$ grounded joints have been virtually floated. Since each forearm is a 4-bar parallelogram that offers one additional axis, plus one for each of the shoulder, elbow and wrist joints, the non-floated stem arm has a total of 4 axes. The other two branches with virtually floated grounded joints, i.e., the two shoulder joints, have totally 3 axes for each arm. However, since the lower tool panel is always keeping horizontal without orientation change, it only needs two points,

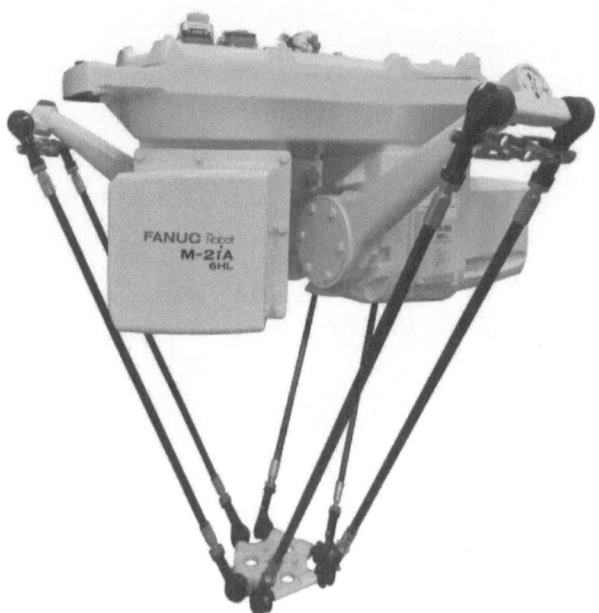

Figure 5.4 A Fanuc M-2iA/6HL industrial Delta parallel/serial hybrid-chain robot. Photo courtesy of Fanuc America Corporation

instead of three points to determine the plane. Thus, the third wrist is constrained by the other two wrists and should not be counted as a free axis. Therefore, the net mobility turns out to be

$$r = \left(\sum_{i=1}^{n-n_g+1} f_i \right) - D_0(n_g - 1) = 4+3+2-6 = 9-6 = 3.$$

This prediction tells us a fact that the tool plate of the Delta robot can be just maneuvered for a 3D translational motion without rotation. Because the three parallelograms on the three forearms ensure to make the tool panel always parallel to the horizontal plane without orientation change, the kinematic output for this Delta robot is only a 3D position vector p_b^p of the tool panel center. Thus, in most cases, a Delta robot is driven and controlled by the three motors mounted on the three shoulders A_b, B_b and C_b.

As we can be further observed from Figure 5.5, the non-floated stem arm has been assigned coordinate frames by exactly following the D-H convention. The parallelogram forearm has only one axis, but we have to assign two axes: z_2 and z_3 for θ_3 and θ_4 joint angles with $\theta_3 = -\theta_4$ at any time due to the parallelogram mechanism. After the coordinate frames are assigned, the D-H table can be created in Table 5.1.

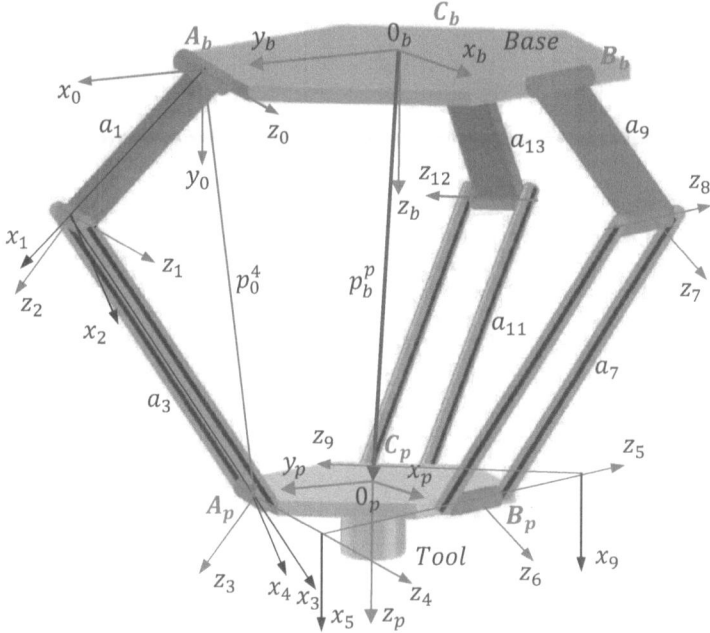

Figure 5.5   Kinematics modeling for a Delta parallel/serial hybrid-chain robot

Joint Angle θ_i	Joint Offset d_i	Twist Angle α_i	Link Length a_i
θ_1	0	0	a_1
θ_2	0	90°	0
θ_3	0	0	a_3
$\theta_4 = -\theta_3$	0	-90°	0

Once the D-H parameter table is prepared, every one-step homogeneous transformation matrix can be readily determined in below:

$$A_0^1 = \begin{pmatrix} c_1 & -s_1 & 0 & a_1 c_1 \\ s_1 & c_1 & 0 & a_1 s_1 \\ 0 & 0 & 1 & 0 \\ 0 & 0 & 0 & 1 \end{pmatrix}, \quad A_1^2 = \begin{pmatrix} c_2 & 0 & s_2 & 0 \\ s_2 & 0 & -c_2 & 0 \\ 0 & 1 & 0 & 0 \\ 0 & 0 & 0 & 1 \end{pmatrix},$$

$$A_2^3 = \begin{pmatrix} c_3 & -s_3 & 0 & a_3 c_3 \\ s_3 & c_3 & 0 & a_3 s_3 \\ 0 & 0 & 1 & 0 \\ 0 & 0 & 0 & 1 \end{pmatrix}, \quad A_3^4 = \begin{pmatrix} c_4 & 0 & -s_4 & 0 \\ s_4 & 0 & c_4 & 0 \\ 0 & -1 & 0 & 0 \\ 0 & 0 & 0 & 1 \end{pmatrix}.$$

Starting link 5, i.e., the lower tool panel, the main stem is split into two branches: one is going along the path of the right arm from B_p to B_b, and the other path is along

θ_i	d_i	α_i	a_i	θ_i	d_i	α_i	a_i
θ_5	d_{51}	$-120°$	0	θ_5	$-d_{52}$	$120°$	0
θ_6	$-d_6$	$-90°$	0	θ_{10}	d_{10}	$-90°$	0
θ_7	0	0	a_7	θ_{11}	0	0	a_{11}
$\theta_8 = -\theta_7$	0	$90°$	0	$\theta_{12} = -\theta_{11}$	0	$90°$	0
θ_9	0	0	a_9	θ_{13}	0	0	a_{13}

Table 5.1

The D-H table for the Delta serial/parallel hybrid-chain system

the left arm from C_p to C_b, as illustrated in Figure 5.5. The one-step homogeneous transformations at the split point are given below:

$$A_4^{51} = \begin{pmatrix} c_5 & \frac{1}{2}s_5 & -\frac{\sqrt{3}}{2}s_5 & 0 \\ s_5 & -\frac{1}{2}c_5 & \frac{\sqrt{3}}{2}c_5 & 0 \\ 0 & -\frac{\sqrt{3}}{2} & -\frac{1}{2} & d_{51} \\ 0 & 0 & 0 & 1 \end{pmatrix}, \quad A_4^{52} = \begin{pmatrix} c_5 & \frac{1}{2}s_5 & \frac{\sqrt{3}}{2}s_5 & 0 \\ s_5 & -\frac{1}{2}c_5 & -\frac{\sqrt{3}}{2}c_5 & 0 \\ 0 & \frac{\sqrt{3}}{2} & -\frac{1}{2} & -d_{52} \\ 0 & 0 & 0 & 1 \end{pmatrix}.$$

Then,

$$A_5^6 = \begin{pmatrix} c_6 & 0 & -s_6 & 0 \\ s_6 & 0 & c_6 & 0 \\ 0 & -1 & 0 & -d_6 \\ 0 & 0 & 0 & 1 \end{pmatrix}, \quad A_6^7 = \begin{pmatrix} c_7 & -s_7 & 0 & a_7c_7 \\ s_7 & c_7 & 0 & a_7s_7 \\ 0 & 0 & 1 & 0 \\ 0 & 0 & 0 & 1 \end{pmatrix},$$

$$A_7^8 = \begin{pmatrix} c_8 & 0 & s_8 & 0 \\ s_8 & 0 & -c_8 & 0 \\ 0 & 1 & 0 & 0 \\ 0 & 0 & 0 & 1 \end{pmatrix}, \quad A_8^9 = \begin{pmatrix} c_9 & -s_9 & 0 & a_9c_9 \\ s_9 & c_9 & 0 & a_9s_9 \\ 0 & 0 & 1 & 0 \\ 0 & 0 & 0 & 1 \end{pmatrix}.$$

$$A_5^{10} = \begin{pmatrix} c_{10} & 0 & -s_{10} & 0 \\ s_{10} & 0 & c_{10} & 0 \\ 0 & -1 & 0 & d_{10} \\ 0 & 0 & 0 & 1 \end{pmatrix}, \quad A_{10}^{11} = \begin{pmatrix} c_{11} & -s_{11} & 0 & a_{11}c_{11} \\ s_{11} & c_{11} & 0 & a_{11}s_{11} \\ 0 & 0 & 1 & 0 \\ 0 & 0 & 0 & 1 \end{pmatrix},$$

$$A_{11}^{12} = \begin{pmatrix} c_{12} & 0 & s_{12} & 0 \\ s_{12} & 0 & -c_{12} & 0 \\ 0 & 1 & 0 & 0 \\ 0 & 0 & 0 & 1 \end{pmatrix}, \quad A_{12}^{13} = \begin{pmatrix} c_{13} & -s_{13} & 0 & a_{13}c_{13} \\ s_{13} & c_{13} & 0 & a_{13}s_{13} \\ 0 & 0 & 1 & 0 \\ 0 & 0 & 0 & 1 \end{pmatrix}.$$

Once all the one-step homogeneous transformation matrices are found, let us first check the validity of kinematics due to the three parallelogram structures. Because

$$A_2^4 = A_2^3 A_3^4 = \begin{pmatrix} c_{34} & 0 & -s_{34} & a_3c_3 \\ s_{34} & 0 & c_{34} & a_3s_3 \\ 0 & -1 & 0 & 0 \\ 0 & 0 & 0 & 1 \end{pmatrix} = \begin{pmatrix} 1 & 0 & 0 & a_3c_3 \\ 0 & 0 & 1 & a_3s_3 \\ 0 & -1 & 0 & 0 \\ 0 & 0 & 0 & 1 \end{pmatrix},$$

due to $\theta_3 + \theta_4 = 0$. This reflects a fact that the parallelogram structure will only change the position vector p_2^4 as a function of θ_3, but will not change the orientation, because R_2^4 as the upper-left 3 by 3 corner of A_2^4 is a constant matrix. Similarly,

$$A_6^8 = \begin{pmatrix} 1 & 0 & 0 & a_7c_7 \\ 0 & 0 & -1 & a_7s_7 \\ 0 & 1 & 0 & 0 \\ 0 & 0 & 0 & 1 \end{pmatrix}, \quad \text{and} \quad A_{10}^{12} = \begin{pmatrix} 1 & 0 & 0 & a_{11}c_{11} \\ 0 & 0 & -1 & a_{11}s_{11} \\ 0 & 1 & 0 & 0 \\ 0 & 0 & 0 & 1 \end{pmatrix},$$

because of $\theta_7 = -\theta_8$ and $\theta_{11} = -\theta_{12}$.

Now, let us take closer look at the homogeneous transformation A_0^4:

$$A_0^4 = A_0^1 A_1^2 A_2^4 = \begin{pmatrix} c_{12} & -s_{12} & 0 & a_1c_1 + a_3c_{12}c_3 \\ s_{12} & c_{12} & 0 & a_1s_1 + a_3s_{12}c_3 \\ 0 & 0 & 1 & a_3s_3 \\ 0 & 0 & 0 & 1 \end{pmatrix}.$$

The upper-left 3 by 3 block of the above A_0^4 is a rotation matrix R_0^4 that represents the orientation of frame 4 with respect to frame 0, which is clearly independent of the parallelogram rotating angle $\theta_3 = -\theta_4$. This implies that θ_3 will not affect the orientation of frame 4. However, the last column of A_0^4 is the position vector p_0^4 that is tailed at the origin A_b of frame 0 and arrow-pointing to the origin A_p of frame 4. It can be directly seen that p_0^4 is not only dependent on θ_1 and θ_2, but also on the parallelogram rotating angle θ_3 in order to keep the tool panel always horizontal. With the same reason, the other two parallelograms will also not interfere the orientations of frame 5 and frame 9, respectively. This is really a smart design to ensure the tool panel orientation invariant at any time.

The purpose of making such a parallelogram structure for each forearm is to ease the IK algorithm as well, in comparison to the 3D 3-leg piston-driven closed parallel-chain system in Figure 3.8. The top platform of the 3D 3-leg system never keeps horizontal so that both the position and orientation are very difficult to find their explicit forms that are related to the three leg displacements. In contrast, the IK algorithm for the Delta robot can be easily deduced without major difficulty.

Let b_a be the vector from the origin of the base frame b to the shoulder A_b on the base ceiling. Also, let β_a be the vector from the origin of frame p to the wrist A_p on the tool panel. Both vectors b_a and β_a are referred to the base frame. Due to no orientation change for the tool panel, both b_a and β_a are constant vectors. If a targeting position vector p_b^p is specified, whose tail is at the origin of the base and arrow-pointing to the origin of frame p, the four vectors p_b^p, b_a, β_a and $R_b^0 p_0^4$ form a quadrilateral, as seen in Figure 5.5. Thus,

$$p_0^4 = R_0^b(p_b^p + \beta_a - b_a) \tag{5.10}$$

can always be determined in terms of the given p_b^p. If at a time instant, $p_0^4 = (x \ y \ z)^T$, then,

$$p_0^4 = \begin{pmatrix} a_1c_1 + a_3c_{12}c_3 \\ a_1s_1 + a_3s_{12}c_3 \\ a_3s_3 \end{pmatrix} = \begin{pmatrix} x \\ y \\ z \end{pmatrix}.$$

After each element of the vectors on both sides is squared, adding them together yields

$$a_1^2 + a_3^2 + 2a_1a_3c_2c_3 = x^2 + y^2 + z^2.$$

On the other hand, $a_3s_3 = z$ so that $s_3 = \frac{z}{a_3}$. Based on the parallelogram structure, $-90° < \theta_3 < 90°$. Within the first and fourth quadrants, $\cos\theta_3 = c_3$ should be always positive. Thus,

$$c_3 = \sqrt{1 - s_3^2} = \sqrt{1 - \left(\frac{z}{a_3}\right)^2}.$$

Therefore,

$$c_2 = \frac{x^2 + y^2 + z^2 - a_1^2 - a_3^2}{2a_1a_3c_3}.$$

In addition, by closely inspecting Figure 5.5, we can discover that θ_2 is rotated about z_1 axis from x_1 to x_2, which is within $0 < \theta_2 < 180°$. In the first and second quadrants, $\sin\theta_2$ is always positive so that $s_2 = \sqrt{1 - c_2^2}$. Thus,

$$\begin{pmatrix} a_1c_2 + a_3c_{12}c_3 \\ a_1s_1 + a_3s_{12}c_3 \end{pmatrix} = \begin{pmatrix} a_1c_1 + a_3(c_1c_2 - s_1s_2)c_3 \\ a_1s_1 + a_3(s_1c_2 + c_1s_2)c_3 \end{pmatrix} = \begin{pmatrix} x \\ y \end{pmatrix}.$$

By rearranging the equation, we have

$$\begin{pmatrix} a_1 + a_3c_2c_3 & -a_3s_2c_3 \\ a_3s_2c_3 & a_1 + a_3c_2c_3 \end{pmatrix} \begin{pmatrix} c_1 \\ s_1 \end{pmatrix} = \begin{pmatrix} x \\ y \end{pmatrix}.$$

Therefore, we can solve this matrix equation for both s_1 and c_1, and $\theta_1 = \text{atan2}(s_1, c_1)$ can be uniquely determined, where $\text{atan2}(\cdot, \cdot)$ is the internal four-quadrant arc tangent function in MATLAB^{TM}. θ_1 is one of the three actuated joints for this Delta robot.

Using the same IK procedure, we can solve the other two actuated joint angles on the shoulders B_b and C_b in terms of the given position vector p_b^p. The entire process is actually a Cartesian motion IK determination. We can also develop an IK algorithm in differential motion by construct a Jacobian matrix for this closed hybrid-chain mechanism.

Previously we have counted the total number of axes after 2 grounded joints at B_b and C_b are virtually floated, i.e.,

$$\sum_{i=1}^{n-n_g+1} f_i = 9,$$

but from the D-H tables, it has 13 θ_i's. The reason is now clear that each of the three parallelogram structures was assigned with an extra axis, plus the wrist axis at C_p is also recovered in the D-H tables. However, the joint position vector of all the independent axes should still be 9-dimensional:

$$q = (\theta_1 \ \theta_2 \ \theta_3 \ \theta_5 \ \theta_6 \ \theta_7 \ \theta_9 \ \theta_{11} \ \theta_{13})^T.$$

Due to the motion in 3D space, $D_0 = 3$, and $D_0(n_g - 1) = 6$, the targeting Jacobian J_0 should be a 6 by 9 matrix that contains an $r = 3$-dimensional null space N^4. Similar to equation (3.27) for the 3D 3-leg closed parallel-chain robotic system in Figure 3.8, we can derive a set of Jacobian matrices for the stem part: J_0^{51} and J_0^{52} by following the D-H tables and one-step homogeneous transformations, as well as the Jacobian matrices for the two branches: J_0^{69} and J_0^{1013}. Then, the overall 6 by 9 Jacobian matrix can be augmented as follows:

$$J_0 = \begin{pmatrix} J_0^{51} & J_0^{69} & O_{3\times 2} \\ J_0^{52} & O_{3\times 3} & J_0^{1013} \end{pmatrix},$$

(5.11)

where each O is the zero matrix.

Since this is actually a closed-chain system, its Jacobian equation should be homogeneous, i.e., $J_0\dot{q} = 0$, which has a nontrivial null solution, i.e., $\dot{q} = (I - J_0^+ J_0)\eta$ with a gradient vector

$$\eta = \frac{\partial p(q)}{\partial q}$$

for some potential function $p(q)$ to drive the system in 3D motion. For instance, if one wishes the position vector p_b^p could follow a desired trajectory $p_b^d(t)$, then

$$p(q) = \frac{1}{2}(p_b^d(t) - p_b^p)^T (p_b^d(t) - p_b^p),$$

which should be minimized. Because we know p_0^4 as a function of θ_1, θ_2 and θ_3 from the homogeneous transformation A_0^4, the gradient vector η can be readily found by

$$\eta = \frac{\partial p(q)}{\partial q} = \left(\frac{\partial p_b^p}{\partial q}\right)^T (p_b^p - p_b^d(t)).$$

A 3D animation study for this Delta closed hybrid-chain robot has been carried out. Figure 5.6 shows the simulated Delta robot being ready to move, while Figure 5.7 demonstrates a realistic motion recording.

In the animation study, the Cartesian motion IK algorithm is adopted. The above stated IK solution is derived only for the stem arm from A_b to A_p to reach the tool panel. For the branch from B_b to B_p, equation (5.10) needs to be modified by premultiplying a rotation matrix $R(z, -120°)$ to create two new constant vectors $b_b = R(z, -120°)b_a$ and $\beta_b = R(z, -120°)\beta_a$. Also, the rotation matrix R_0^b should be postmultiplied by $R(z, 120°)$ to project the frame from the base to the new frame 0 at B_b. By replacing θ_1 through θ_5 in the previous IK algorithm by θ_6 through θ_{10}, respectively, follow the exact same procedure to find each of θ_6 through θ_{10} in terms of the new rotated output vector p_0^9, i.e.,

$$p_0^9 = R_0^b R(z, 120°)(p_b^p + \beta_b - b_b).$$

(5.12)

In the same way, let θ_1 through θ_5 be replaced by θ_{11} through θ_{15} for the 3rd arm from C_b to C_p. Then, solve the IK problem in terms of another new rotated output vector p_0^{14}, i.e.,

$$p_0^{14} = R_0^b R(z, -120°)(p_b^p + \beta_c - b_c),$$

(5.13)

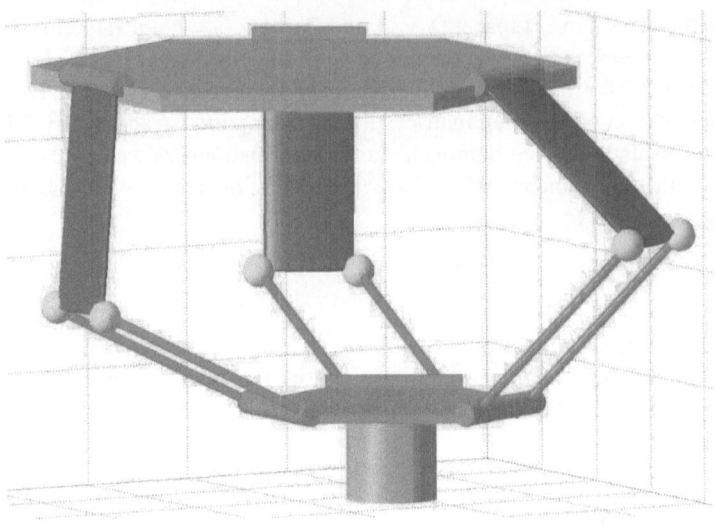

Figure 5.6 The Delta parallel/serial hybrid-chain robot is ready for animation study

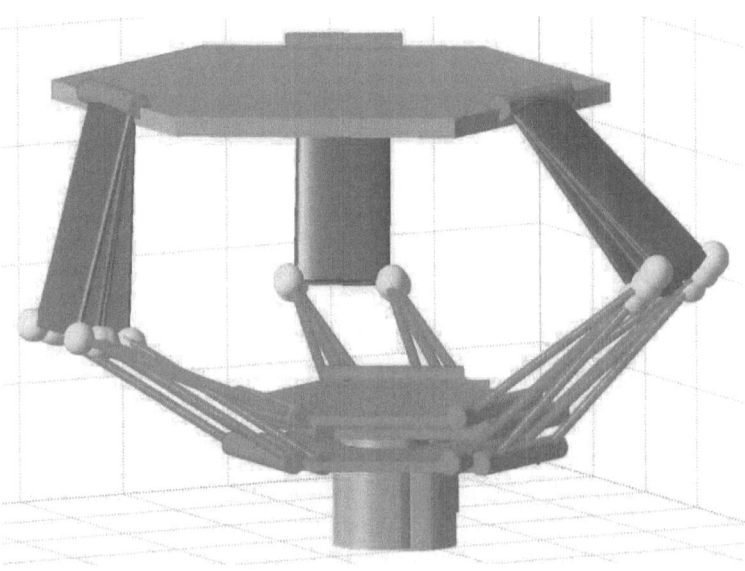

Figure 5.7 The Delta parallel/serial hybrid-chain robot is realistically moving in 3D animation space

where $b_c = R(z, 120)b_a$ and $\beta_c = R(z, 120°)\beta_a$. All the 15 joint angles can now determine an instantaneous posture of the entire Delta robot at each sampling point.

One can also employ the Jacobian matrix (5.11) and its null solution along with the above gradient vector η to perform a differential motion simulation study. Since the IK solution in Cartesian motion is so easy for this particular Delta robotic system, the animation study in both Figure 5.6 and Figure 5.7 has adopted this IK version developed in Cartesian motion. However, if one wishes to develop a dynamic model based on the floating grounded joints virtual model, the D-H tables in Table 5.1, and Jacobian matrix in (5.11) as well as every sub-Jacobian matrix may be all required to go through.

5.3 DUALITY BETWEEN OPEN SERIAL-CHAIN AND CLOSED PARALLEL-CHAIN SYSTEMS

Duality is a general and broad concept that has manifested in almost every area of mathematics, physics and science. While it constantly plays an important and even significant role in the development of mathematical and physical theories, there is no single definition that can cover all instances of the phenomenon. Duality in mathematics is not a theorem, but a "principle". It has a simple origin, but is very powerful and useful with a long history. Over the hundreds of years it has been continuously developed and expanded, and until now we can still use it in our new research and development. It reveals an intrinsic connection between different phenomena in nature and appears in many subjects in almost every branch of mathematics, such as geometry, topology, algebra, and analysis, as well as in physics and other scientific fields. Fundamentally, duality is to view the same object in two different angles. In other words, duality translates concepts, theorems or mathematical structures into opposite concepts, theorems or structures, in a one-to-one correspondence fashion, not by accident or incident, but by a natural involution: if A is the dual of B, then B is the dual of A [10–12].

In the science of engineering, static electric and magnetic fields are tied with a clear duality relation. An electrical displacement vector obeys Guass's Law $\nabla \cdot \vec{D} = \rho_v$, while the magnetic field intensity vector satisfies Ampere's Circuital Law $\nabla \times \vec{H} = \vec{J}$. Thus, $\vec{D}$ and $\vec{H}$ are dual counterparts to each other. Similarly, because $\nabla \times \vec{E} = 0$ and $\nabla \cdot \vec{B} = 0$ in a static field case, the electric field intensity $\vec{E}$ is a dual counterpart of the magnetic flux density $\vec{B}$, and vice versa. Controllability and observability in a linear time-invariant control system are also a typical pair of duality. If a system is given by

$$\begin{cases} \dot{x} = Ax + Bu \\ y = Cx, \end{cases}$$

then the system

$$\begin{cases} \dot{x} = A^T x + C^T u \\ y = B^T x, \end{cases}$$

is the dual system of the primary system. If the primary system is controllable (observable), then the dual system is observable (controllable). Similarly, for a constant control gain matrix K, the input $u = -Kx$ is called a state-feedback control such that it can make the state equation $\dot{x} = Ax + Bu = (A - BK)x$. While a dual state-feedback input $u = -L^T x$ can make the dual system $\dot{x} = A^T x + C^T u = (A - LC)^T x$, which is the foundation of a system *observer* to best estimate the state $\hat{x}$ with a converging estimation error $e = x - \hat{x} \to 0$ as $t \to \infty$. In fact, the best estimated state obeys the following equation:

$$\dot{\hat{x}} = A\hat{x} + Bu + L(y - C\hat{x}).$$

After the above estimated state equation is subtracted from the primary state equation, it reaches an estimation error equation $\dot{e} = (A - LC)e$, where $e = x - \hat{x}$ can asymptotically converge to zero, provided that the matrix $(A - LC)$ has all the eigenvalues in the left-half of complex plane (LHP).

Many duality relations are theoretically elegant to offer a principle, an insight, and a guideline of directing for further conceptual developments, but they may not always be an indispensable tool to substantially solve the problem. However, one of the most useful duality relations in engineering science is the duality between electric and magnetic circuits. Let us look at an example of 3-coil electromagnetic circuit system given in Figure 5.8. Suppose that the magnetic core material is ideal, i.e., its relative permeability $\mu_r \to \infty$, and it has 3 air-gaps, one at each leg. The reluctance for each air-gap is given by

$$\mathscr{R}_i = \frac{l_{gi}}{\mu_0 A_i}, \quad \text{for } i = 1, 2, 3,$$

where l_{gi} is the air-gap length and A_i is the cross-section area of gap i. Also, N_i is the number of turns for coil i in Figure 5.8 for $i = 1, 2, 3$.

A Three-Airgap Electromagnetic Circuit

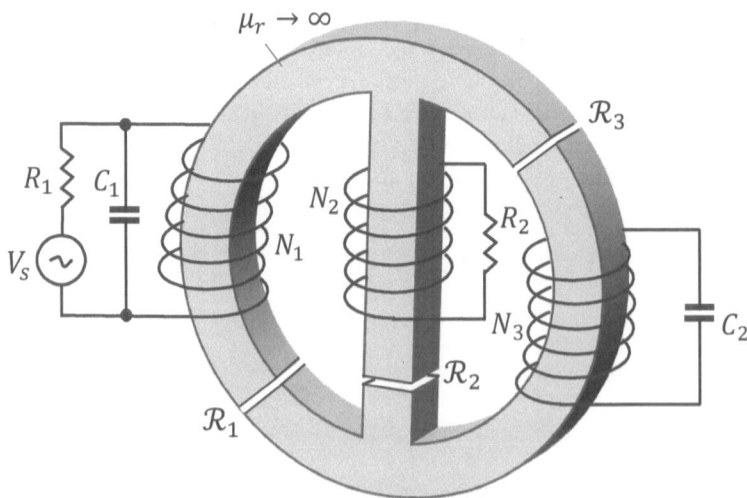

Figure 5.8 A three-coil electromagnetic circuit system

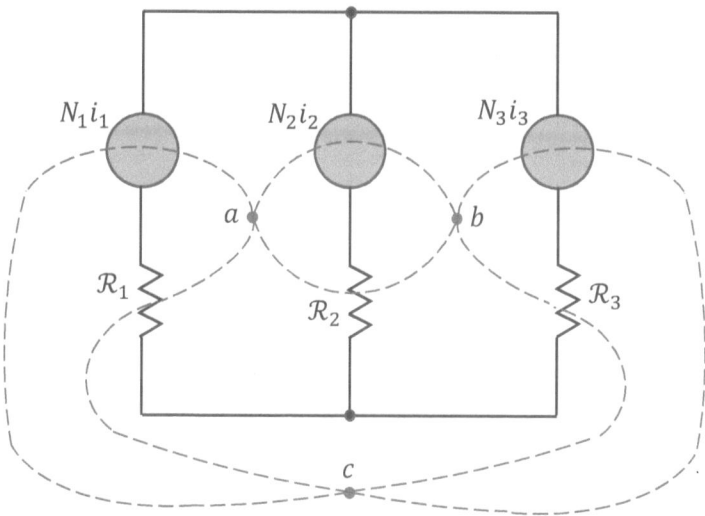

Figure 5.9 The equivalent magnetic circuit and its duality

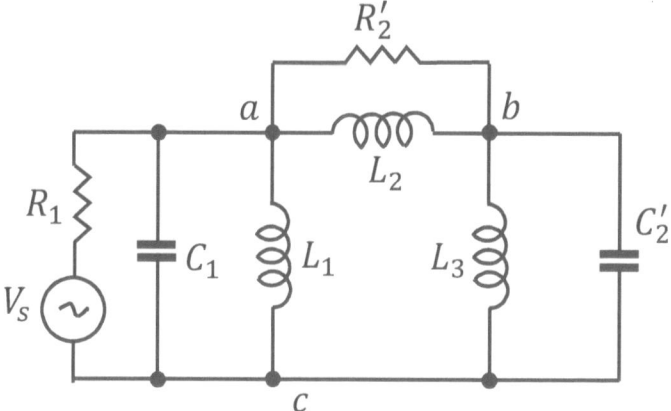

Figure 5.10 The final equivalent electric circuit

Then, the equivalent magnetic circuit for this 3-coil electromagnetic circuit system can be determined in Figure 5.9. Assume that we pick the circuit that is connected to coil 1 as a reference coil. Let us now apply *the principle of duality* between the electric and magnetic circuits to find an equivalent electric circuit towards the solution. Since the equivalent magnetic circuit, as shown in Figure 5.9, has 3 meshes a, b and c, including the background mesh c, their dual counterparts are the 3 nodes a, b and c in its dual electric circuit. Between each adjacent pair of the meshes, such as between meshes a and b, it has an *mmf* (magnetomotive force) $N_2 i_2$ and a reluctance $\mathcal{R}_2$, both of which are connected in series. Based on the duality, the dual counterpart of $N_2 i_2$ is the current i_2 flowing into the circuit port connecting to the corresponding resistance R'_2 along with a dual counterpart of the reluctance $\mathscr{R}_2$, i.e., an inductance L_2 in parallel, as shown in Figure 5.10. Because we pick coil 1 as the reference, the resistance R'_2 and capacitance C'_2 as well as each inductance, as a dual counterpart of the corresponding reluctance, should be

$$L_i = \frac{N_1^2}{\mathscr{R}_i}, \quad \text{for } i = 1,2,3, \quad \text{and} \quad R'_2 = \left(\frac{N_1}{N_2}\right)^2 R_2, \quad C'_2 = \left(\frac{N_3}{N_1}\right)^2 C_2.$$

Under the duality principle, the final equivalent electric circuit for the entire electromagnetic circuit system is given in Figure 5.10. One can evidently see that it is intricate and even impossible to find an equivalent electric circuit without this particular duality-based conversion procedure. Now, with such powerful duality principle, only three steps can directly reach the solution. In summary, the duality relation between electric and magnetic circuits can be outlined in the following table:

Categories	Electric Circuit	Magnetic Circuit
Components	In Parallel	In Series
Connection	In Series	In Parallel
Sources	Current Source	Ni, (mmf)
Circuit	Mesh	Node
Topology	Node	Mesh
Circuit	Open	Short
Branches or Ports	Short	Open
E-M Bridge	Inductance $L = \frac{N^2}{\mathscr{R}}$	Reluctance $\mathscr{R}$
Kirchhoff's Law	KCL $\sum i_{in} = \sum i_{out}$	KVL $\sum Ni = \sum Hl$
KVL & KCL	KVL $\sum v_{rise} = \sum v_{drop}$	KCL $\sum \phi_{in} = \sum \phi_{out}$

In addition to this useful duality relation between electric and magnetic circuits, the duality relation between open serial-chain and closed parallel-chain robotic systems is also remarkable and applicable. Table 5.2 demonstrates a convincing evidence of duality between each pair of the dual counterparts, such as their forward and inverse kinematics and their kinematics and statics. On the other hand, through a further duality study in robotics, by referring to Table 5.2, we make the following remarks:

1. For the kinematics in Cartesian motion, the forward kinematics $(R_0^p \ p_0^p) = f^{-1}(q)$ for a closed parallel-chain robot has almost no hope to find an explicit form of $f^{-1}(q)$. Namely, given 6 prismatic piston lengths $q = (d_1 \cdots d_6)^T$ as an input, it is almost unable to determine what the top panel output position p_0^p and orientation R_0^p will be, in contrast to its inverse kinematics that is almost exceptionally straightforward. The difficult or even impossible FK solution may lead to a kinematics-based isometric embedding $z = \zeta(q)$ without an explicit full

Categories	Open Serial-Chain Robots	Closed Parallel-Chain Robots
Kinematics	$(R_0^t \; p_0^t) = f(q)$	$q = f(R_0^p \; p_0^p)$
(Cartesian Motion)	$q = f^{-1}(R_0^t \; p_0^t)$	$(R_0^p \; p_0^p) = f^{-1}(q)$
Kinematics	$V = J_0 \dot{q}$	$\dot{q} = J_c^T V$
(Differential Motion)	$\dot{q} = J_0^{-1} V$	$V = J_c^{T-1} \dot{q}$
Statics	$\tau = J_0^T F$	$F = J_c \tau$
Kinematics (Embedding)	$\dot{z} = J \dot{q}$	$\dot{q} = J_z^T \dot{z}$
Dynamics	$\tau + \tau_g = J^T \ddot{z}$	$\ddot{z} = J_z(\tau + \tau_g)$
Arguments of z and J	Joint variables q	Cartesian variables R_0^p & p_0^p

Table 5.2

A table on the principle of duality between open serial and closed parallel-chain robotic systems

function form of $\zeta(\cdot)$, which will result in an arduous determination of its genuine mathematical Jacobian matrix J_z if one keeps using the open serial-chain approach.

2. However, the FK in differential motion for the closed parallel-chain robotic systems has its explicit form that is solvable in contrast to the FK in Cartesian motion.

3. For the statics, the validity of $F = J_c \tau$ for a closed parallel-chain robot as a dual form of the open serial-chain counterpart is under the condition that every non-actuated passive axis must have no static friction or no additional force/torque imposed on. Otherwise, this equation may not be accurate.

4. In the dynamics category, the dynamic equations in the isometric embedding version between two counterparts are very similar to their respective statics equations, as can be clearly seen from Table 5.2. It is important that no matter how different from one to the other between two dynamic models, the two dual counterparts are unified to meet the common energy conservation law. In other words, the kinetic energy in an open serial-chain system and in a closed parallel-chain system is always determined in a common way: $K = \frac{1}{2} \dot{z}^T \dot{z}$.

5. For the two very different dual equations in the dynamics category in Table 5.2, which one is more suitable for a given closed parallel/serial hybrid-chain robotic system should fully depend on which dual category between $\dot{z} = J \dot{q}$ and $\dot{q} = J_z^T \dot{z}$ the kinematics belongs to.

6. In general speaking, an isometric embedding and its Jacobian matrix in the open serial-chain or open hybrid-chain case can always be an explicit function of the joint variables q, while the isometric embedding and its Jacobian matrix in a closed parallel-chain or closed hybrid-chain case are often a direct function of the Cartesian variables, such as the positions and orientations, instead of the joint variables q.

5.4 ISOMETRIC EMBEDDING BASED DYNAMICS MODELING FOR PARALLEL AND HYBRID-CHAIN ROBOTS

By recalling Theorem 4.1, a robotic system can be dynamically modeled with a reduced formula given in equation (4.2). To apply this, it must satisfy the following two key conditions:

1. The mapping $\zeta : M^n \to \mathbb{R}^m$ should be an isometric embedding that can send the configuration manifold (C-manifold) M^n of the robotic system into a spacious enough ambient Euclidean space $\mathbb{R}^m$ with the Riemannian metric preserved;
2. The Jacobian matrix J must be a genuine mathematical Jacobian, i.e., $J = \frac{\partial z}{\partial q}$.

Under these two key conditions, the isometric embedding based dynamics modeling procedure should more specifically meet the following requirements:

1. Since the formation of an isometric embedding completely relies on the kinematic model for open hybrid-chain robotic systems, both the position and orientation of the kinematic model should be an explicit function of the joint or axis variables q in order to determine its genuine mathematical Jacobian matrix.
2. If the dominant link or platform is only in 3 d.o.f. translational motion, then, its isometric embedding can be defined as $z = \zeta(q) = \sqrt{m} p_0^p(q)$, the position vector with respect to the fixed base as an explicit function form of the joint position vector q.
3. If the dominant link or platform of a robot, in addition to the 3 d.o.f. translation, also has orientation changes, then the isometric embedding $z = \zeta(q)$ has to be augmented by its rotation part, such as given by equation (4.4) or (4.6). But they must also be the explicit functions of q.
4. However, in many closed parallel or hybrid-chain robot cases, such as the Stewart platform, the isometric embedding $z = \zeta(\cdot)$ is often an explicit function of Cartesian variables, such as the position and orientation. Then, both the kinematic and dynamic equations should follow the column of the "Closed Parallel-Chain Robots" of the duality Table 5.2.
5. Therefore, for dynamics modeling, there are two dual formulations to make a choice, one is $\tau + \tau_g = J^T \ddot{z}$, and the other one is $\ddot{z} = J_z(\tau + \tau_g)$. To determine which one is suitable, the given robotic system must first be identified to see which kinematics category it belongs to before continuing its dynamics modeling. In other words, if the duality category is obvious, just keep using the matched dynamic formula. If the category is temporarily unclear, such as a hybrid-chain mechanism or as the degree of mobility is less than 6, then one has to check its kinematic equation to see which one it can fit between $\dot{z} = J\dot{q}$ and $\dot{q} = J_z^T \dot{z}$, as given by Table 5.2. If the kinematics is in the former type, then the dynamics modeling follows the open serial-chain dynamic formulation, and $z = \zeta(q)$ should be an explicit function of q. If the kinematics is in the latter type, the dynamics modeling should follow the closed parallel-chain dynamic formulation, and both z and its Jacobian J_z should be the function of Cartesian positions p_0^p and orientations R_0^p.

6. After the isometric embedding is confirmed, $z = \zeta(\cdot)$ must obey

$$\frac{1}{2}\dot{z}^T\dot{z} = K,$$

where K is the kinetic energy of the dominant link or platform of the robotic system.

We now focus on the following five robotic systems as typical examples to explore and develop both their kinematic and dynamic models.

5.4.1 THE STEWART PLATFORM

Let us first study the 6-leg closed parallel-chain Stewart platform again. As we can see from Figure 5.1, if the top 6 spherical joints that connect the 6 prismatic pistons to the top panel can be distributed to form a perfect symmetric hexagon with equal sides, or every pair of the adjacent spherical joints are merged in one and all the 3 pairs form an equilateral triangle, and so are the bottom 6 universal joints, then

$$\sum_{i=1}^{6} b_p^i = 0, \quad \text{and} \quad \sum_{i=1}^{6} a_0^i = 0.$$

Based on the inverse kinematics (IK) equation given in (5.1), we obtain

$$\sum_{i=1}^{6} d_i l_i^o = \sum_{i=1}^{6} (p_0^p + R_0^p b_p^i - a_0^i) = 6p_0^p + R_0^p \sum_{i=1}^{6} b_p^i - \sum_{i=1}^{6} a_0^i = 6p_0^p.$$

If the mass center position p_0^c of the top panel, as a dominant body, is the same as its centroid position p_0^p, then

$$p_0^c = p_0^p = \frac{1}{6}\sum_{i=1}^{6} d_i l_i^o. \tag{5.14}$$

At the first glance, this equation is a perfect explicit form of the function of all the prismatic displacements d_i's. However, inside the summation, it is also related to the unit vector l_i^o of each leg, which may also depend on almost all the active and passive axes, including the 6 prismatic piston displacements. Therefore, the above p_0^c in (5.14) is failed to be part of the isometric embedding $z = \zeta(q)$ as an explicit function of all the piston displacements d_i's due to the unavailability of the FK solution for such a typical closed parallel-chain Stewart platform.

Therefore, the Stewart platform has been justified that it does not fit to the open serial-chain category. Now, let us in turn try the dual dynamic formula in the closed parallel-chain category, i.e.,

$$\ddot{z} = J_z(\tau + \tau_g), \tag{5.15}$$

which is based on equation (5.9) and given from Table 5.2. However, since the Jacobian matrix J_z is a 12 by 6 tall matrix, based on linear algebra, equation (5.15) may not have a compatible solution. Regardless the compatibility, the minimum norm and least square error solution to (5.15) can always be found in terms of the pseudo-inverse J_z^+ of the Jacobian matrix J_z, i.e.,

$$\tau + \tau_g = J_z^+ \ddot{z} = (J_z^T J_z)^{-1} J_z^T \ddot{z}.$$

Actually, the isometric embedding in equation (5.5) and its Jacobian matrix in (5.7) have been well-developed in the previous kinematics study section for the Stewart platform, and let us rewrite them here again,

$$z = \begin{pmatrix} \sqrt{m}p_0^p \\ \beta_1 r_x \\ \beta_2 r_y \\ \beta_3 r_z \end{pmatrix}, \quad J_z = \begin{pmatrix} (l_1^o \ l_2^o \ l_3^o \ l_4^o \ l_5^o \ l_6^o)A \\ (l_1^o \ l_2^o \ l_3^o \ l_4^o \ l_5^o \ l_6^o)B_1 \\ (l_1^o \ l_2^o \ l_3^o \ l_4^o \ l_5^o \ l_6^o)B_2 \\ (l_1^o \ l_2^o \ l_3^o \ l_4^o \ l_5^o \ l_6^o)B_3 \end{pmatrix},$$

where $R_0^p = (r_x \ r_y \ r_z)$. Each unit directional vector l_i^o that is required in the Jacobian matrix J_z can be directly found in terms of the given p_0^p and R_0^p by the IK equation (5.1), i.e.,

$$l_i^o = \frac{1}{d_i}(p_0^p + R_0^p b_p^i - a_0^i).$$

Then, both z and J_z can be determined at each sampling point in terms of the Cartesian motion input. Therefore, the dynamic equation for the Stewart platform such a typical closed parallel-chain system can be formulated by (5.15). If a trajectory-tracking control scheme given in equation (4.9) is adopted, then the dynamic control model can be expressed as

$$\ddot{z} = \ddot{z}^d + K_2(\dot{z}^d - \dot{z}) + K_1(z^d - z) = J_z(\tau + \tau_g), \qquad (5.16)$$

where $z^d = z^d(t)$ is a desired trajectory, and $K_1 > 0$ and $K_2 > 0$ are two positive control gain constants.

5.4.2   THE 3D 3-LEG HYBRID-CHAIN ROBOTIC SYSTEM

This 3D 3-leg robotic system seems to belong to the closed parallel-chain category as well, but it can only offer 3 d.o.f. of mobility. We have to first test its kinematics in order to determine which dual category of dynamic formulation this kind of robot should belong to. As a comparative study, the major differences between this robot and the Stewart platform are outlined as follows:

1. Although each leg is of the same prismatic piston type in comparison to the Stewart platform, this robot has only 3 legs;
2. The degree of mobility for this robot is only $r = 3$;
3. While the position of the top platform with respect to the base can be specified as a required kinematic output, the orientation is unknown and is actually a by-product of meeting the specified position requirement.

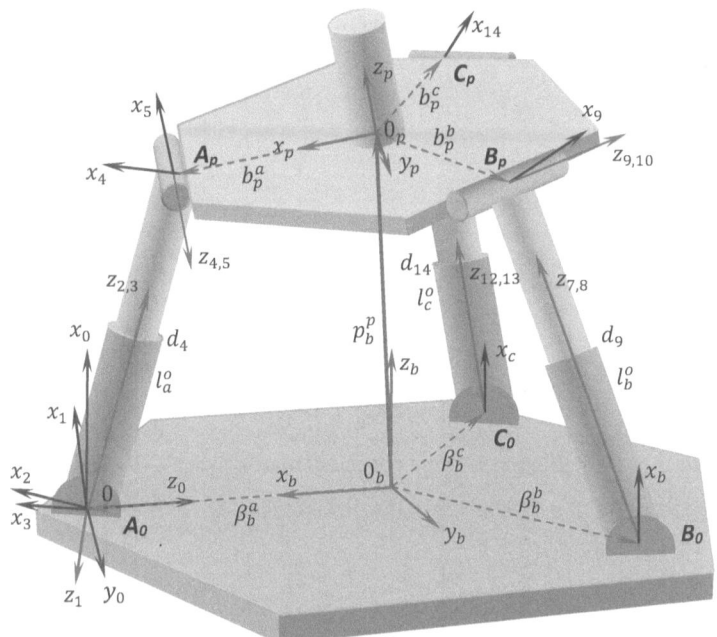

Figure 5.11 Dynamic analysis of the 3D 3-leg hybrid-chain robotic system

According to Figure 5.11, it also contains a vector quadrilateral between the top platform and the bottom base. Obviously it consists of a given required position vector p_b^p, the directional vector $d_4 l_a^o$ of leg a, and two constant vectors b_p^a on the top platform and β_b^a on the bottom base. They should satisfy the following vector relation that is similar to equation (5.1) for the Stewart platform:

$$d_4 l_a^o = p_b^p + R_b^p b_p^a - \beta_b^a.$$

While the orientation R_b^p is an unknown by-product, we may temporarily assume that it is also given. Then, in a similar way, the time-derivative of the piston displacement for leg a can be found as

$$\dot{d}_4 = (l_a^o)^T \dot{p}_b^p + (l_a^o)^T \dot{R}_b^p b_p^a.$$

Similarly, for leg b and leg c,

$$d_9 l_b^o = p_b^p + R_b^p b_p^b - \beta_b^b, \quad \text{and} \quad \dot{d}_9 = (l_b^o)^T \dot{p}_b^p + (l_b^o)^T \dot{R}_b^p b_p^b,$$

$$d_{14} l_c^o = p_b^p + R_b^p b_p^c - \beta_b^c, \quad \text{and} \quad \dot{d}_{14} = (l_c^o)^T \dot{p}_b^p + (l_c^o)^T \dot{R}_b^p b_p^c,$$

By following the same procedure as in the Stewart platform kinematics modeling, an isometric embedding and its time-derivative can be constructed by

$$
z = \begin{pmatrix} \sqrt{m}p_b^p \\ h_1 r_x \\ h_2 r_y \\ h_3 r_z \end{pmatrix}, \quad \dot{z} = \begin{pmatrix} \sqrt{m}\dot{p}_b^p \\ h_1 \dot{r}_x \\ h_2 \dot{r}_y \\ h_3 \dot{r}_z \end{pmatrix},
$$

where $(r_x \ r_y \ r_z) = R_b^p$.

Let a new 12 by 3 Jacobian matrix be defined as

$$
J_z = \begin{pmatrix} (l_a^o \ l_b^o \ l_c^o)A \\ (l_a^o \ l_b^o \ l_c^o)B_1 \\ (l_a^o \ l_b^o \ l_c^o)B_2 \\ (l_a^o \ l_b^o \ l_c^o)B_3 \end{pmatrix},
$$

where all the 3 by 3 diagonal constant coefficient matrices A, B_1, B_2, and B_3 are given by

$$
A = \begin{pmatrix} \frac{1}{\sqrt{m}} & 0 & 0 \\ 0 & \frac{1}{\sqrt{m}} & 0 \\ 0 & 0 & \frac{1}{\sqrt{m}} \end{pmatrix}, \quad B_1 = \begin{pmatrix} \frac{b_{px}^a}{h_1} & 0 & 0 \\ 0 & \frac{b_{px}^b}{h_1} & 0 \\ 0 & 0 & \frac{b_{px}^c}{h_1} \end{pmatrix},
$$

$$
B_2 = \begin{pmatrix} \frac{b_{py}^a}{h_2} & 0 & 0 \\ 0 & \frac{b_{py}^b}{h_2} & 0 \\ 0 & 0 & \frac{b_{py}^c}{h_2} \end{pmatrix}, \quad B_3 = \begin{pmatrix} \frac{b_{pz}^a}{h_3} & 0 & 0 \\ 0 & \frac{b_{pz}^b}{h_3} & 0 \\ 0 & 0 & \frac{b_{pz}^c}{h_3} \end{pmatrix}.
$$

Thus, the 3D 3-leg closed hybrid-chain robot should obey the same Jacobian equation as that for the Stewart platform:

$$
\dot{q} = J_z^T \dot{z}.
$$

Because of this kinematics modeling outcome, its dynamic formulation should follow equation (5.15), instead of equation (4.2). However, the orientation R_b^p is a by-product that is actually unknown so that both z and J_z cannot be thoroughly determined. This difficulty is consistent to its tough IK solution issue, as addressed in Chapter 3, which becomes an obstacle for further dynamics modeling development. To tackle this tough issue, let us further try to develop a solution to finding the passive by-product orientation for the top platform of the 3D 3-leg robotic system.

In fact, if the position of the tip point with a height t_4 above the top platform is given by p_b^t, we can take advantage of the revolute joint mechanical constraint at A_p, B_p and C_p to resolve the by-product orientation R_b^p for this 3D 3-leg closed parallel-chain system, as depicted in Figure 5.12. Clearly, if $t_4 = 0$, the tip point meets the mass center of the top platform.

From Figure 5.12, it is obvious that the vector that is tailed at A_0 and arrow-pointing to the tip point is $\eta_a = p_b^t - \beta_b^a$. By the same token, $\eta_b = p_b^t - \beta_b^b$ and $\eta_c = p_b^t - \beta_b^c$ at B_0 and C_0, respectively. Therefore, under a given tip-position vector

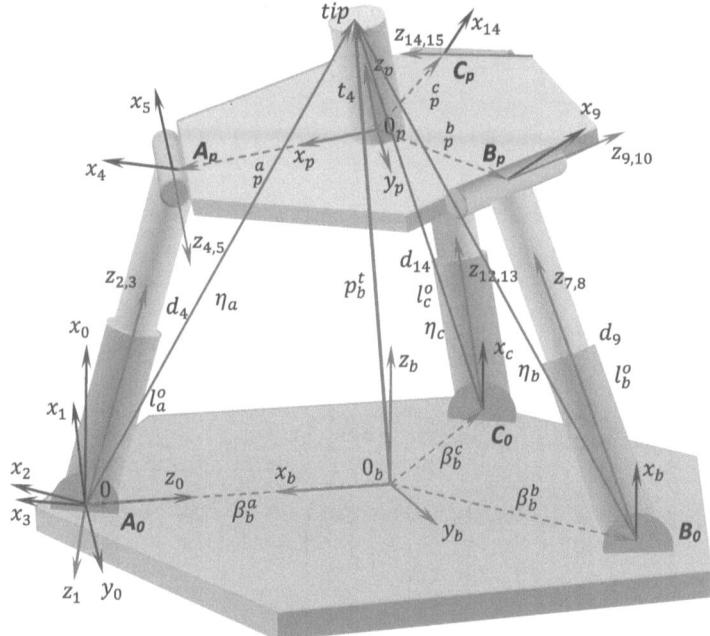

Figure 5.12 To solve the by-product orientation R_b^p for the 3D 3-leg closed-chain robotic system

p_b^t, all the three vectors η_a, η_b and η_c can be uniquely determined due to the fact that the three vectors β_b^a, β_b^b and β_b^c fixed on the base panel are all known with respect to the base frame. Furthermore, the vector η_a is always perpendicular to the axis z_4 of the revolute joint at A_p, i.e., $\eta_a \perp z_4$, and also, $\eta_b \perp z_9$ and $\eta_c \perp z_{14}$ at any time instant.

Moreover, the unit axes z_4, z_9 and z_{14} can directly be projected on frame p that is fixed on the top platform. If they form an equilateral triangle, as depicted in Figure 5.12, then by referring to frame p,

$$z_4 = \begin{pmatrix} 0 \\ 1 \\ 0 \end{pmatrix}, \quad z_9 = \begin{pmatrix} -\frac{\sqrt{3}}{2} \\ -\frac{1}{2} \\ 0 \end{pmatrix}, \quad z_{14} = \begin{pmatrix} \frac{\sqrt{3}}{2} \\ -\frac{1}{2} \\ 0 \end{pmatrix}.$$

Alternatively, if they form a right-isosceles triangle, or a rectangle, please refer to the literature [8]. Therefore,

$$\eta_a^T R_b^p z_4 = 0, \quad \eta_b^T R_b^p z_9 = 0, \quad \text{and} \quad \eta_c^T R_b^p z_{14} = 0. \tag{5.17}$$

Intuitively, when the 3 legs of this closed parallel-chain system are sliding individually to meet the position p_b^t requirement, the passive orientation change for the top platform should be an only 2 d.o.f. rotation without spinning. Based on this intuition, let us define the by-product orientation R_b^p of the top platform as a rotation about the x_b and y_b axes of the base only, but no spin about the z_b-axis, i.e.,

$$R_b^p = R(x,\phi)R(y,\psi) = \begin{pmatrix} 1 & 0 & 0 \\ 0 & c\phi & -s\phi \\ 0 & s\phi & c\phi \end{pmatrix} \begin{pmatrix} c\psi & 0 & s\psi \\ 0 & 1 & 0 \\ -s\psi & 0 & c\psi \end{pmatrix},$$

where $s\phi$ and $c\phi$ are the abbreviations of $\sin\phi$ and $\cos\phi$, respectively, and so are $s\psi$ and $c\psi$ as the short notation for $\sin\psi$ and $\cos\psi$. Therefore, solving for both two angles ϕ and ψ can uniquely determine the passive orientation R_b^p of the top platform.

Now by substituting $R_b^p = R(x,\phi)R(y,\psi)$ into the first equation of (5.17), we can have

$$\eta_a^T R_b^p z_4 = \eta_a^T \begin{pmatrix} 1 & 0 & 0 \\ 0 & c\phi & -s\phi \\ 0 & s\phi & c\phi \end{pmatrix} \begin{pmatrix} c\psi & 0 & s\psi \\ 0 & 1 & 0 \\ -s\psi & 0 & c\psi \end{pmatrix} \begin{pmatrix} 0 \\ 1 \\ 0 \end{pmatrix}$$

$$= \eta_a^T \begin{pmatrix} 1 & 0 & 0 \\ 0 & c\phi & -s\phi \\ 0 & s\phi & c\phi \end{pmatrix} \begin{pmatrix} 0 \\ 1 \\ 0 \end{pmatrix} = \eta_a^T \begin{pmatrix} 0 \\ c\phi \\ s\phi \end{pmatrix} = 0.$$

Thus,

$$\eta_a(2)c\phi + \eta_a(3)s\phi = 0 \quad \text{so that} \quad \phi = -\tan^{-1}\left(\frac{\eta_a(2)}{\eta_a(3)}\right), \qquad (5.18)$$

where we confine ϕ as well as ψ being rotated within $(-90°, 90°)$, which is a quite reasonable and practical constraint in this particular case.

Once ϕ is solved, the first rotation matrix $R(x,\phi)$ is well-determined. Now, based on the second and third equations of (5.17),

$$\eta_b^T R(x,\phi) \begin{pmatrix} c\psi & 0 & s\psi \\ 0 & 1 & 0 \\ -s\psi & 0 & c\psi \end{pmatrix} \begin{pmatrix} -\frac{\sqrt{3}}{2} \\ -\frac{1}{2} \\ 0 \end{pmatrix} = \eta_b^T R(x,\phi) \begin{pmatrix} -\frac{\sqrt{3}}{2}c\psi \\ -\frac{1}{2} \\ \frac{\sqrt{3}}{2}s\psi \end{pmatrix} = 0,$$

and also,

$$\eta_c^T R(x,\phi) \begin{pmatrix} c\psi & 0 & s\psi \\ 0 & 1 & 0 \\ -s\psi & 0 & c\psi \end{pmatrix} \begin{pmatrix} \frac{\sqrt{3}}{2} \\ -\frac{1}{2} \\ 0 \end{pmatrix} = \eta_c^T R(x,\phi) \begin{pmatrix} \frac{\sqrt{3}}{2}c\psi \\ -\frac{1}{2} \\ -\frac{\sqrt{3}}{2}s\psi \end{pmatrix} = 0..$$

However, the above two equations can easily fall into a linearly dependent dilemma, especially near the home position with both ϕ and ψ close to zero due to the clear fact that all the three revolute joint axes z_4, z_9 and z_{14} are on the same 2D plane of the top platform with the common zero z-projection. Therefore, to avoid any possible singularity, we have to rely on only one of them in solving the angle ψ. Let us pick the first one. Let $h_b = \eta_b^T R(x,\phi)$. Then,

$$-\frac{\sqrt{3}}{2}h_b(1)c\psi - \frac{1}{2}h_b(2) + \frac{\sqrt{3}}{2}h_b(3)s\psi = 0.$$

Shifting the first cosine term to the right-hand side and then squaring both sides yield

$$\frac{3}{4}h_b(3)^2 s^2 \psi + \frac{1}{4}h_b(2)^2 - \frac{\sqrt{3}}{2}h_b(2)h_b(3)s\psi = \frac{3}{4}h_b(1)^2 c^2 \psi = \frac{3}{4}h_b(1)^2(1-s^2\psi).$$

Thus, $s\psi$ satisfies the following quadratic equation:

$$\frac{3}{4}(h_b(1)^2 + h_b(3)^2)s^2 \psi - \frac{\sqrt{3}}{2}h_b(2)h_b(3)s\psi + \frac{1}{4}h_b(2)^2 - \frac{3}{4}h_b(1)^2 = 0.$$

By solving this quadratic equation, we obtain

$$s\psi = \frac{h_b(2)h_b(3) + \sqrt{3h_b(1)^4 + 3h_b(1)^2 h_b(3)^2 - h_b(1)^2 h_b(2)^2}}{\sqrt{3}(h_b(1)^2 + h_b(3)^2)}. \tag{5.19}$$

Finally, due to the constraint $-90° < \psi < 90°$, the angle ψ can be resolved by the arc sine function, i.e., $\psi = \mathrm{asin}(s\psi)$.

The passive orientation change in order to meet the given position p_b^t requirement can now be determined as $R_b^p = R(x, \phi)R(y, \psi)$. As a numerical example, if the position vector p_b^t and the three fixed vectors β_b^a, β_b^b and β_b^c for the universal joints on the base panel with respect to the base frame are given by

$$p_b^t = \begin{pmatrix} 0.6000 \\ 1.6400 \\ 1.8918 \end{pmatrix}, \quad \beta_b^a = \begin{pmatrix} 1.2000 \\ 0 \\ 0.0800 \end{pmatrix}, \quad \beta_b^b = \begin{pmatrix} -0.6000 \\ 1.0392 \\ 0.0800 \end{pmatrix}, \quad \beta_b^c = \begin{pmatrix} -0.6000 \\ -1.0392 \\ 0.0800 \end{pmatrix},$$

then,

$$\eta_a = p_b^t - \beta_b^a = \begin{pmatrix} -0.6000 \\ 1.6200 \\ 1.8462 \end{pmatrix}, \quad \eta_b = p_b^t - \beta_b^b = \begin{pmatrix} 1.2000 \\ 0.5808 \\ 1.8462 \end{pmatrix},$$

and

$$\eta_c = p_b^t - \beta_b^c = \begin{pmatrix} 1.2000 \\ 2.6592 \\ 1.8462 \end{pmatrix}.$$

Based on equation (5.18), $\phi = -41.2665°$. Then, using equation (5.19), $\psi = 21.9535°$. The passive orientation is now equal to

$$R_b^p = R(x, \phi)R(y, \psi) = \begin{pmatrix} 0.9275 & 0 & 0.3739 \\ -0.2466 & 0.7516 & 0.6117 \\ -0.2810 & -0.6596 & 0.6971 \end{pmatrix}.$$

Figure 5.13 illustrates the instantaneous posture with the above numerical data in an animation study for this 3D 3-leg based 3+3 parallel/serial-chain hybrid robot as it is drawing a sine wave in the 3D space. Through the above procedure of finding the passive by-product orientation R_b^p in order to meet the position p_b^t requirement, we can now completely define an isometric embedding that augments p_b^t along with

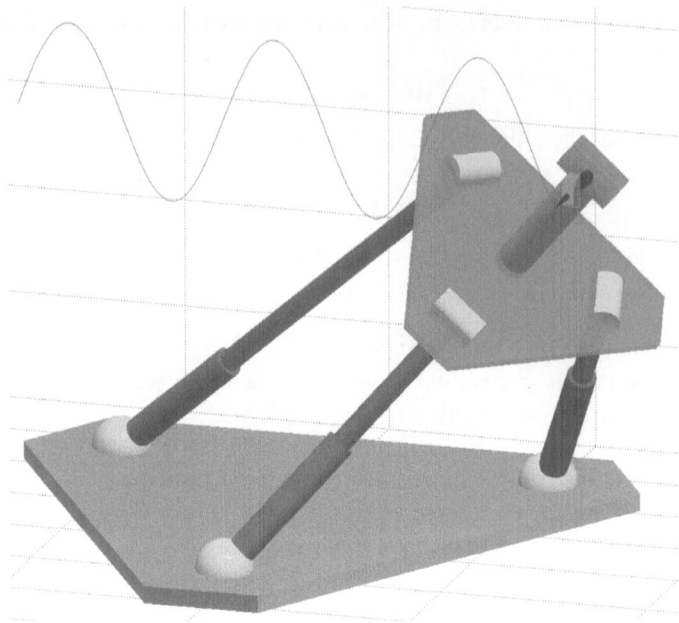

Figure 5.13 The instantaneous posture of the 3D 3-leg closed-chain robotic system in the numerical example

the three columns of the passive rotation matrix R_b^p towards the dynamic formulation $\ddot{z} = J_z(\tau + \tau_g)$ in the closed-chain version, as given in Table 5.2. It is observable that the arguments of the isometric embedding z and its Jacobian matrix J_z are the Cartesian position p_b^t and the passive orientation R_b^p, instead of the joint variables.

Since this particular robotic system looks like a mechanism in-between the two dual counterparts, it may be worthwhile to alternatively explore the dynamics modeling development in joint space. Let us now return back to try an open serial-chain dynamic formulation $\tau + \tau_g = J^T \ddot{z}$. To do so, let us re-assign the coordinate frames for this closed 3D 3-leg parallel-chain system, as illustrated in Figure 5.11. This time each of the three legs starts from their respective grounded universal joints A_0, B_0 and C_0 to assign coordinate frames, but using different axis angles. Leg a starts A_0 with the axis angles θ_1 through θ_5. Leg b starts B_0 with the axis angles θ_6 through θ_{10}, and leg c starts C_0 with the axis angles θ_{11} through θ_{15}. Because leg b and leg c can follow the same D-H tables as leg a, they share the same forms of homogeneous transformation matrices but just in terms of the different axis angles.

According to Example 3.7 in Chapter 3, this 3D 3-leg system has a complete set of homogeneous transformations in the floating grounded joint model. We can use its D-H table to find the homogeneous transformation from frame 0 to frame 4:

$$A_0^4 = \begin{pmatrix} c_1c_2c_3 - s_1s_3 & -c_1s_2 & -c_1c_2s_3 - s_1c_3 & d_4c_1s_2 \\ s_1c_2c_3 + c_1s_3 & -s_1s_2 & -s_1c_2s_3 + c_1c_3 & d_4s_1s_2 \\ -s_2c_3 & -c_2 & s_2s_3 & d_4c_2 \\ 0 & 0 & 0 & 1 \end{pmatrix}.$$

The upper left 3 by 3 block of A_0^4 is the orientation of frame 4, while the 4-th column is the position vector p_0^4 that is representing the position $l_a = d_4 l_a^o$ of leg a, where l_a^o is the unit vector of l_a. Similarly, A_0^9 and A_0^{14} can have the same form of homogeneous transformation matrices as A_0^4 for leg b and leg c, respectively, but each 0 subscript is different, which only indicates each to start from A_0, B_0 and C_0. Likewise, the last column of A_0^9 is the position vector $l_b = d_9 l_b^o$ for leg b, and the last column of A_0^{14} is the position vector $l_c = d_{14} l_c^o$ for leg c.

One can also see from Figure 5.11 that the four vectors l_a, b_p^a, β_b^a and p_b^p form a quadrilateral so that

$$l_a = d_4 l_a^o = p_b^p + R_b^p b_p^a - \beta_b^a.$$

Similarly, for leg b and leg c,

$$l_b = d_9 l_b^o = p_b^p + R_b^p b_p^b - \beta_b^b, \quad l_c = d_{14} l_c^o = p_b^p + R_b^p b_p^c - \beta_b^c.$$

Since A_0, B_0 and C_0 form an equilateral triangle, and A_p, B_p and C_p also form an equilateral triangle,

$$\beta_b^a + \beta_b^b + \beta_b^c = 0, \quad \text{and} \quad b_p^a + b_p^b + b_p^c = 0.$$

Thus,

$$d_4 l_a^o + d_9 l_b^o + d_{14} l_c^o = 3 p_b^p.$$

Therefore, if p_b^p is the mass center position vector of the top panel, as a dominant body, with respect to the base, then

$$p_b^c = p_b^p = \frac{1}{3}(d_4 l_a^o + d_9 l_b^o + d_{14} l_c^o),$$

where the three unit directional vectors can now be determined in terms of the related axis angles:

$$l_a^o = \begin{pmatrix} c_1s_2 \\ s_1s_2 \\ c_2 \end{pmatrix}, \quad l_b^o = \begin{pmatrix} c_6s_7 \\ s_6s_7 \\ c_7 \end{pmatrix}, \quad l_c^o = \begin{pmatrix} c_{11}s_{12} \\ s_{11}s_{12} \\ c_{12} \end{pmatrix}.$$

Actually, the orientation of the top platform, as a by-product, has only a 2 d.o.f. rotation, i.e., to rotate about the x_p and y_p axes without spinning rotation about the z_p axis as can be determined by the preceding algorithm in equations (5.18) and (5.19). We can now alternatively find this only 2-dimensional rotation matrix R_b^p by the above three equations of l_a, l_b and l_c. Thus, an isometric embedding can be defined by

$$z = \begin{pmatrix} \sqrt{m}p_b^c \\ h_1 r_x \\ h_2 r_y \\ h_3 r_z \end{pmatrix},$$

where $R_b^p = (r_x \ r_y \ r_z)$. To find its Jacobian matrix, we may just take derivative for z with respect only to d_4, d_9 and d_{14} so that

$$J_d = \frac{\partial z}{\partial q_d} = \begin{pmatrix} \frac{\sqrt{m}}{3}(l_a^o \ l_b^o \ l_c^o) \\ h_1 \frac{\partial r_x}{\partial q_d} \\ h_2 \frac{\partial r_y}{\partial q_d} \\ h_3 \frac{\partial r_z}{\partial q_d} \end{pmatrix},$$

where $q_d = (d_4 \ d_9 \ d_{14})^T$.

After the above preparation, the dynamics for this 3D 3-leg hybrid-chain robotic system can be formulated by $\tau_d + \tau_{gd} = J_d^T \ddot{z}$, where the joint torque vector τ_d is only 3-dimensional to reflect the three piston driving forces. Since the Jacobian matrix depends also on the axis angles other than q_d, to implement this dynamic equation, every passive and active axis position has to be measured at each sampling point. If the rest passive axes angles other than q_d are augmented and defined as a new joint position vector q_r, then,

$$J_r = \frac{\partial z}{\partial q_r}.$$

The corresponding part of dynamic equation becomes

$$\tau_r + \tau_{gr} = \tau_{gr} = J_r^T \ddot{z},$$

due to $\tau_r = 0$ for all the passive non-actuated axes, but they are still acted by τ_{gr} due to gravity. This is actually an internal dynamic constraint, which is hard to be further determined explicitly. Therefore, because of this new obstacle, we may have to return back for this kind of robot to the closed parallel-chain category.

5.4.3   THE DELTA CLOSED HYBRID-CHAIN ROBOT

The Delta robot is actually a closed parallel/serial hybrid-chain manipulator [9]. Because it is designed to use a four-bar parallelogram mechanism for each of the three forearm, the tool panel can keep a 3 d.o.f. translational motion without orientation change. Such a relatively simple kinematics makes its IK solution more straightforward. For its dynamics modeling, an isometric embedding for the tool panel, as a dominant body, can be formed using only the mass center position vector with respect to the base frame, i.e.,

$$z = \zeta(q) = \sqrt{m}p_b^c,$$

and it is only 3-dimensional to send its 3-dimensional configuration manifold (C-manifold) M^3 into the Euclidean 3-space $\mathbb{R}^3$.

In order to find the mass center position vector p_b^c for the lower tool panel, let us first assume that it matches and is equal to the centroid position vector p_b^p. Since on the base ceiling, $b_a + b_b + b_c = 0$, and also on the lower tool panel, $\beta_a + \beta_b + \beta_c = 0$, due to the three arms jointed with the base as well as the tool panel to form an equilateral triangle on each, as shown in Figure 5.5. Thus,

$$R_b^0 p_0^4 + R(z, -120°)R_b^0 p_0^9 + R(z, 120°)R_b^0 p_0^{14} = 3p_b^p.$$

Therefore,

$$p_b^c = p_b^p = \frac{1}{3}[R_b^0 p_0^4 + R(z, -120°)R_b^0 p_0^9 + R(z, 120°)R_b^0 p_0^{14}],$$

where R_b^0 and $R(z, \pm 120°)$ are constant rotation matrices.

Recalling equations (5.10), (5.12) and (5.13) from the last section,

$$p_0^4 = \begin{pmatrix} a_1 c_1 + a_3 c_{12} c_3 \\ a_1 s_1 + a_3 s_{12} c_3 \\ a_3 s_3 \end{pmatrix},$$

and also,

$$p_0^9 = \begin{pmatrix} a_1 c_6 + a_3 c_{67} c_8 \\ a_1 s_6 + a_3 s_{67} c_8 \\ a_3 s_8 \end{pmatrix}, \quad \text{and} \quad p_0^{14} = \begin{pmatrix} a_1 c_{11} + a_3 c_{1112} c_{13} \\ a_1 s_{11} + a_3 s_{1112} c_{13} \\ a_3 s_{13} \end{pmatrix}.$$

Then, the mass center position $p_b^c = p_b^p$ can be determined as an explicit function of all the joint angles.

Because the motor drives are installed at the 3 joints θ_1, θ_6 and θ_{11} that are all connected to the base ceiling of the robot, a special Jacobian matrix J_d for the isometric embedding $z = \zeta(q)$, which only covers the three actuated joints can be determined by taking derivative with respect to just 3 driving joint angles. Let $q_d = (\theta_1 \ \theta_6 \ \theta_{11})^T$ and the remaining joint angles $q_r = (\theta_2 \ \theta_3 \ \theta_7 \ \theta_8 \ \theta_{12} \ \theta_{13})^T$. Also, by referring to Figure 5.5,

$$R_b^0 = \begin{pmatrix} 0 & 0 & 1 \\ 1 & 0 & 0 \\ 0 & 1 & 0 \end{pmatrix}, \quad \text{and} \quad R(z, \pm 120°) = \begin{pmatrix} -\frac{1}{2} & \mp\frac{\sqrt{3}}{2} & 0 \\ \pm\frac{\sqrt{3}}{2} & -\frac{1}{2} & 0 \\ 0 & 0 & 1 \end{pmatrix}.$$

Then, the special 3 by 3 Jacobian matrix $J_d = \frac{\partial \zeta(q)}{\partial q_d} =$

$$\frac{\sqrt{m}}{3} \begin{pmatrix} 0 & -\frac{\sqrt{3}}{2}(a_1 s_6 + a_3 s_{67} c_8) & \frac{\sqrt{3}}{2}(a_1 s_{11} + a_3 s_{1112} c_{13}) \\ -a_1 s_1 - a_3 s_{12} c_3 & \frac{1}{2}(a_1 s_6 + a_3 s_{67} c_8) & \frac{1}{2}(a_1 s_{11} + a_3 s_{1112} c_{13}) \\ a_1 c_1 + a_3 c_{12} c_3 & a_1 c_6 + a_3 c_{67} c_8 & a_1 c_{11} + a_3 c_{1112} c_{13} \end{pmatrix},$$

and the remaining 3 by 6 Jacobian matrix becomes

$$J_r = \frac{\partial \zeta(q)}{\partial q_r}.$$

Since the torque due to gravity is a gradient vector of the gravitational potential energy,

$$\tau_g = -\frac{\partial P_g}{\partial q} = \begin{pmatrix} \tau_{gd} \\ \tau_{gr} \end{pmatrix} = -\begin{pmatrix} \frac{\partial P_g}{\partial q_d} \\ \frac{\partial P_g}{\partial q_r} \end{pmatrix},$$

the z-component of p_b^p is now the sum of every z-component of $R_b^0 p_0^4$, $R_b^0 p_0^9$ and $R_b^0 p_0^{14}$, i.e.,

$$P_g = mgh = \frac{1}{3}mg(a_1 s_1 + a_3 s_{12} c_3 + a_1 s_6 + a_3 s_{67} c_8 + a_1 s_{11} + a_3 s_{1112} c_{13}).$$

Therefore, the dynamic equation for the dominant tool panel of this Delta closed parallel/serial hybrid-chain robot can be formulated as

$$\begin{cases} \tau_d = J_d^T \ddot{z} - \tau_{gd} \\ 0 = J_r^T \ddot{z} - \tau_{gr}. \end{cases} \tag{5.20}$$

As we can clearly see from (5.20), the first equation is to determine the 3 actu-ated joint torques $\tau_d = (\tau_1 \ \tau_6 \ \tau_{11})^T$, while the second equation becomes an internal dynamic constraint, which is difficult to be further determined. Therefore, we have to go back to explore both the kinematic and dynamic models for this Delta robot by using the dual counterpart formulation.

To do so, let three directional vectors be defined by

$$l_a = R_b^0 p_0^4, \quad l_b = R(z, -120°) R_b^0 p_0^9, \quad l_c = R(z, 120°) R_b^0 p_0^{14},$$

each of which represents the vector from A_b to A_p, the vector from B_b to B_p, and the vector from C_b to C_p, respectively. Since we have already had a complete IK solution for the Delta robot, given a directional vector l_a, the joint angles θ_1, θ_2 and θ_3 can be solved, and so are the rest joint angles if l_b and l_c are given.

Similar to the 3D 3-leg robotic system, by defining an isometric embedding for this Delta robot, $z = \sqrt{m}p_b^c$, we can derive its time-derivative and Jacobian matrix:

$$\dot{z} = \sqrt{m}\dot{p}_b^c, \quad \text{and} \quad J_z = \frac{1}{\sqrt{m}}(l_a^o \ l_b^o \ l_c^o), \tag{5.21}$$

where the Jacobian J_z is a 3 by 3 square matrix, and l_a^o, l_b^o and l_c^o are the unit di-rectional vectors of l_a, l_b and l_c, respectively. If we define the lengths of the three directional vectors as d_a, d_b and d_c such that $l_a = d_a l_a^o$, $l_b = d_b l_b^o$, and $l_c = d_c l_c^o$, then the following kinematic equation in differential motion is valid:

$$\dot{q} = \begin{pmatrix} \dot{d}_a \\ \dot{d}_b \\ \dot{d}_c \end{pmatrix} = J_z^T \dot{z}.$$

With this kinematic model, we achieve its dynamic model in the following form:

$$\ddot{z} = J_z(\tau + \tau_g) \quad \text{so that} \quad \tau + \tau_g = J_z^{-1}\ddot{z}. \tag{5.22}$$

Therefore, based on equations (5.10), (5.12) and (5.13), each directional vector is directly related to the given position vector $p_b^c = p_b^p$, i.e.,

$$l_a = R_b^0 p_0^4 = p_b^p + \beta_a - b_a,$$

and

$$l_b = R(z, -120°)R_b^0 p_0^9 = p_b^p + \beta_b - b_b, \quad l_c = R(z, 120°)R_b^0 p_0^{14} = p_b^p + \beta_c - b_c.$$

In other words, given a required position p_b^p, the three directional vectors can be calculated so that the isometric embedding z and its Jacobian matrix J_z in equation (5.21) can be determined in terms of p_b^p. Thus, by using the dynamic equation in (5.22), the joint torque/force τ can be found as well.

However, the 3-dimensional joint torque vector τ found from equation (5.22) is $\tau = (f_{da} \ f_{db} \ f_{dc})^T$, instead of a joint torque τ_i about the z_{i-1} axis. This means that the resultant f_{da} is a virtual force along p_0^4, f_{db} is a virtual force along p_0^9 and f_{dc} is another virtual force along p_0^{14}. Therefore, we need to convert each of them to the actuated joint torques by using the statics equation. Namely, let us first find each kinematic Jacobian matrix, such as

$$J_0^4 = \frac{\partial p_0^4}{\partial q_a} = \begin{pmatrix} -a_1 s_1 - a_3 s_{12} c_3 & -a_3 s_{12} c_3 & -a_3 c_{12} s_3 \\ a_1 c_1 + a_3 c_{12} c_3 & a_3 c_{12} c_3 & -a_3 s_{12} s_3 \\ 0 & 0 & a_3 c_3 \end{pmatrix}. \tag{5.23}$$

The other two Jacobian matrices J_0^9 and J_0^{14} can also be found in the same format but different joint angles. Then, let each kinematic Jacobian matrix be projected onto the base, i.e., $J_b^4 = R_b^0 J_0^4$, $J_b^9 = R_b^0 J_0^9$, and $J_b^{14} = R_b^0 J_0^{14}$. The joint torques can thus be found based on the statics equation, i.e.,

$$\tau_a = (J_b^4)^T f_{da} l_a^o, \quad \tau_b = (J_b^9)^T f_{db} l_b^o, \quad \tau_c = (J_b^{14})^T f_{dc} l_c^o,$$

where $\tau_a = (\tau_1 \ \tau_2 \ \tau_3)^T$, $\tau_b = (\tau_6 \ \tau_7 \ \tau_8)^T$, and $\tau_c = (\tau_{11} \ \tau_{12} \ \tau_{13})^T$.

However, each of the three joint torque vectors τ_a has three components, and we cannot implement all the nine joint torques for the Delta robot, which has only three actuated joints. Therefore, let the virtual force f_{da} calculated from the dynamic equation be further decomposed into two vector components:

$$\vec{f}_{da} = \vec{f}_{d1} + \vec{f}_{d2},$$

one is $\vec{f}_{d1}$ that is parallel to the tool panel as a pulling/pushing force to be balanced and absorbed by the tool panel structure, the other one is $\vec{f}_{d2}$ that is along the forearm of leg a, as illustrated in Figure 5.14. Obviously, only the virtual force f_{d2} calculated from the dynamic equation will be required for the actuated joint θ_1. In fact, due

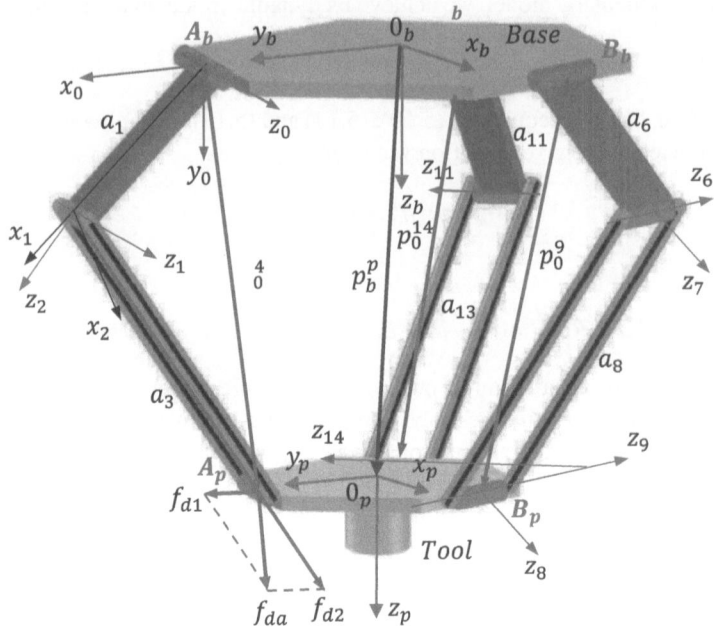

Figure 5.14 Dynamic torque determination for the Delta closed hybrid-chain robot

to $\vec{f}_{d2}$ along the forearm parallelogram link, the new torque vector $\tau_a = (J_b^4)^T \vec{f}_{d2} = (\tau_1 \ 0 \ 0)^T$. Namely, only τ_1 is nonzero, and the rest two are all zero. In the same way, the other two legs can also find the nonzero actuated joint torques τ_6 and τ_{11}, and all the other joint torques vanish. Finally, the three actuated joint torques can be well-determined based on the dynamic equation given by (5.22) through the above decomposition and statics procedures for such a Delta closed hybrid-chain robot manipulator.

To verify the decomposition and statics procedures, let the joint angles at a time instant for leg a be $\theta_1 = 60°$, $\theta_2 = 80°$ and $\theta_3 = 5°$, and let the link lengths $a_1 = 0.8$ and $a_3 = 1$ meters. According to equations (5.10) and (5.23) as well as

$$R_b^0 = \begin{pmatrix} 1 & 0 & 0 \\ 0 & 0 & 1 \\ 0 & 1 & 0 \end{pmatrix},$$

we obtain the following numerical results:

$$J_b^4 = R_b^0 J_0^4 = \begin{pmatrix} 0 & 0 & 0.9962 \\ -1.3332 & -0.6403 & 0.0668 \\ -0.3631 & -0.7631 & -0.0560 \end{pmatrix},$$

and

$$p_b^4 = R_b^0 p_0^4 = \begin{pmatrix} 0.0872 \\ -0.3631 \\ 1.3332 \end{pmatrix}, \quad p_{2(b)}^4 = R_b^0 p_{2(0)}^4 = \begin{pmatrix} 0.0872 \\ -0.7631 \\ 0.6403 \end{pmatrix}.$$

Suppose that the magnitude of the dynamic force along p_0^4 is $f_{da} = 2$ Newtons. In order to decompose the virtual dynamic force $\vec{f}_{da}$ into $\vec{f}_{d1}$ that is horizontal and $\vec{f}_{d2}$ that is along the forearm $p_{2(b)}^4$, the z-components of $\vec{f}_{da}$ and $\vec{f}_{d2}$ must be equal to each other due to the horizontal tool panel. Thus, let

$$\vec{f}_{da} = f_{da} \cdot p_b^4 / \|p_b^4\| = \begin{pmatrix} 0.1259 \\ -0.5246 \\ 1.9259 \end{pmatrix}.$$

Also, let the ratio

$$k = \frac{\vec{f}_{da}(3)}{p_{2(b)}^4(3)} = 3.0076$$

multiply on $p_{2(b)}^4$ to find the decomposed force vector

$$\vec{f}_{d2} = k \cdot p_{2(b)}^4 = \begin{pmatrix} 0.2621 \\ -2.2952 \\ 1.9259 \end{pmatrix}.$$

We can now clearly see that the z-components of $\vec{f}_{d2}$ and $\vec{f}_{da}$ have the same value. Then, this $\vec{f}_{d2}$ can be guaranteed to line up with the forearm $p_{2(b)}^4$. Finally, the statics equation gives rise to

$$\tau_a = (J_b^4)^T \vec{f}_{d2} = \begin{pmatrix} 2.3605 \\ 0 \\ 0 \end{pmatrix}.$$

The first nonzero element is $\tau_1 = 2.3605$ Newton-meters as an actuated dynamic torque at θ_1. Using the same procedure, we can find τ_6 and τ_{11} for the other two actuated joints. Therefore, choosing the closed parallel-chain category to develop both the kinematics and dynamics modeling for the Delta robot becomes quite successful.

5.4.4 DYNAMICS MODELING FOR LEGGED ROBOTS

Legged robots are a type of mobile robotic systems, which use articulated limbs, such as leg mechanisms, to provide locomotion. They are more versatile than the wheeled robots and can traverse many different terrains, though these advantages require increased complexity and power consumption. Legged robots often imitate legged animals, such as humans, dogs, cats or insects, as a typical biomimic mechanism [13–16].

One of the most representative legged robots was created by Boston Dynamics, called Spot, which has only 55 lb of weight. A quadrupedal legged robot model is

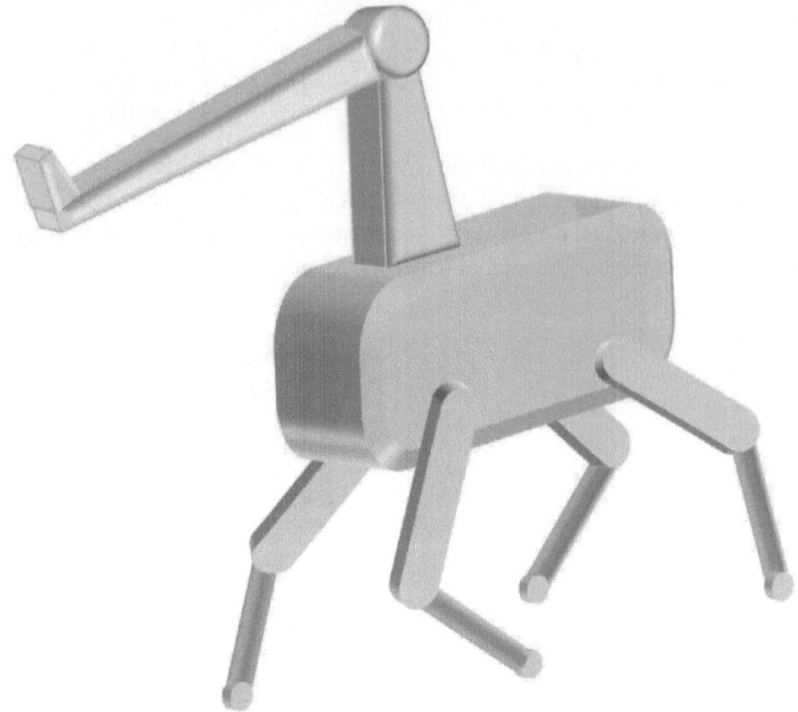

Figure 5.15 A quadrupedal legged robot model

given in Figure 5.15. In order to develop a dynamic model for such a quadrupedal mobile robot, like modeling a humanoid robot in biped motion, we need a complete kinematic model first. Example 2.3 in Chapter 2 has exhibited a complete D-H convention-based kinematic model for the 40-joint humanoid robot. The first six joints are all virtual to represent the entire body translation along the x_0, y_0 and z_0 base coordinate axes and body rotation: spin, fall over and fall aside. This treatment is very common for most mobile robotic systems. However, to model a quadrupedal mobile robot, let us focus only on an instantaneous dynamic motion at a time instant so that the first six virtual axes can be skipped over to reset the joint axis numbering. Figure 5.16 is a schematic drawing to reflect the coordinate system re-assignment.

Based on the D-H convention, leg a of this quadrupedal mobile robot model has the following D-H table:

Joint Angle θ_i	Joint Offset d_i	Twist Angle α_i	Link Length a_i
θ_1	0	$-90°$	0
θ_2	0	$90°$	a_2
θ_3	0	0	$-a_3$

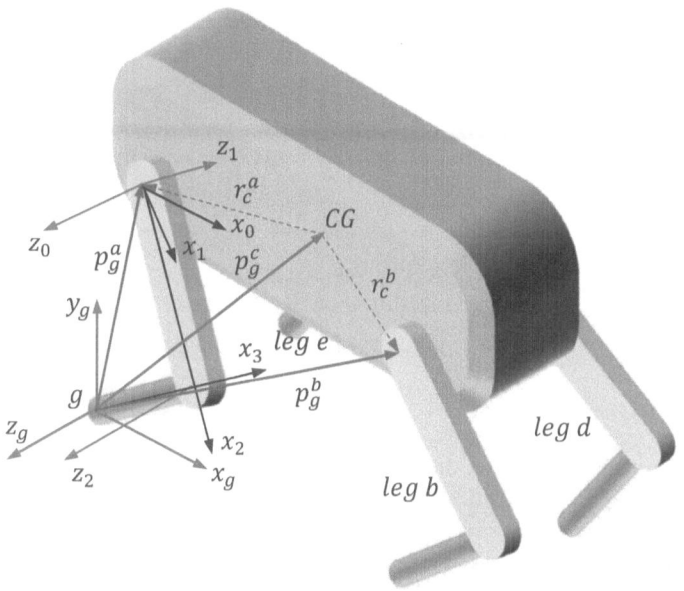

Figure 5.16 Schematic diagram for the quadrupedal legged robot

It has only three joints, and the tip ending point of the leg is actually touching down on ground at the origin g of the ground frame, which is the same origin of frame 3, as depicted in Figure 5.16. All the three one-step homogeneous transformations can be derived based on the D-H table:

$$A_0^1 = \begin{pmatrix} c_1 & 0 & -s_1 & 0 \\ s_1 & 0 & c_1 & 0 \\ 0 & -1 & 0 & 0 \\ 0 & 0 & 0 & 1 \end{pmatrix}, \quad A_1^2 = \begin{pmatrix} c_2 & 0 & s_2 & a_2 c_2 \\ s_2 & 0 & -c_2 & a_2 s_2 \\ 0 & 1 & 0 & 0 \\ 0 & 0 & 0 & 1 \end{pmatrix},$$

$$\text{and} \quad A_2^3 = \begin{pmatrix} c_3 & -s_3 & 0 & -a_3 c_3 \\ s_3 & c_3 & 0 & -a_3 s_3 \\ 0 & 0 & 1 & 0 \\ 0 & 0 & 0 & 1 \end{pmatrix}.$$

We can now readily calculate the total homogeneous transformation for leg a from frame 0 to frame 3:

$$A_0^3 = \begin{pmatrix} c_1 c_2 c_3 - s_1 s_3 & -c_1 c_2 s_3 - s_1 c_3 & c_1 s_2 & a_2 c_1 c_2 - a_3 c_1 c_2 c_3 + a_3 s_1 s_3 \\ s_1 c_2 c_3 + c_1 s_3 & -s_1 c_2 s_3 + c_1 c_3 & s_1 s_2 & a_2 s_1 c_2 - a_3 s_1 c_2 c_3 - a_3 c_1 s_3 \\ -s_2 c_3 & s_2 s_3 & c_2 & -a_2 s_2 + a_3 s_2 c_3 \\ 0 & 0 & 0 & 1 \end{pmatrix}.$$

The last column of the above matrix A_0^3 is the position vector p_0^3. Its inverse vector $p_3^0 = -R_3^0 p_0^3$ is tailed at the origin of frame 3 and arrow-pointing to the origin of frame 0 for leg a. However, since frame g is defined to be the exact same orientation as frame 0, the inverse position vector $p_g^0 = -R_g^0 p_0^3 = -p_0^3$ due to $R_g^0 = R_0^g = I$, the identity. Therefore,

$$p_g^a = p_g^0 = -p_0^3 = \begin{pmatrix} -a_2c_1c_2 + a_3c_1c_2c_3 - a_3s_1s_3 \\ -a_2s_1c_2 + a_3s_1c_2c_3 + a_3c_1s_3 \\ a_2s_2 - a_3s_2c_3 \end{pmatrix}.$$

Because θ_2 is specifically designed for a possible abducting or adducting action on each leg in case the robot needs to be balanced against gravity, but $\theta_2 = 0$ in most normal cases. If the abduction/adduction angle $\theta_2 = 0$, then p_g^a is reduced to a simple form:

$$p_g^a = \begin{pmatrix} a_3c_{13} - a_2c_1 \\ a_3s_{13} - a_2s_1 \\ 0 \end{pmatrix}, \quad \text{if } \theta_2 = 0.$$

Using the same procedure, the other three legs b, d and e can also have their position vectors p_g^b, p_g^d and p_g^e derived as functions of their respective joint angles. However, they must be referred to the same ground frame. Since for a rigid body, its position and orientation can be uniquely determined by three unaligned points, called the **three-point model**, which has been adopted in finding an isometric embedding for a robotic system configuration manifold (C-manifold) in Chapter 4. Although three of the four position vectors are sufficient enough to determine an embedding $z = \zeta(q)$, it must also be isometric, i.e., $\frac{1}{2}\dot{z}^T\dot{z} = K$ should hold, where K is the kinetic energy of the system.

Therefore, instead of directly augmenting three of the four position vectors, we have to find the position vector for the mass center first, or the center of gravity (CG) of the quadrupedal legged robot body p_g^c. In fact, p_g^c can be a linear combination of the three position vectors:

$$p_g^c = \alpha p_g^a + \beta p_g^b + \gamma p_g^e,$$

for three constants α, β and γ.

Once p_g^c is determined, we need then to find three radial vectors that are tailed at the mass center. Namely,

$$r_{c(g)}^a = p_g^a - p_g^c, \quad r_{c(g)}^b = p_g^b - p_g^c, \quad r_{c(g)}^e = p_g^e - p_g^c.$$

After that, an isometric embedding $z = \zeta(q)$ can be formed by

$$z = \zeta(q) = \begin{pmatrix} a p_g^c \\ b_1 r_{c(g)}^a \\ b_2 r_{c(g)}^b \\ b_3 r_{c(g)}^e \end{pmatrix},$$

with some constants a, b_1, b_2 and b_3. The dynamics for this quadrupedal legged robot can be modeled as

$$\tau = J^T \ddot{z} - \tau_g = \left(\frac{\partial \zeta}{\partial q} \right)^T \ddot{z} + \frac{\partial P_g}{\partial q},$$

where P_g is the gravitational potential energy, and q is the joint position vector. Since each leg has 3 joints, q is 9-dimensional. However, if each $\theta_2 = 0$, then q is only 6-dimensional.

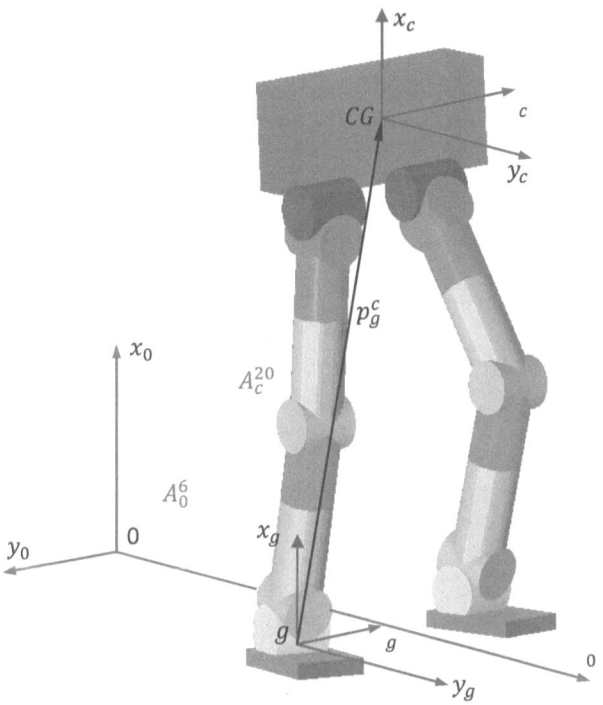

Figure 5.17 A biped robot as a lower body of the humanoid robot

We now turn to study a biped robot, which is actually the lower body of the humanoid robot in Figure 2.7. Let the upper body of the humanoid robot form a single rigid body, as a dominant object of this biped robot. If the center of gravity (CG) is defined, and the right foot is touching down on ground at a time instant, as shown in Figure 5.17, the position vector p_g^c that is tailed at the origin of frame g and arrow-pointing to the CG can be found by the kinematic model. Using the same D-H coordinate frame assignment numbering as modeling for the humanoid robot in Figure 2.7, we have developed the homogeneous transformation A_6^{20}. Frame 6 is originated at the H-triangle center, and we purposely define all the three frames: the ground frame g, the CG frame c and frame 6 have the same initial orientation, as depicted in Figure 5.17. Then, $R_6^{20} = I$, the identity so that $p_g^c = -p_6^{20}$ if the origin of frame 6 is matched with the CG. Even if they have a distance away, we can add a shift into p_6^{20}.

Based on the kinematics modeling for the humanoid robot in Chapter 2, the right leg has 7 joints from θ_{14} through θ_{20}, and they are the hip extension/flexion, hip abduction/adduction, hip medial/lateral, knee extension/flexion, knee medial/lateral, ankle dorsiflexion/plantar, and the toe extension/flexion. The position vector $p_g^c = -p_6^{20}$ can now be an explicit function of all the 7 joint angles. In addition, the orientation of frame c for the dominant upper body can also be determined by $(R_6^{20})^T = R_g^c$ that is the explicit function of all the 7 joint angles as well. Therefore, an isometric embedding for the dominant body can be easily defined by

$$
z = \zeta(q) = \begin{pmatrix} \sqrt{m} p_g^c \\ b_1 r_x \\ b_2 r_y \\ b_3 r_z \end{pmatrix},
$$

where r_x, r_y and r_z are the three columns of R_g^c, i.e., $R_g^c = (r_x \; r_y \; r_z)$. Then, it is also straightforward to find its 12 by 7 Jacobian matrix $J = \frac{\partial \zeta}{\partial q}$. Its dynamic equation for the dominant body of the biped robot at the time instant as the right foot touches down on ground can be formulated by

$$
\tau = J^T \ddot{z} - \tau_g,
$$

for all the 7 instantaneous joint torques.

If the biped robot switches its stance, the left foot is touching down on ground, then, $p_g^c = -p_6^{13}$ and R_6^{13} will take over the kinematics. The left leg has also 7 joints, from θ_7 through θ_{13}. The new isometric embedding z can be redefined with the same format but in terms of these 7 joint angles for the left leg. Both its Jacobian matrix J and the dynamic equation also have the same form with different joint angles, which will produce 7 joint torques for the left leg at the new stance.

5.4.5 A SUMMARY OF DYNAMICS MODELING

Systems	Mobility	Mechanism	Outputs	Kinematics	Dynamics
Stewart Platform	6	Closed Parallel	Both p and R	$\dot{q} = J_z^T \dot{z}$	$\ddot{z} = J_z(\tau + \tau_g)$
3D 3-Leg Robot	3	Closed Parallel	p	$\dot{q} = J_z^T \dot{z}$	$\ddot{z} = J_z(\tau + \tau_g)$
Delta Robot	3	Closed Hybrid	p	$\dot{q} = J_z^T \dot{z}$	$\ddot{z} = J_z(\tau + \tau_g)$
Quadrupedal	6	Open Hybrid	Both p and R	$\dot{z} = J\dot{q}$	$J^T \ddot{z} = \tau + \tau_g$
Bipedal Robot	6	Open Hybrid	Both p and R	$\dot{z} = J\dot{q}$	$J^T \ddot{z} = \tau + \tau_g$

Table 5.3

A comparative table on open hybrid and closed parallel or hybrid-chain robotic systems

Dynamics modeling for closed parallel and open parallel/serial hybrid chain robotic systems has been explored and investigated. Table 5.3 summarizes the outcomes and findings from the detailed study on the most typical robot manipulators in such challenging mechanisms. Due to the remarkable principle of duality between open serial and closed parallel-chain robotic systems, we can always have two dual counterparts for making choice in both the kinematics and dynamics modelings. While most open serial-chain robots and the Stewart platform stand on the two dual extremes, many open or closed hybrid-chain systems are situated in-between. Which category of dynamics an in-between system should fit to now becomes clear that it fully depends on what type of kinematics is modeled. In other words, each of those in-between systems can choose a feasible kinematic model from the two dual counterparts, and then its dynamics modeling has to follow up within the same category.

REFERENCES

1. Stewart, D., (1965–1966) A Platform with Six Degrees of Freedom. *Proceedings of the Institution of Mechanical Engineers*. 180 (1, No.15): pp. 371-386.
2. Dasgupta, B. and Mruthyunjaya, T.S., (2000) The Stewart platform manipulator: a review, *Mechanism and Machine Theory*, Vol.35, Issue 1, January, pp. 15-40.
3. Merlet, J.P., (2008) Parallel Robots, 2nd Edition. Springer. ISBN 978-1-4020-4132-7.
4. Gogu, G., (2008). Structural Synthesis of Parallel Robots, Part 1: Methodology. Springer. ISBN 978-1-4020-5102-9.
5. Gogu, G., (2009). Structural Synthesis of Parallel Robots, Part 2: Translational topologies with Two and Three Degrees of Freedom. Springer. ISBN 978-1-4020-9793-5.
6. Kong, X. and Gosselin, C., (2007). Type Synthesis of Parallel Mechanisms. Springer. ISBN 978-3-540-71989-2.
7. Gallardo-Alvarado, J., (2016). Kinematic Analysis of Parallel Manipulators by Algebraic Screw Theory. Springer. ISBN 978-3-319-31124-1.
8. Gu, Edward Y.L., (2013) A Journey from Robot to Digital Human. Springer, Heidelberg, New York.
9. Clavel, R., (1988) Delta: a fast robot with parallel geometry, *Proc. 18th Int. Symp. Ind. Robots*, Sydney, Australia, pp. 91-100.
10. Atiyah, M., (2007), Duality in Mathematics and Physics, *Lecture notes from the Institut de Matematica de la Universitat de Barcelona (IMUB)*.
11. Kostrikin, A.I., (2001) Duality, *Encyclopedia of Mathematics*, EMS Press.
12. Gowers, Timothy, (2008) III.19 Duality, *The Princeton Companion to Mathematics*, Princeton University Press, pp. 187-190.
13. Bekey, George A., (2005). Autonomous robots: from biological inspiration to implementation and control. Cambridge, Massachusetts: MIT Press. ISBN 978-0-262-02578-2.
14. Raibert, M.H., (1986) Legged Robots That Balance. Cambridge, MA: MIT Press.
15. Wang, Lingfeng, Tan, K.C. and Chew, C.M., (2006) Evolutionary robotics: from algorithms to implementations. Hackensack, N.J.: World Scientific Pub. ISBN 978-981-256-870-0.
16. McCarthy, J. Michael, (2019). Kinematic Synthesis of Mechanisms: A project based approach. MDA Press, Chicago, Illinois.

6 Nonlinear Control Theories

6.1 LYAPUNOV STABILITY THEORIES AND CONTROL STRATE-GIES

A nonlinear system can be represented by a general state-space equation:

$$\dot{x} = f(t,x,u), \quad \text{or} \quad \dot{x} = f(x,u), \tag{6.1}$$

depending on whether each parameter is time-varying in the first form, or is time-invariant in the second form, where the state $x \in \mathbb{R}^n$ is a vector field, the input vector $u \in \mathbb{R}^r$, and $f(\cdot, \cdot)$ is interpreted as a tangent vector field. In special cases, if the state x is solved from the equation and its solution is a trajectory $x(t)$ as $t \geq 0$ with an initial state $x(0)$, then (6.1) can be more specifically written as

$$\dot{x}(t) = f(t,x(t),u(t)), \quad \text{or} \quad \dot{x}(t) = f(x(t),u(t)).$$

The **equilibrium points** for a system given by (6.1) are defined for those points such that $\dot{x} = 0$. Thus, all the equilibrium points are the roots of the algebraic equation $f(t,x,0) = 0$ in a time-varying (or non-autonomous) case or $f(x,0) = 0$ in a time-invariant (or autonomous) case under the zero input $u = 0$.

It can be easily seen that a nonlinear system often has multiple *isolated equilibrium points*, because $f(t,x) = 0$ or $f(x) = 0$ can have more than one solution. In contrast, for a linear time-invariant system, i.e., $f(x) = Ax$ under $u = 0$, the solution to $Ax = 0$ is only one at $x = 0$ if the constant matrix A is non-singular. However, if A is singular, the equation $Ax = 0$ has an infinite number of solutions as internal points in the null space of A, but they are continuous, not isolated from each other.

Furthermore, since any coordinates shift will not affect the nonlinear system form $\dot{x} = f(x,u)$, without loss of generality, we can always assume an equilibrium point of interest at the origin of the state space, i.e., $x_e = 0$. For instance, let a 1-dimensional system be $\dot{x} = x - x^3$. It has three isolated equilibrium points: $x = 0$ and $x = \pm 1$ that are the roots of $x - x^3 = 0$. When one of the three equilibrium point $x = 1$ is under study, we may define a new state $z = x - 1$ such that $x = z + 1$. Substituting $x = z + 1$ into the original state-space equation yields $\dot{z} = (z+1) - (z+1)^3 = -z(z+1)(z+2)$, which obviously has an equilibrium point at the origin $z = 0$ as seen in the z-coordinate system.

Now, a formal definition of stability with respect to an equilibrium point $x_e = 0$ for a general time-varying system in the form of (6.1) can be stated as follows:

Definition 6.1. The system (6.1) at its equilibrium point $x_e = 0$ is said to be stable at $t = t_0$ if for each $\varepsilon > 0$, there exists $\delta(t_0, \varepsilon) > 0$, such that

$$\|x(t_0)\| < \delta(t_0, \varepsilon) \Rightarrow \|x(t)\| < \varepsilon, \quad \text{for all } t \geq t_0.$$

The above definition of systems stability with respect to an equilibrium point can be interpreted as a system that is able to transit from an initial state to any destination close to the equilibrium point. This formal definition is conventionally called **stability in the Lyapunov sense**, which can be thought of as a bottom line for a system to be stable [1–3]. Aleksandr Lyapunov (1857–1918) was a great Russian mathematician and physicist. He is well-known for his development of the stability theory for a dynamical system, as well as for his many contributions to mathematical physics and probability theory.

For many dynamic systems, however, only to be stable in the Lyapunov sense at an equilibrium point is not satisfactory enough. Instead, it is also desired to converge to the equilibrium point. This leads to the concept of **asymptotic stability**.

Definition 6.2. The system (6.1) with respect to the equilibrium point $x_e = 0$ is said to be asymptotically stable at $t = t_0$ if

1. it is stable at $t = t_0$, and
2. there exists a $\delta_1(t_0) > 0$ such that

$$\|x(t_0)\| < \delta_1(t_0) \Rightarrow \|x(t)\| \to 0 \text{ as } t \to \infty.$$

Naturally, to determine if a system is stable at an equilibrium point can be done theoretically by directly solving the system equation to see if the solution is close to or converges to the equilibrium point. However, to find an explicit form of the solution is often a difficult task, and even impossible for many nonlinear systems. We are now facing a challenging question that can we determine the stability for a given system without need of solving the equation? The answer is yes, and there is an effective approach to testing and determining the stability for a system, called the **Lyapunov direct method** [1–3].

The most typical example of the stability test is a steel ball that is rolling on the inner concave wall of a basin under gravitational field. Intuitively, the steel ball leaving from an initial point on the inner wall will always keep its total energy (kinetic energy plus potential energy) unchanged under a friction-free condition, or the total energy is reduced gradually until the ball stops at the bottom point of the basin if there is a rolling friction. As a counter situation, if the ball is placed on the top point of a smooth convex surface under gravity, then it will no longer be back to the initial top point again once the ball is leaving from the top point and sliding down the hill. In this case, the ball will keep rolling down the hill and its kinetic energy will be unbounded. This simple example implies that every dynamic system commonly tends to the minimum energy state in order to stabilize itself. In other words, the lower the energy level is, the more stable the system will be. Therefore, we are motivated to explore an *energy-like function* associated with the system equation as a dynamic constraint for the stability testing.

To effectively determine a system stability, an energy-like scalar function must be formalized under a number of conditions. For instance, consider a two-dimensional system with an equilibrium point at $x = 0$. Without loss of generality, we assume that the two state variables x_1 and x_2 of the system span a horizontal x_1-x_2 phase plane, while the vertical V-axis that is perpendicular to the phase plane represents the value of the energy-like scalar function $V(x)$. Then, the scalar function $V(x)$ can geometrically be created like a bowl shape with its bottom point touched at the origin of the x_1-x_2 phase plane, while all the other points on the surface are above the phase plane, as shown in Figure 6.1. Every phase trajectory of the system on the horizontal phase plane can now be thought of as the projection of a curve on the bowl surface. For the sake of clear discussion, we refer to such a curve on the bowl surface as a *V-lifted curve* for the corresponding phase trajectory.

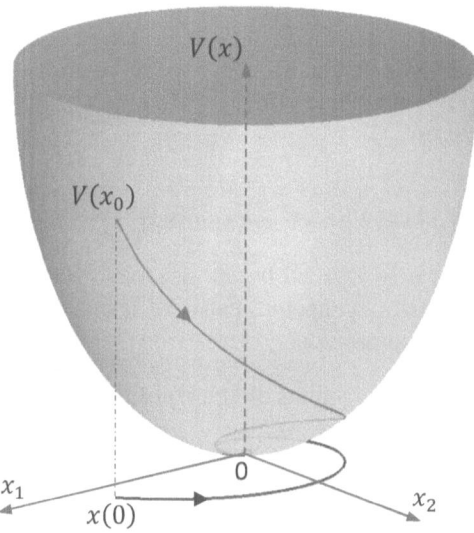

Figure 6.1   An energy-like function $V(x)$ and a V-lifted trajectory

Suppose that a phase trajectory that satisfies the system equation is specified by starting from a non-zero initial point and ending at the origin of the phase plane, i.e., the equilibrium point of the system. Then, the V-lifted curve can be viewed, at least in an open neighborhood of the origin, to slide down along the bowl surface until reaching the bottom point, as shown in Figure 6.1. In this case, the equilibrium point, i.e., the origin of the x_1-x_2 plane, is asymptotically stable for the system.

Motivated by the above geometrical imagination and physical interpretation, we now formally introduce an energy-like scalar function associated with a given system equation to determine the stability at an equilibrium point. First, we realize that the energy-like function is not just a positive function, but it should also satisfy some additional conditions to form a bowl-like geometrical surface. Such a special positive scalar function is called a *positive-definite function*, or a *p.d.f.* in abbreviation. Here we will focus our discussions only on the time-invariant cases, i.e., the functions do not contain time explicitly, or do not contain any time-varying parameter. The discussions on time-explicit positive-definite functions can be found in the literature [1–3].

Definition 6.3. A scalar continuous function $V(x)$ of $x \in \mathbb{R}^n$ is said to be a local positive-definite function (*l.p.d.f.*) if in some open ball $B(\delta) = \{x \mid \|x\| < \delta\}$, $V(0) = 0$ and $V(x) > 0$ for $x \neq 0$. Furthermore, $V(x)$ is said to be a positive-definite function (p.d.f.) if the ball limit $B(\delta)$ can be extended to the entire state space $\mathbb{R}^n$. If a p.d.f. $V(x) \to \infty$ as $\|x\| \to \infty$, then this p.d.f. $V(x)$ is further called a strong positive-definite function (s.p.d.f.).

It can be seen from the above formal definition that a positive scalar function is different from an *l.p.d.f.* or a p.d.f., and the latter requires stronger conditions than the former. For instance, if $x \in \mathbb{R}^1$, the scalar function $\exp(-x^2)$ (Gaussian) is always positive. However, it is neither an *l.p.d.f.* nor a p.d.f. The reason is quite obvious that this Gaussian function does not vanish at $x = 0$. In contrast to this, the function $V(x) = 1 - \exp(-x^2)$ is a p.d.f., due to $V(0) = 0$ and $V(x) > 0$ for any $x \neq 0$, but it is not an s.p.d.f., because $V(x) \to 1 < \infty$ as $x \to \infty$.

Lemma 6.1. *A quadratic form $V(x) = x^T M x$ for a vector field $x \in \mathbb{R}^n$ is s.p.d.f. if and only if the n by n symmetric coefficient matrix M is positive-definite.*

This lemma can be easily justified by the quadratic form property from linear algebra. To test if an n by n symmetric matrix M is positive-definite, denoted as $M > 0$, we can have two approaches:

1. All the n leading principal minors of M are positive; or
2. All the n eigenvalues of M are positive.

For example, let a quadratic form be $V_1(x) = 3x_1^2 + 2x_2^2 + x_3^2 - 2x_1x_2 - 4x_2x_3$. Clearly, its coefficient matrix is

$$M_1 = \begin{pmatrix} 3 & -1 & 0 \\ -1 & 2 & -2 \\ 0 & -2 & 1 \end{pmatrix}.$$

The three *leading principal minors* of M_1 are given by

$$\det(3) = 3, \quad \det\begin{pmatrix} 3 & -1 \\ -1 & 2 \end{pmatrix} = 5, \quad \det\begin{pmatrix} 3 & -1 & 0 \\ -1 & 2 & -2 \\ 0 & -2 & 1 \end{pmatrix} = -7.$$

On the other hand, the eigenvalues of M_1 are $\lambda_1 = -0.6691$, $\lambda_2 = 2.5240$, and $\lambda_3 = 4.1451$, where one of the three is negative. Therefore, M_1 is not positive-definite, and based on Lemma 6.1, the quadratic form $V_1(x)$ is not a p.d.f. In fact, if $x_1 = 0$, $x_2 = 1$ and $x_3 = 1$, $x^T M_1 x = -1 < 0$.

If the above quadratic form is modified to $V_2(x) = 3x_1^2 + 2x_2^2 + x_3^2 - 2x_1x_2 - 2x_2x_3$, its coefficient matrix becomes

$$M_2 = \begin{pmatrix} 3 & -1 & 0 \\ -1 & 2 & -1 \\ 0 & -1 & 1 \end{pmatrix}.$$

The three new leading principal minors of M_2 turn out to be

$$\det(3) = 3, \quad \det \begin{pmatrix} 3 & -1 \\ -1 & 2 \end{pmatrix} = 5, \quad \det \begin{pmatrix} 3 & -1 & 0 \\ -1 & 2 & -1 \\ 0 & -1 & 1 \end{pmatrix} = 2.$$

Thus, M_2 is a positive-definite matrix, and any x_1, x_2 and x_3 will keep $V_2(x) > 0$ always. On the other hand, the eigenvalues of M_2 are $\lambda_1 = 0.2679$, $\lambda_2 = 2.0000$, and $\lambda_3 = 3.7321$, and they are all positive. Therefore, based on Lemma 6.1, $V_2(x)$ is not just a p.d.f., but is an s.p.d.f., i.e., $V_2(x) = 0$ at $x = 0$, $V_2(x) > 0$ for $x \neq 0$ and $V_2(x) \to \infty$ as $x \to \infty$.

Another example is $V(x) = x_1^2 + \sin^2 x_2$, where $x \in \mathbb{R}^2$. It can be justified that $V(x)$ is an l.p.d.f., but not a p.d.f. First, $V(0) = 0$, and second, one can define an open ball $B(\delta)$ around $x = 0$ with a radius $\delta = \pi$. Then, for any $0 \neq x \in B(\delta)$, $V(x) > 0$. However, when one extends x to the entire space, i.e., $\delta \to \infty$, this scalar function $V(x)$ can have multiple zeros along the x_2-axis as $x_1 = 0$ at every $x_2 = k\pi$ for $k = \pm 1, \pm 2, \cdots$. Therefore, one of the p.d.f. conditions in Definition 6.3 $V(x) > 0$ for $x \neq 0$ does not hold for such a scalar function $V(x)$.

The third example is $V(x) = x_1^2$, where $x = (x_1 \ x_2)^T \in \mathbb{R}^2$. It looks like a p.d.f., because $V(0) = 0$ and $V(x) = x_1^2 > 0$ when $x_1 \neq 0$. Actually, because this scalar function only contains x_1 and misses x_2, at every point on the straight line $x_1 = 0$ in the 2D state space, any non-zero value of x_2 will make the function value vanish. For instance, if $x = (0 \ 4)^T \neq 0$, then $V(x) = 0$. This violates the conditions for an l.p.d.f. or p.d.f., which require $V(x) = 0$ only at the single point $x = 0$ in the entire state space. Therefore, $V(x) = x_1^2$ in 2D state space is neither an l.p.d.f. nor a p.d.f. This example implies an important fact that if a scalar function $V(x)$ has one or more states excluded, it will no longer be an l.p.d.f, nor a p.d.f.

After introducing the formal definitions of the l.p.d.f., p.d.f. and s.p.d.f., we now state the following stability theorem in the Lyapunov sense:

Theorem 6.1. *The equilibrium point $x = 0$ of a system $\dot{x} = f(x)$ is stable if there exists an l.p.d.f. $V(x)$ such that its time-derivative along any trajectory of the system is non-positive, i.e., $\dot{V}(x) \leq 0$ for each $x \in B(\delta) \subset \mathbb{R}^n$.*

This is commonly called the **Lyapunov Direct Method** of stability. In the above theorem, "along any trajectory" means that $f(x)$ should be substituted into every term containing $\dot{x}$ after taking the time-derivative on $V(x)$.

A more important question, which sometimes may not be so straightforward when this theorem is being applied, is how do we find such an l.p.d.f. $V(x)$ for a given system? The theorem does not directly tell how. However, the system would be guaranteed to be stable in the Lyapunov sense if $V(x)$ could be found and satisfy all the conditions in the theorem. On the other hand, if one couldn't find a qualified $V(x)$, the system may still possibly be stable, because being unable to find a qualified function $V(x)$ does not imply that it does not exist! Therefore, using the Lyapunov direct method to determine a system stability requires effort and skill to find a right function

$V(x)$, often called a *Lyapunov candidate*, such that all the conditions in Theorem 6.1 hold.

Note that Theorem 6.1 only provides a criterion for a system to be *stable*. In order to promote it to a local or global asymptotic stability, some of the conditions need to be made stronger. In addition to Theorem 6.1, if

1. the condition "$\dot{V}(x) \leq 0$" is replaced by "$-\dot{V}(x)$ is an *l.p.d.f.* in an open ball $B(\delta)$", then the system at $x = 0$ is also *locally asymptotically stable*; and if
2. the condition "an *l.p.d.f.* $V(x)$" is replaced by "an s.p.d.f. $V(x)$" and the condition "$\dot{V}(x) \leq 0$" is replaced by "$-\dot{V}(x)$ is an s.p.d.f.", then the system at $x = 0$ is *globally asymptotically stable*.

For example, consider a 1-dimensional system $\dot{x} = -x$ with its equilibrium point at $x = 0$. Let $V(x) = x^2$ be a Lyapunov candidate. Obviously, $V(x)$ is an s.p.d.f in the 1D state space. Then, $\dot{V}(x) = 2x\dot{x} = -2x^2$, which is not only non-positive, but also $-\dot{V}(x) = 2x^2$ is an s.p.d.f. Therefore, the system is globally asymptotically stable with respect to the equilibrium point $x = 0$. In fact, the solution of $\dot{x} = -x$ is $x(t) = x(0)e^{-t}$. Thus, for any finite $x(0)$ as a starting state, $x(t) \to 0$ as $t \to \infty$.

As a second example, let a 2-dimensional system be given by

$$\begin{cases} \dot{x}_1 = x_1(x_1^2 + x_2^2 - 1) - x_2 \\ \dot{x}_2 = x_1 + x_2(x_1^2 + x_2^2 - 1). \end{cases} \tag{6.2}$$

This system has an equilibrium point at $x_1 = 0$ and $x_2 = 0$. Let $V(x) = x_1^2 + x_2^2$ that is obviously an s.p.d.f. in the 2D phase plane. Since

$$\dot{V}(x) = 2x_1\dot{x}_1 + 2x_2\dot{x}_2 = 2(x_1^2 + x_2^2)(x_1^2 + x_2^2 - 1),$$

$-\dot{V}(x)$ is clearly an *l.p.d.f.* in the open unity ball $B(1)$, i.e., $x_1^2 + x_2^2 < 1$. Thus, the system is locally asymptotically stable at $x = (x_1 \ x_2)^T = 0$.

A well-known simple-pendulum mechanical system can be modeled by

$$\begin{cases} \dot{x}_1 = x_2 \\ \dot{x}_2 = -bx_2 - k\sin x_1, \end{cases} \tag{6.3}$$

where $x_1 = \theta$ is the angular displacement of the pendulum and $x_2 = \dot{\theta}$ is its angular velocity. In such a second-order mechanical system, only picking the kinetic energy $K = \frac{1}{2}x_2^2$ may not be qualified as an *l.p.d.f.* or a p.d.f., because K contains only one component of $x = (x_1 \ x_2)^T$. If we add a gravitational potential energy term $k(1 - \cos x_1)$ to form a Lyapunov candidate:

$$V(x) = k(1 - \cos x_1) + \frac{1}{2}x_2^2, \tag{6.4}$$

then $V(x)$ can be an *l.p.d.f.* as the angle $x_1 = \theta$ is confined within $(-\pi, \pi)$.

Since

$$\dot{V}(x) = k\sin x_1 \dot{x}_1 + x_2 \dot{x}_2 = k\sin x_1 x_2 - bx_2^2 - kx_2 \sin x_1 = -bx_2^2 \leq 0, \qquad (6.5)$$

the system is stable at $x = 0$ when $b > 0$. In this case, $-\dot{V}(x) = bx_2^2 > 0$ is just positive, but is not an *l.p.d.f.* Therefore, the simple-pendulum system currently cannot be locally asymptotically stable by using Theorem 6.1.

However, intuitively this system is supposed to be also locally asymptotically stable at $x = 0$ if $b > 0$. The reason is clear that with a positive viscous friction coefficient $b > 0$, the pendulum should converge to the equilibrium point $x = 0$ from any starting point within $-\pi < x_1 < \pi$ under gravity.

La Salle had first discovered this insufficiency [1, 3], and showed the following important theorem to relax the conditions for an asymptotic stability:

Theorem 6.2. (La Salle) – *For an autonomous (time-invariant) system $\dot{x} = f(x)$, suppose that $V : \mathbb{R}^n \to \mathbb{R}$ is a continuously differentiable l.p.d.f. (s.p.d.f.), $\dot{V}(x) \leq 0$ for $x \in B(\delta)$ ($x \in \mathbb{R}^n$), and the set $S = \{x \in \mathbb{R}^n \mid \dot{V}(x) \equiv 0\}$ contains no non-trivial trajectories of the system. Then, $x = 0$ is a locally (globally) asymptotically stable equilibrium point.*

This Theorem shows that only the condition $\dot{V}(x) \leq 0$ may already imply an asymptotic stability, provided that the set S has only a trivial trajectory, i.e., the equilibrium point $x = 0$. Applying the La Salle's Theorem to the simple-pendulum system in equation (6.3), one may, at first glance, observe that the set S contains only $x_2 \equiv 0$ due to (6.5). However, by substituting $x_2 \equiv 0$ and also $\dot{x}_2 = 0$ into the two state equations in (6.3), we obtain $\dot{x}_1 = 0$ and $\sin x_1 = 0$. Since x_1 is confined to $(-\pi, \pi)$, it suffices to show that $x_1 \equiv 0$, too. Thus, the set S contains only the equilibrium point $x = 0$ that is trivial. Therefore, at $x = 0$, the simple-pendulum system is not only stable, but is also locally asymptotically stable if the damping coefficient $b > 0$, which is consistent to the physical meaning.

In general, if a mechanical system can be modeled as

$$\begin{cases} \dot{x}_1 = x_2 \\ \dot{x}_2 = -f(x_2) - g(x_1), \end{cases} \qquad (6.6)$$

and the following conditions hold:

1. $f(\cdot)$ and $g(\cdot)$ are continuous;
2. $f(0) = g(0) = 0$, and $\sigma f(\sigma) > 0$ and $\sigma g(\sigma) > 0$ for each $\sigma \neq 0$;
3. $\int_0^{\sigma} g(\xi)d\xi \to \infty$ as $|\sigma| \to \infty$;

then, a Lyapunov candidate can be defined by

$$V(x) = \frac{x_2^2}{2} + \int_0^{x_1} g(\xi)d\xi. \qquad (6.7)$$

It can be shown first that the system has the equilibrium point at $x = 0$. Second, $V(x)$ is an s.p.d.f., and $\dot{V}(x) = x_2\dot{x}_2 + g(x_1)\dot{x}_1 = -x_2 f(x_2) \le 0$. Therefore, the general mechanical system (6.6) is at least stable at $x = 0$ if all the above three conditions hold. This general case also implies that in most mechanical systems, only a kinetic energy term may not be enough to define a Lyapunov candidate, and a potential energy term is thus needed to form an energy-like function. In fact, the second integral term of (6.7) is just a general form of the potential energy for many second-order mechanical systems.

Let us look at a 3rd-order system example. Consider a nonlinear system $\dot{x} = f(x)$, where

$$f(x) = \begin{pmatrix} -x_2 - (2x_1 + x_3)^3 \\ x_1 \\ x_2 \end{pmatrix}.$$

Obviously, $x = 0$ is the equilibrium point. In order to justify its stability at $x = 0$, let a Lyapunov candidate be defined as $V(x) = x_1^2 + \frac{1}{2}x_2^2 + \frac{1}{2}x_3^2 + x_1 x_3$. Since this quadratic form $x^T M x$ has the following coefficient matrix:

$$M = \begin{pmatrix} 1 & 0 & \frac{1}{2} \\ 0 & \frac{1}{2} & 0 \\ \frac{1}{2} & 0 & \frac{1}{2} \end{pmatrix},$$

by checking all the three leading principal minors to see if they are positive, based on Lemma 6.1, $V(x)$ is an s.p.d.f.

Now, taking time-derivative for $V(x)$ yields

$$\dot{V}(x) = 2x_1\dot{x}_1 + x_2\dot{x}_2 + x_3\dot{x}_3 + \dot{x}_1 x_3 + x_1\dot{x}_3$$

$$= -2x_1[x_2 + (2x_1 + x_3)^3] + x_1 x_2 + x_2 x_3 - x_2 x_3 - (2x_1 + x_3)^3 x_3 + x_1 x_2$$

$$= -2x_1(2x_1 + x_3)^3 - x_3(2x_1 + x_3)^3 = -(2x_1 + x_3)^4 \le 0,$$

which is non-positive, but $-\dot{V}(x)$ is not a p.d.f., because every point on the $x_3 = -2x_1$ plane can make $\dot{V}(x) = 0$. Based on Theorem 6.1, the system is just stable. We turn to test it using the La Salle Theorem 6.2. Since the zero points of $\dot{V}(x)$ are on the plane $x_3 = -2x_1$, this leads to the system equation simplified to

$$\begin{cases} \dot{x}_1 = -x_2 \\ \dot{x}_2 = x_1 \\ \dot{x}_3 = x_2. \end{cases}$$

Since $\dot{x}_3 = -2\dot{x}_1 = 2x_2$, but $\dot{x}_3 = x_2$ from the 3rd equation, it forces $x_2 \equiv 0$ so that $\dot{x}_2 = x_1 = 0$. Due to $x_1 = 0$, $x_3 = -2x_1 = 0$. Thus, the set S contains only the equilibrium point $x = 0$ and nothing else. Based on the La Salle Theorem, the 3rd-order system is globally asymptotically stable.

The second example is a realistic robotic system. In almost every industrial application, an independent joint control of the robot manipulator has been demonstrated to be effective in operating a set-point task. Now, we are going to take this example to show mathematically that an independent PD (proportion-derivative) control law on each joint of the robot can asymptotically stabilize the robot to reach a desired destination. The major difference between a set-point task and a trajectory-tracking task lies in the fact that if $q^d(t)$ is a given desired joint position vector, then $\dot{q}^d = 0$ for the set-point task while $\dot{q}^d \neq 0$ for the trajectory-tracking. In other words, $q^d(t)$ represents a fixed destination point in the former case, while it describes a desired trajectory in the latter case. This also implies that a set-point task operation can only control the robot to reach a given desired destination point, but cannot guarantee what kind of actual trajectory it travels.

Let $\Delta q = q^d - q$ be a joint error vector for an n-joint robot manipulator in a set-point task operation. Then, the time-derivative of the error becomes $\Delta \dot{q} = \dot{q}^d - \dot{q} = -\dot{q}$. Let an independent joint PD control law be defined as

$$u = \tau = K_p \Delta q + K_d \Delta \dot{q} - \tau_g = K_p \Delta q - K_d \dot{q} - \tau_g, \tag{6.8}$$

where both K_p and K_d are n by n diagonal positive-definite matrices that represent the proportional and derivative gain constants, respectively. The input $u \in \mathbb{R}^n$ is a control signal to excite each joint of the robot as an n-dimensional joint torque vector $\tau = u$.

Let $x = \begin{pmatrix} q \\ \dot{q} \end{pmatrix} \in \mathbb{R}^{2n}$ be the state vector of the robotic system. According to the robot dynamic model developed in equation (3.22), we can rewrite it in the following state-space equation form:

$$\dot{x} = f(x) + G(x)u,$$

where

$$f(x) = \begin{pmatrix} \dot{q} \\ W^{-1} \left[\tau_g - \left(W_d^T - \frac{1}{2} W_d \right) \dot{q} \right] \end{pmatrix}, \quad \text{and} \quad G(x) = \begin{pmatrix} O \\ W^{-1} \end{pmatrix}$$

with the n by n zero matrix O in $G(x)$. The control objective can now be stated to make the norm of the state-error

$$\Delta x = \begin{pmatrix} \Delta q \\ \Delta \dot{q} \end{pmatrix} = \begin{pmatrix} \Delta q \\ -\dot{q} \end{pmatrix}$$

go to 0 as $t \to \infty$.

To achieve this objective, let us define a following *Lyapunov candidate* that is a positive-definite scalar function of the state error Δx as well as the kinetic energy of the robotic system:

$$V(x) = \frac{1}{2} \dot{q}^T W \dot{q} + \frac{1}{2} \Delta q^T K_p \Delta q. \tag{6.9}$$

Since the dynamic equation for a robotic system, based on (3.22), is given by

$$W \ddot{q} + \frac{1}{2} \dot{W} \dot{q} + \frac{1}{2} (W_d^T - W_d) \dot{q} - \tau_g = \tau,$$

If $V(x)$ is	and $\dot{V}(x)$ is	Type of Stability
l.p.d.f.	$\dot{V}(x) \leq 0$ for $x \in B(\delta)$	Locally Stable
s.p.d.f.	$\dot{V}(x) \leq 0$ for $x \in \mathbb{R}^n$	Globally Stable
l.p.d.f.	$-\dot{V}(x)$ is l.p.d.f. or	Locally Asymptotically
	$\dot{V}(x) \leq 0$ & only $x = 0$ in the local set S	Stable
s.p.d.f.	$-\dot{V}(x)$ is p.d.f. or	Globally Asymptotically
	$\dot{V}(x) \leq 0$ & only $x = 0$ in the global set S	Stable

Table 6.1

Stability determination for nonlinear systems

multiplying $\dot{q}^T$ to both sides will lead to

$$\dot{q}^T W \ddot{q} + \frac{1}{2}\dot{q}^T \dot{W} \dot{q} + \frac{1}{2}\dot{q}^T (W_d^T - W_d)\dot{q} - \dot{q}^T \tau_g = \dot{q}^T \tau.$$

Because $W_d^T - W_d$ is a skew-symmetric matrix, $\dot{q}^T (W_d^T - W_d)\dot{q} \equiv 0$ so that

$$\dot{q}^T W \ddot{q} + \frac{1}{2}\dot{q}^T \dot{W} \dot{q} = \dot{q}^T \tau + \dot{q}^T \tau_g.$$

Now, taking time-derivative for (6.9) yields

$$\dot{V}(x) = \dot{q}^T W \ddot{q} + \frac{1}{2}\dot{q}^T \dot{W} \dot{q} + \Delta\dot{q}^T K_p \Delta q.$$

Substituting the above equation and invoking the PD control law (6.8), we obtain

$$\begin{aligned} \dot{V} &= \dot{q}^T (\tau + \tau_g) - \dot{q}^T K_p \Delta q && (6.10) \\ &= \dot{q}^T (K_p \Delta q - K_d \dot{q}) - \dot{q}^T K_p \Delta q = -\dot{q}^T K_d \dot{q}. \end{aligned}$$

Since $K_d > 0$, $\dot{V}$ is shown to be at least non-positive, but $-\dot{V}(x)$ is not a p.d.f. due to missing Δq in $\dot{V}(x)$. However, we can further justify that if $\Delta\dot{q} = -\dot{q} = 0$ so that $\dot{V} = 0$, then $\Delta q = 0$ as well by using the PD control law (6.8) and the robot state equation (3.22). Therefore, based on the La Salle's Theorem, $\|\Delta x\| \to 0$ asymptotically as $t \to \infty$. In other words, $q \to q^d$ and $\dot{q} \to 0$ as $t \to \infty$. This also shows that *an independent-joint PD control law given by (6.8) with knowledge of the joint torque τ_g due to gravity can perfectly drive the robotic system to perform any set-point task.*

In summary, using the Lyapunov Direct method in Theorem 6.1 and La Salle's Theorem 6.2, we can determine the stability for a system at the equilibrium point $x = 0$ using the testing procedure in Table 6.1. Therefore, this table provides **a quick reference in stability determination** for nonlinear systems based on the Lyapunov Direct Method and La Salle's Theorem.

6.1.1 THE LOCAL LINEARIZATION PROCEDURE

For a nonlinear autonomous (time-invariant) control system given in the second equation of (6.1), the equilibrium points can be determined by solving the algebraic equation $f(x,0) = 0$ with the zero input. Based on the coordinates parallel shift procedure, any nonzero equilibrium point $x = x_e \neq 0$ can always be shifted by defining $z = x - x_e$ to the origin of the z-coordinate system with the state-space equation transformed to $\dot{z} = \bar{f}(z, u)$. Therefore, without loss of generality, every nonlinear system can assume the origin of the state space as its equilibrium point as a default.

When we study a system characteristic, test its stability, or design a control law for the system, a neighboring region around the equilibrium point $x = 0$ is the most important domain to be considered. Thus, we can use the first-order truncation of the Taylor expansion around $x = 0$ to locally linearize the entire control system (6.1). The Taylor series of the function $f(x, u)$ around $x = 0$ and $u = 0$ is given by

$$f(x,u) = f(0,0) + \left(\frac{\partial f}{\partial x}\right)_{0,0} x + \left(\frac{\partial f}{\partial u}\right)_{0,0} u + o(x,u),$$

where $o(x,u)$ is the amount of higher-order residue terms. With the first-order approximation, and noticing that $f(0,0) = 0$ due to the equilibrium point definition, we can thus represent the system in a neighborhood of the equilibrium point $x = 0$ with $u = 0$ by

$$\dot{x} = f(x,u) \approx Ax + Bu, \tag{6.11}$$

where

$$A = \left(\frac{\partial f}{\partial x}\right)_{0,0}, \quad \text{and} \quad B = \left(\frac{\partial f}{\partial u}\right)_{0,0}.$$

Clearly, A and B are n by n and n by r mathematical *Jacobian matrices*, respectively, both evaluated at $x = 0$ and $u = 0$, and thus, they are two constant matrices. The nonlinear control system (6.1) is therefore locally linearized to a time-invariant system in (6.11) within a neighborhood around the equilibrium point $x = 0$ and the zero input $u = 0$.

It can be seen that such a **local linearization** procedure requires the computation of two Jacobian matrices $\partial f/\partial x$ and $\partial f/\partial u$ symbolically before substituting $x = 0$ and $u = 0$ to determine the matrices A and B. This procedure, though convenient, is valid only in a small neighboring region of the equilibrium point and the zero input. With the increasing of nonlinearity and system dimensions, the local linearization approach may experience more limitations in general analysis of nonlinear systems.

To demonstrate the local linearization procedure, let us look at a following nonlinear autonomous control system:

$$\begin{cases} \dot{x}_1 = 4x_1 + x_2^2 - sat(2x_2 + u) \\ \dot{x}_2 = 2\sin x_1 - x_2 + u, \end{cases} \tag{6.12}$$

where $sat(\sigma)$ is a saturated-ramp function defined by

$$sat(\sigma) = \begin{cases} \sigma, & \text{if } |\sigma| \le 1 \\ +1, & \text{if } \sigma > 1 \\ -1, & \text{if } \sigma < 1. \end{cases} \tag{6.13}$$

Obviously, $x_1 = 0$ and $x_2 = 0$ is an isolated equilibrium point under $u = 0$. In a small neighboring ball $B(\delta_x, \delta_u)$ around the equilibrium point and under the zero input, let us determine the matrices A and B. Since both x_2 and u in $B(\delta_x, \delta_u)$ are very small, the saturation term $sat(2x_2 + u) = 2x_2 + u$ based on the definition (6.13). Therefore,

$$\frac{\partial f}{\partial x} = \begin{pmatrix} 4 & 2x_2 - 2 \\ 2\cos x_1 & -1 \end{pmatrix}, \quad \text{and} \quad \frac{\partial f}{\partial u} = \begin{pmatrix} -1 \\ 1 \end{pmatrix}.$$

By substituting $x_1 = 0$ and $x_2 = 0$ into the above Jacobian matrices, we obtain

$$A = \begin{pmatrix} 4 & -2 \\ 2 & -1 \end{pmatrix}, \quad \text{and} \quad B = \begin{pmatrix} -1 \\ 1 \end{pmatrix}.$$

Finally, the equation $\dot{x} = Ax + Bu$ with the above A and B represents a locally linearized control system valid only in a small ball $B(\delta_x, \delta_u)$ around the origin $x = 0$ and $u = 0$ for the original nonlinear control system (6.12).

6.1.2 INDIRECT METHOD OF SYSTEMS STABILITY TEST

In addition to the Lyapunov direct method for testing systems stability, there is another method available for the same purpose. Every nonlinear system can be approximately linearized around the equilibrium point $x = 0$ by the first-order truncation of the Taylor expansion in equation (6.11). Specifically, for a nonlinear zero-input system:

$$\dot{x} = f(x), \tag{6.14}$$

its locally linearized system can be written by

$$\dot{x} = Ax \quad \text{with} \quad A = \left(\frac{\partial f}{\partial x}\right)_{x=0}. \tag{6.15}$$

Now, an interesting and also important question is that can the stability of the locally linearized system (6.15) predict and imply the stability of the original nonlinear system (6.14)? Since testing stability for a linear system is relatively simple, the answer to the above question is practically useful. In fact, there has been a theorem, called the **Lyapunov Indirect Method**, developed for this purpose. Let us formally state this theorem without giving proof:

Theorem 6.3. *For a nonlinear zero-input system (6.14),*

1. *If its locally linearized system (6.15) is asymptotically stable, i.e., all eigenvalues of A are strictly located in the left-half of complex plane (LHP), or A is Hurwitz, then $x = 0$ for the original nonlinear system (6.14) is a locally asymptotically stable equilibrium point;*
2. *If (6.15) is unstable, i.e., at least one eigenvalue of A is strictly located in the right-half of complex plane (RHP), then $x = 0$ is also unstable for (6.14);*
3. *If (6.15) is critically stable, i.e., at least an eigenvalue or one pair of complex-conjugate eigenvalues of A is on the imaginary axis (distinctly if more than one) and all others are strictly in the LHP, then one cannot conclude any type of stability at $x = 0$ for (6.14).*

This Lyapunov indirect method of stability is seen to be convenient and useful except when the last case is encountered. Therefore, it is suggested that to test *local stability* for a nonlinear system at $x = 0$, one can first locally linearize it, and then apply the indirect method. If the third case is encountered, then the testing process has to switch to the direct method. This suggestion is very helpful for many nonlinear systems with relatively lower dimensions, for which finding a Lyapunov candidate is more difficult than locally linearizing the system. However, for high-dimensional systems with higher nonlinearity, since it is more difficult to manipulate the Jacobian matrix $\frac{\partial f}{\partial x}$ and eigenvalues at $x = 0$, the direct method may become more suitable in such cases. Furthermore, if it is desired to test a nonlinear system for *global stability*, we must rely on the direct method, because the indirect method is only capable of testing systems for local stability.

As example, a 2D nonlinear system, called the Rayleigh's equation is given by

$$\begin{cases} \dot{x}_1 = x_2 \\ \dot{x}_2 = -x_1 + \delta(x_2 - \frac{x_2^3}{3}). \end{cases} \tag{6.16}$$

In order to test its stability at its equilibrium point $x_1 = 0$ and $x_2 = 0$, let us first try the indirect method. Since the Jacobian matrix at $x = 0$ is

$$A = \left(\frac{\partial f}{\partial x} \right)_0 = \begin{pmatrix} 0 & 1 \\ -1 & \delta(1 - x_2^2) \end{pmatrix}_0 = \begin{pmatrix} 0 & 1 \\ -1 & \delta \end{pmatrix},$$

the characteristic equation of A is given by $\det(\lambda I - A) = \lambda^2 - \delta\lambda + 1 = 0$, where I is the 2 by 2 identity. Solving this quadratic equation yields

$$\lambda_{1,2} = \frac{\delta \pm \sqrt{\delta^2 - 4}}{2}.$$

Therefore, if $\delta < 0$, then both roots are either negative or have negative real parts. Based on the Lyapunov indirect method in Theorem 6.3, the Rayleigh's system at $x = 0$ is (at least) locally asymptotically stable. However, if $\delta > 0$, then both roots are either positive or have positive real parts; thus $x = 0$ is unstable for the

Rayleigh's system. Finally, for $\delta = 0$, $\lambda_{1,2} = \pm j$, then there is no basis to determine the stability by the indirect method. Now, trying the direct method for the case $\delta = 0$, the Rayleigh's equation is reduced to $\dot{x}_1 = x_2$ and $\dot{x}_2 = -x_1$. Let us define a Lyapunov candidate $V(x) = x_1^2 + x_2^2$, which is obviously an s.p.d.f. Because $\dot{V}(x) = 2x_1\dot{x}_1 + 2x_2\dot{x}_2 = 2x_1x_2 - 2x_2x_1 = 0$, based on the direct method given by Theorem 6.1, $x = 0$ is a stable equilibrium point for the Rayleigh's system under $\delta = 0$, but is not asymptotically stable.

6.1.3   A THEOREM FOR DETERMINATION OF SYSTEM INSTABILITY

The Lyapunov direct method of stability is effective to determine whether a system is either stable, or locally or globally asymptotically stable, provided that a Lyapunov candidate can be found. However, Theorem 6.1 only provides a sufficient but not necessary condition; in other words, one is unable to prove by this theorem that a system is unstable at an equilibrium point. If one could not find a proper Lyapunov candidate to justify the stability of a system, it does not imply that the system is unstable. This may be thought of as a major drawback of the direct method. In contrast to the direct method, the Lyapunov indirect method can be used to determine the instability for a nonlinear system if its locally linearized system is unstable. However, when the locally linearized system is critically stable, the determination of the system instability is still unanswered.

   In this section, we introduce an extended theorem which reverses the Lyapunov direct method with some conditions adjusted for the determination of system instability.

Theorem 6.4. *The equilibrium point $x = 0$ for the system (6.14) is unstable if there exist a continuous scalar function $V(x)$ such that*

   1. *$\dot{V}(x)$ is an l.p.d.f., and*
   2. *$V(0) = 0$ and $V(x) > 0$ for some x arbitrarily close to 0.*

   This instability theorem can be interpreted by reversing the Lyapunov direct method of stability. In the stability theorem, the stable equilibrium point requires $\dot{V}(x) \leq 0$. Whereas the instability theorem requires $\dot{V}(x)$ to be at least an *l.p.d.f.*, but it relaxes the condition for $V(x)$, i.e., it requires only $V(0) = 0$ and $V(x) > 0$ for *some* point $x \neq 0$ close to the equilibrium point, but not necessarily for *all* $x \neq 0$ in a neighborhood ball $B(\delta)$ of the equilibrium point $x = 0$.

   As an example, let us test the following nonlinear system:

$$\begin{cases} \dot{x}_1 = x_1 - x_2 + x_1x_2 \\ \dot{x}_2 = -x_2 - x_2^2. \end{cases} \tag{6.17}$$

We try $V(x) = (2x_1 - x_2)^2 - x_2^2$ and clearly, $V(0) = 0$. There exist some points $x \neq 0$, for instance, let $x_2 = 0$ and x_1 be any small value such that $V(x) > 0$. Since

$$\dot{V}(x) = 2(2x_1 - x_2)(2\dot{x}_1 - \dot{x}_2) - 2x_2\dot{x}_2 = [(2x_1 - x_2)^2 + x_2^2](1 + x_2),$$

it can be easily seen that this $\dot{V}(x)$ is an l.p.d.f. in a ball $B(\delta)$ for any $\delta < 1$. Therefore, the system is unstable at $x = 0$.

6.1.4 STABILIZATION OF NONLINEAR CONTROL SYSTEMS

For a control system $\dot{x} = f(x, u)$, subject to the existence of input u, we may take advantage of defining an appropriate input $u = u(x)$ to stabilize the entire system $\dot{x} = f(x, u(x))$. Specifically, for the originally unstable or critically stable zero-input system $\dot{x} = f(x, 0)$, there can be a chance to stabilize the system or improve the stability for the system. Since the input u can be defined as a function of the state x that is retrieved to update the input, we call such an input function $u = u(x)$ the **state-feedback control**.

As a typical example, let us revisit the system given by (6.12),

$$\begin{cases} \dot{x}_1 = 4x_1 + x_2^2 - sat(2x_2 + u) \\ \dot{x}_2 = 2\sin x_1 - x_2 + u, \end{cases} \tag{6.18}$$

where, again, $sat(\sigma)$ is a saturated-ramp function, as defined in (6.13). It has been shown in the previous section that the locally linearized system of (6.18) around the origin $x = 0$ is $\dot{x} = Ax + Bu$, where

$$A = \begin{pmatrix} 4 & -2 \\ 2 & -1 \end{pmatrix}, \quad \text{and} \quad B = \begin{pmatrix} -1 \\ 1 \end{pmatrix}.$$

The characteristic equation of A is given by

$$\det(\lambda I - A) = \lambda^2 - 3\lambda = 0,$$

and its roots are $\lambda_1 = 0$ and $\lambda_2 = 3$. Therefore, according to the Lyapunov Indirect Method in Theorem 6.3, the nonlinear system (6.18) with zero input is unstable. Now, define $u = -Kx$ with a 1 by 2 constant matrix K such that the matrix $A - BK$ is Hurwitz, i.e., all the eigenvalues are strictly in the LHP. For instance, let $K = (-5 \ 1)$, then, the locally linearized control system becomes

$$\dot{x} = Ax - BKx = (A - BK)x = \begin{pmatrix} -1 & -1 \\ 7 & -2 \end{pmatrix} x. \tag{6.19}$$

The characteristic equation of $A - BK$ is $\lambda^2 + 3\lambda + 9 = 0$, which obviously has two complex conjugate roots with negative real parts. Thus, the new linearized system (6.19) is asymptotically stable at $x = 0$.

However, the above stabilization is done only for the locally linearized system, instead of the original nonlinear system (6.18). An important question now naturally arises: is there any relationship between the stabilization of a nonlinear control system and its locally linearized one? Let the following theorem answer the question. Suppose that the nonlinear control system is given by $\dot{x} = f(x, u)$, where $x \in \mathbb{R}^n$ and $u \in \mathbb{R}^r$, and its locally linearized system at the equilibrium point $x = 0$ is found to be $\dot{x} = Ax + Bu$.

Theorem 6.5. (Feedback Stabilization) – *Under the following assumptions:*

1. $f(\cdot,\cdot)$ *is continuous and differentiable, and* $f(0,0) = 0$, *and*
2. *the locally linearized system is controllable.*

Also, suppose that there exists a constant matrix $K \in \mathbb{R}^{r \times n}$ *such that the matrix* $A - BK$ *is Hurwitz, then* $u = -Kx$ *is a state-feedback control such that* $x = 0$ *is a locally asymptotically stable equilibrium point for the resultant nonlinear system* $\dot{x} = f(x, -Kx)$.

Note that it has been discussed in almost every literature on the linear control systems that a linear time-invariant system $\dot{x} = Ax + Bu$ is controllable if $\text{rank}(B \ AB \ \cdots \ A^{n-1}B) = n$, which is just required by the second assumption of the above theorem. The controllability issue for nonlinear systems will be further addressed in next sections.

Theorem 6.5 shows an important fact that to stabilize a nonlinear control system, one can design a state-feedback control to stabilize its locally linearized system, and then the nonlinear system stability will also be guaranteed by this state-feedback control, provided that the above two assumptions hold. This is referred to as the **Lyapunov Indirect Stabilization Method**.

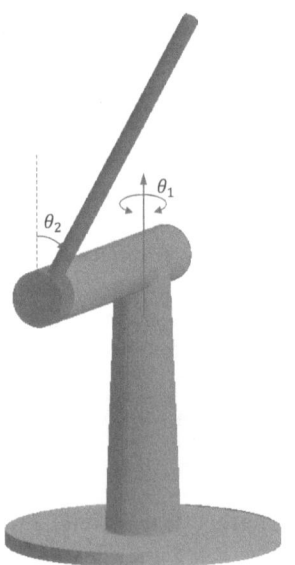

Figure 6.2   A rotating inverted pendulum system

As a practical example, let us look at a rotating inverted pendulum system, which is different from the well-known sliding inverted pendulum, as shown in Figure 3.1 and dynamically modeled in Chapter 3. Instead of sliding the cart to stabilize the top inverted pendulum, this new system is applying a torque $\tau = u$, as a single system input, to rotate and control the arm angle θ_1 about the shoulder axis for balancing the inverted pendulum, as illustrated in Figure 6.2. In order to simulate this system towards a realistic animation, we need to develop its dynamic model first. Through the dynamics modeling procedure as well-developed in Chapter 3, we can formulate its complete dynamic equation below:

$$W\ddot{q} + c(q,\dot{q}) - \tau_g = \begin{pmatrix} u \\ 0 \end{pmatrix},$$

where the angular displacement vector is $q = \begin{pmatrix} \theta_1 \\ \theta_2 \end{pmatrix}$, the inertial matrix is

$$W = \begin{pmatrix} m_1\frac{l_1^2}{4} + I_1 + m_2l_1^2 & \frac{1}{2}m_2l_1l_2\cos\theta_2 \\ \frac{1}{2}m_2l_1l_2\cos\theta_2 & m_2\frac{l_2^2}{4} + I_2 \end{pmatrix},$$

and the centrifugal and Coriolis force term is

$$c(q,\dot{q}) = \left(W_d^T - \frac{1}{2}W_d\right)\dot{q} = \begin{pmatrix} -\frac{1}{2}m_2l_1l_2\dot{\theta}_2^2\sin\theta_2 \\ 0 \end{pmatrix}.$$

Since the gravitational potential energy can be found as $P = m_2g\frac{l_2}{2}\cos\theta_2$, the arm required rotating torque due to gravity becomes

$$\tau_g = -\frac{\partial P}{\partial q} = \begin{pmatrix} 0 \\ m_2g\frac{l_2}{2}\sin\theta_2 \end{pmatrix}.$$

To form a standard state-space equation, let the 4-dimensional state vector be defined as $x = \begin{pmatrix} q \\ \dot{q} \end{pmatrix}$. Similar to equation (6.1), we can combine $f(x)$ and $G(x)u$ to merge them into a single tangent vector:

$$\dot{x} = f(x,u) = \begin{pmatrix} \dot{q} \\ W^{-1}[\tau_g - c(q,\dot{q})] + W^{-1}\begin{pmatrix} u \\ 0 \end{pmatrix} \end{pmatrix}.$$

Since the inverse of the inertial matrix W is involved in $f(x,u)$, we encounter a difficulty to symbolically derive the Jacobian matrices for both $\frac{\partial f}{\partial x}$ and $\frac{\partial f}{\partial u}$ before evaluating them at the equilibrium point $x = 0$ and $u = 0$, as required by the local linearization equation in (6.11). However, in most robot modeling cases, we can go through an alternative but equivalent way. Namely, let every second or higher order term of state variables approximate to zero, and set $\cos x_i \approx 1$ and $\sin x_i \approx x_i$ for each i before inverting the inertial matrix W. If doing this way, the related locally linearized matrices become

$$\hat{W} = \begin{pmatrix} m_1\frac{l_1^2}{4} + I_1 + m_2l_1^2 & \frac{1}{2}m_2l_1l_2 \\ \frac{1}{2}m_2l_1l_2 & m_2\frac{l_2^2}{4} + I_2 \end{pmatrix},$$

and $\hat{c}(q, \dot{q}) = 0$, while

$$\hat{t}_g = \begin{pmatrix} 0 \\ m_2 g \frac{l_1}{2} \theta_2 \end{pmatrix}.$$

Then, the locally linearized dynamic model can be directly found without going through the Jacobian matrix procedure:

$$\ddot{q} = \hat{W}^{-1} \hat{t}_g + \hat{W}^{-1} \begin{pmatrix} u \\ 0 \end{pmatrix}.$$

The matrices A and B for the locally linearized system can thus be determined as follows:

$$A = \begin{pmatrix} O_{2 \times 2} & I_{2 \times 2} \\ (O_{2 \times 1} \quad m_2 g \frac{l_2}{2} \xi_2) & O_{2 \times 2} \end{pmatrix}, \quad B = \begin{pmatrix} O_{2 \times 1} \\ \xi_1 \end{pmatrix},$$

where ξ_1 and ξ_2 are the first and second columns of $\hat{W}^{-1}$, respectively, and $I_{2 \times 2}$ is the 2 by 2 identity, and $O_{2 \times 1}$ and $O_{2 \times 2}$ are all the zero matrices.

As a numerical solution, let the arm length be $l_1 = 0.8$ and let the inverted pendulum length be $l_2 = 1.2$, all in meters. Let the arm mass be $m_1 = 2$ and let the inverted pendulum mass be $m_2 = 0.75$, all in Kg. Furthermore, let the moments of inertia for the arm and inverted pendulum be $I_1 = 0.4$ and $I_2 = 0.1$, respectively, all in Kg-m^2. The gravitational acceleration is $g = 9.806$ m/sec^2. Then, from MATLAB^{TM}, we obtain the following numerical values of both the matrices A and B:

$$A = \begin{pmatrix} 0 & 0 & 1.0000 & 0 \\ 0 & 0 & 0 & 1.0000 \\ 0 & -5.0527 & 0 & 0 \\ 0 & 16.8424 & 0 & 0 \end{pmatrix}, \quad \text{and} \quad B = \begin{pmatrix} 0 \\ 0 \\ 1.1768 \\ -1.1450 \end{pmatrix}.$$

Since the controllability matrix can be calculated as

$$(B \ AB \ A^2 B \ A^3 B) = \begin{pmatrix} 0 & 1.1768 & 0 & 5.7855 \\ 0 & -1.1450 & 0 & -19.2852 \\ 1.1768 & 0 & 5.7855 & 0 \\ -1.1450 & 0 & -19.2852 & 0 \end{pmatrix},$$

which is nonsingular, the locally linearized system is controllable. However, because the eigenvalues of A are $\lambda_{1,2} = 0$, and $\lambda_{3,4} = \pm 4.1039$, one of which is in the RHP, based on Theorem 6.3, the original nonlinear system is unstable. In order to stabilized this unstable but controllable rotating inverted pendulum system, let us place the new eigenvalues to

$$P = (-1 + j2, \ -1 - j2, \ -2, \ -3).$$

By calling the MATLAB^{TM} internal function $place(A, B, P)$, we will immediately be responded with a state feedback control gain

$$K = (-2.1375 \ -35.2458 \ -2.6362 \ -8.8228)$$

such that $A - BK$ has the eigenvalues exactly at P. Therefore, the entire nonlinear rotating inverted pendulum system can now be controlled by the state-feedback $u = -Kx$ to asymptotically converge to the equilibrium point $x_e = 0$, provided that the initial deviation of $x(0)$ from $x_e = 0$ is not too large. In fact, the simulation study shows that it diverges if the initial condition $x(0)$ is relatively large, which just reflects the nature of *local stabilization*.

This practical example with its simulation and 3D graphical animation demonstrates that the Lyapunov indirect stabilization method is quite useful and effective for many robotic systems or non-robot dynamic systems as long as the initial condition is relatively small.

6.2   CONTROLLABILITY AND OBSERVABILITY

Controllability is a key issue that must be addressed before starting to design a control law for a given system. Consider a class of nonlinear control systems with the inputs appearing "linearly", i.e. affine in the inputs:

$$\dot{x} = f(x) + \sum_{i=1}^{r} g_i(x)u_i = f(x) + G(x)u, \quad x \in M \subset \mathbb{R}^n, \tag{6.20}$$

where $u = (u_1 \cdots u_r)^T \in U \subset \mathbb{R}^r$ is the input vector, and $G(x) = (g_1(x) \cdots g_r(x))$ is an n by r coefficient matrix of u. The system is controllable if there exists an admissible input vector $u(t)$ such that the system state $x(t)$ can travel from an initial point $x(t_0) = x_0 \in M$ to its equilibrium point $x(t_f) = x_e \in M$ within a finite time interval $T = t_f - t_0 < \infty$. The controllability reveals whether the control system has a complete set of "healthy" input channels, through which an input can excite the states effectively to reach the destination x_e. Based on this interpretation, the controllability should clearly depend upon the function forms of all vector fields $f(x)$ and $g_i(x)$'s in (6.20).

Observability, on the other hand, is another parallel key issue for a system with an output equation, i.e.,

$$\begin{cases} \dot{x} = f(x) \\ y = h(x), \end{cases} \tag{6.21}$$

where $y \in Y \subset \mathbb{R}^m$ is an output vector. The above system is said to be observable if for each pair of distinct states x_1 and x_2, the corresponding outputs y_1 and y_2 are also distinguishable. Clearly, the observability can be interpreted as a testing criterion to see if the entire system has sufficient output channels to measure (or observe) each internal state value. Intuitively, the observability should depend upon the function forms of both $f(x)$ and $h(x)$.

Before we formally discuss the two important concepts for nonlinear systems, let us first introduce some mathematical preliminaries, and then deduce their testing criteria along with the necessary physical interpretations.

6.2.1   CONTROL LIE ALGEBRA AND CONTROLLABILITY

Consider a linear time-invariant control system $\dot{x} = Ax + Bu$, where x is an n-dimensional state vector and u is an r-dimensional input vector, and A and B are n by n and n by r constant matrices, respectively. As well known from the linear control theory, the system is controllable if

$$\text{rank}(B \ AB \ A^2B \ \cdots \ A^{n-1}B) = n. \tag{6.22}$$

However, for a nonlinear control system described by (6.20), there are no matrices A and B, and such a testing criterion (6.22) is no longer usable. We need now to develop a more general approach to testing the controllability for both linear and nonlinear control systems.

Let us recall a special Lie algebra $\mathscr{L}$ that collects all n-dimensional differentiable vector fields in $\mathbb{R}^n$ along with a commutative derivative relation, as introduced in Chapter 2. Namely, for any two vector fields f, $g \in \mathbb{R}^n$, both of which are smooth functions of $x \in \mathbb{R}^n$, the Lie bracket is defined as

$$[f,g] = \frac{\partial g}{\partial x}f - \frac{\partial f}{\partial x}g.$$

In order to extend the above Lie bracket between two vector fields to higher-order derivatives, a more compact notation may be adopted here, called an **adjoint operator**, i.e., $[f,g] = ad_f g$. This new notation treats the Lie bracket $[f,g]$ as a vector field g operated by an adjoint operator $ad_f = [f, \cdot]$. Therefore, an nth order Lie bracket $(n > 1)$ can be expressed as

$$\underbrace{[f, \cdots [f,g] \cdots]}_{n \qquad n} = ad_f^n g.$$

For a nonlinear control system given by (6.20), we now specifically define a so-called **Control Lie Algebra** $\mathscr{L}_c$ that is spanned by all of up to $(n-1)$st-order Lie brackets among f, and g_1 through g_r, i.e.,

$$\mathscr{L}_c = span\{g_1, \cdots, g_r, ad_f g_1, \cdots, ad_f g_r, \cdots, ad_f^{n-1}g_1, \cdots, ad_f^{n-1}g_r\}. \tag{6.23}$$

With this control Lie algebra concept, we may show that the following theorem is true and is also a general effective testing criterion for system controllability [5, 6]:

Theorem 6.6. *The control system (6.20) is controllable if and only if*

$$\dim(\mathscr{L}_c) = \dim(M) = n.$$

Note that since each element in $\mathscr{L}_c$ is a function of x, the dimension of $\mathscr{L}_c$ may be different from one point to the others. Thus, if the above condition of dimension is valid only in a neighborhood of a point in $M \subset \mathbb{R}^n$, we say that the system is *locally controllable*. If the condition of dimension can cover all the region M, then it is *globally controllable*.

As the first example, let us test a linear time-invariant control system $\dot{x} = Ax + Bu$, where $x \in M \subset \mathbb{R}^n$ and $u \in U \subset \mathbb{R}^r$. Comparing this linear equation with (6.20), we have $f(x) = Ax$ and $G(x) = (g_1(x) \cdots g_r(x)) = B$. Since in this linear case, each column b_i of B is a constant vector, $ad_f b_i = -(\frac{\partial f}{\partial x})b_i = -Ab_i$, and $ad_f^2 b_i = ad_f(ad_f b_i) = ad_f(-Ab_i) = A^2 b_i$. Thus, we conclude that in a linear time-invariant control system, the control Lie algebra is spanned by all columns of B and all $A^j b_i$'s, where $1 \le i \le r$ and $1 \le j \le n-1$. This conclusion clearly shows that the well-known linear system controllability condition (6.22) is just a special case of the above controllability theorem for general nonlinear control systems.

Let us now look at a disease control system given by

$$\begin{cases} \dot{x}_1 = -ax_1 + bx_1x_2 + u \\ \dot{x}_2 = -bx_1x_2. \end{cases} \tag{6.24}$$

The story behind the equation is that suppose there is a fatal disease spreading across a village. If any one is infected, then this person will either die, or cure with permanent immune from the disease [1]. To model this system, we may define the number of infected people in the village as the first state component x_1, while the number of non-infected healthy people is defined as the second state component x_2. To test its controllability, let a control Lie algebra $\mathcal{L}_c$ be constructed by the basis $\{g, ad_f g\}$. Thus, for equation (6.24),

$$g = \begin{pmatrix} 1 \\ 0 \end{pmatrix}, \text{ and } ad_f g = \frac{\partial g}{\partial x}f - \frac{\partial f}{\partial x}g = -\begin{pmatrix} -a+bx_2 & bx_1 \\ -bx_2 & -bx_1 \end{pmatrix}\begin{pmatrix} 1 \\ 0 \end{pmatrix} = \begin{pmatrix} a-bx_2 \\ bx_2 \end{pmatrix}.$$

It shows that the matrix augmented by g and $ad_f g$ is

$$(g \ ad_f g) = \begin{pmatrix} 1 & a-bx_2 \\ 0 & bx_2 \end{pmatrix},$$

and its determinant is bx_2, which is non-zero if $x_2 \ne 0$. Thus, the disease control system (6.24) is controllable except when $x_2 = 0$, i.e., on the x_1-axis. In fact, the x_2 is the number of healthy people. The system becomes uncontrollable if everyone in the village is infected by the disease.

If we add one more input into the disease control system in (6.24), i.e.,

$$\begin{cases} \dot{x}_1 = -ax_1 + bx_1x_2 + u_1 \\ \dot{x}_2 = -bx_1x_2 + u_2. \end{cases} \tag{6.25}$$

then, $g_1 = \begin{pmatrix} 1 \\ 0 \end{pmatrix}$ and $g_2 = \begin{pmatrix} 0 \\ 1 \end{pmatrix}$. Without need of calculating the adjoint terms $ad_f g_1$ and $ad_f g_2$, the system has already been controllable. It might be a trivial case, because for a system having two state variables $n = 2$, there would be too many inputs $r = 2$ in the control system model. In regular cases, $r < n$.

The next example is given by equation (6.18) from the last section. The system model is given by

$$\begin{cases} \dot{x}_1 = 4x_1 + x_2^2 - sat(2x_2 + u) \\ \dot{x}_2 = 2\sin x_1 - x_2 + u. \end{cases} \quad (6.26)$$

To discuss its controllability, let us first consider a near-equilibrium case. Because the equilibrium point for the system with $u = 0$ is $x = 0$, in the near-equilibrium point region, we have $|2x_2 + u| < 1$ so that $sat(2x_2 + u) = 2x_2 + u$. The equation becomes

$$\begin{cases} \dot{x}_1 = 4x_1 + x_2^2 - 2x_2 - u \\ \dot{x}_2 = 2\sin x_1 - x_2 + u. \end{cases} \quad (6.27)$$

Clearly, in (6.27),

$$f(x) = \begin{pmatrix} 4x_1 + x_2^2 - 2x_2 \\ 2\sin x_1 - x_2 \end{pmatrix} \quad \text{and} \quad g(x) = \begin{pmatrix} -1 \\ 1 \end{pmatrix}.$$

Thus,

$$ad_f g = -\begin{pmatrix} 4 & 2x_2 - 2 \\ 2\cos x_1 & -1 \end{pmatrix}\begin{pmatrix} -1 \\ 1 \end{pmatrix} = \begin{pmatrix} -2x_2 + 6 \\ 2\cos x_1 + 1 \end{pmatrix}.$$

Since the matrix augmented by g and $ad_f g$ has the determinant $2x_2 - 2\cos x_1 - 7$, which should be non-zero within the near-equilibrium region, the system is locally controllable.

In contrast to the near-equilibrium case, consider $2x_2 + u > 1$ so that $sat(2x_2 + u) = 1$. In this saturated case, $f(x)$ and $g(x)$ become

$$f(x) = \begin{pmatrix} 4x_1 + x_2^2 - 1 \\ 2\sin x_1 - x_2 \end{pmatrix} \quad \text{and} \quad g(x) = \begin{pmatrix} 0 \\ 1 \end{pmatrix},$$

and

$$ad_f g = -\begin{pmatrix} 4 & 0 \\ 2\cos x_1 & -1 \end{pmatrix}\begin{pmatrix} 0 \\ 1 \end{pmatrix} = \begin{pmatrix} 0 \\ 1 \end{pmatrix}.$$

It becomes obvious that the two components g and $ad_f g$ in the basis of the control Lie algebra $\mathscr{L}_c$ are parallel to each other. Therefore, $\dim \mathscr{L}_c = 1 < 2$, and the system when $|2x_2 + u| > 1$ in the saturated case is uncontrollable. The physical meaning can be explained that in the saturated case under $|2x_2 + u| > 1$, the input channel is narrowed down.

6.2.2   OBSERVATION SPACE AND OBSERVABILITY

Likewise, for a linear time-invariant system along with a linear output equation, $\dot{x} = Ax$ and $y = Cx$, where y is an m-dimensional output vector, and A and C are constant matrices, the system is observable if

$$
\text{rank}\begin{pmatrix} C \\ CA \\ \vdots \\ CA^{n-1} \end{pmatrix} = n. \tag{6.28}
$$

The notion behind the above criterion is that since the dimension of y is always less than the dimension of x, i.e., $m < n$. Only $y = Cx$ cannot uniquely determine each component of x. Then, we add $\dot{y} = C\dot{x} = CAx$ and $\ddot{y} = CA\dot{x} = CA^2x$ until

$$
\begin{pmatrix} y \\ \dot{y} \\ \vdots \\ y^{(n-1)} \end{pmatrix} = \begin{pmatrix} C \\ CA \\ \vdots \\ CA^{n-1} \end{pmatrix} x
$$

can resolve each component of x in terms of the output y and its time-derivatives as a collection of measurement. Therefore, the linear system is observable if and only if (6.28) is full-ranked.

Since a nonlinear system with a nonlinear output equation given in (6.21) does not have any constant coefficient matrices A and C, the observability criterion (6.28) cannot directly be applied to the nonlinear system. A generalized criterion of testing observability for both linear and nonlinear systems via the Lie algebra mathematical tool is therefore needed.

First, let us introduce a **Lie derivative** that is virtually a *directional derivative* for a scalar field $\mu(x)$ as a function of $x \in \mathbb{R}^n$ along the direction of any n-dimensional vector field $f(x)$. The mathematical expression is given as:

$$
L_f\mu(x) = \frac{\partial\mu(x)}{\partial x}f(x). \tag{6.29}
$$

Since $\frac{\partial\mu(x)}{\partial x}$ is a 1 by n gradient vector of the scalar $\mu(x)$, the norm of a gradient vector often represents the maximum rate of function value changes, and the inner product between the gradient and the vector field $f(x)$ in (6.29) results in a directional derivative of $\mu(x)$ along $f(x)$. Therefore, the Lie derivative of a scalar field defined by (6.29) is also a scalar field.

If each component of a vector field $h(x) \in \mathbb{R}^m$ is considered to take its Lie derivative along $f(x) \in \mathbb{R}^n$, then, all components can be acted concurrently, and the result is a vector field that has the same dimension as $h(x)$, while its i-th element is the Lie derivative of the i-th component of $h(x)$. Namely, if $h(x) = (h_1(x) \cdots h_m(x))^T$ and each component $h_i(x)$ is a scalar field, then the Lie derivative of the vector field $h(x)$ is defined by

$$
L_fh(x) = \begin{pmatrix} L_fh_1(x) \\ \vdots \\ L_fh_m(x) \end{pmatrix}. \tag{6.30}
$$

With the Lie derivative concept, we now define a so-called **Observation Space** $\mathcal{O}$ over $\mathbb{R}^m$ as

$$\mathcal{O} = span\{h(x), L_f h(x), \cdots, L_f^{n-1} h(x)\}. \tag{6.31}$$

In other words, this space is spanned by all up to $(n-1)$st-order Lie derivatives of the output function $h(x)$ along the tangent vector field $f(x)$. Then, motivated by the notion of the linear observability, let us further define an *Observability Distribution*, denoted by $d\mathcal{O}$, which is spanned by all the gradient vectors of every component in $\mathcal{O}$. Namely,

$$d\mathcal{O} = span\left\{ \frac{\partial \phi}{\partial x} \;\middle|\; \phi \in \mathcal{O} \right\}. \tag{6.32}$$

With these two consecutive definitions, we can now present the following theorem for testing the system observability [6]:

Theorem 6.7. *The system (6.21) is observable if and only if* $\dim(d\mathcal{O}) = n$.

Likewise, the above testing criterion also has *locally observable* and *globally observable* cases, depending on whether the above condition of dimension in the theorem is valid only in a neighborhood of a point or over the entire state space.

For example, a linear time-invariant system $\dot{x} = Ax$ and $y = Cx$ has $f(x) = Ax$ and $h(x) = Cx$. Based on the above observation space definition, $\mathcal{O}$ should be spanned by $\{Cx, CAx, \cdots, CA^{n-1}x\}$. Then, its observability distribution becomes the following vector collection:

$$d\mathcal{O} = \{C, CA, \cdots, CA^{n-1}\}.$$

Obviously, the result based on the above observability theorem is consistent to (6.28) that is well-known in the linear systems theory.

Consider now a following nonlinear system:

$$\begin{cases} \dot{x}_1 = x_2 \\ \dot{x}_2 = -\sin x_2. \end{cases} \tag{6.33}$$

If we define an output equation $y = h(x) = x_1$, then $L_f h(x) = (1 \; 0) \begin{pmatrix} x_2 \\ -\sin x_2 \end{pmatrix} = x_2$, and the observation space $\mathcal{O}$ is spanned by x_1 and x_2. Thus, the observability distribution $d\mathcal{O}$ is formed by two row vectors and they are the gradients of x_1 and x_2, i.e., $(1 \; 0)$ and $(0 \; 1)$. Since the two gradients are augmented to yield the 2 by 2 identity matrix that is obviously full-ranked, the system (6.33) with $y = x_1$ is observable.

If $y = h(x) = x_2$ is defined as the output function, then $L_f h(x) = (0 \; 1) f(x) = -\sin x_2$. The observation space is now spanned by x_2 and $-\sin x_2$, which leads to the observability distribution $d\mathcal{O} = span\{(0 \; 1), (0 \; -\cos x_2)\}$. Therefore, the dimension of $d\mathcal{O}$ is only one, and the system (6.33) with $y = x_2$ is unobservable.

The above two cases of the same system but different outputs typically demonstrate that the observability characterizes how sufficient the output channels are in order to observe (or measure) each internal state of a system through them. In the first case, the first state x_1 of the system is directly measured through the output channel $y = x_1$, and its time-derivative $\dot{y} = \dot{x}_1$ can also be used to determine the second state x_2 via the first equation of (6.33). Whereas in the second case, although x_2 can directly be observed by $y = x_2$, the first state x_1 cannot be uniquely determined by the first equation of (6.33), because the integral of $\dot{x}_1 = x_2$ requires the knowledge of additional initial condition that is unavailable. Therefore, the second case is unobservable.

6.3 INPUT-STATE AND INPUT-OUTPUT STATE-FEEDBACK LINEARIZATION

In mathematics, a distribution Δ of dimension k on a smooth n-dimensional manifold M^n $(k \leq n)$ is a smooth collection of k-dimensional subspaces $\Delta_p \subset T_p M$ with $\Delta = \bigcup_{p \in M} \Delta_p$. The distribution Δ is said to be **integrable** if an immersed submanifold N can be found at every point $p \in M^n$ such that $\Delta_p = T_p N$. Furthermore, Δ is said to be **completely integrable** if an integral manifold of Δ exists through each point of M^n.

Suppose that there are k smooth n-dimensional tangent vector fields $f_1, \cdots, f_k$, where $k \leq n$. A **distribution** Δ is therefore a vector subspace that is spanned by all these tangent vector fields, i.e.,

$$\Delta = span\{f_1, \cdots, f_k\}.$$

After taking the Lie bracket for each pair of the vector fields in the distribution Δ, if every resultant vector field is still contained in Δ, we say the distribution is **involutive**. Formally, a distribution Δ of dimension k is said to be involutive if for each pair $f_i \in \Delta$ and $f_j \in \Delta$, $(1 \leq i, j \leq k)$, $[f_i, f_j] \in \Delta$, where the Lie bracket is defined by

$$[f_i, f_j] = ad_{f_i} f_j = \frac{\partial f_j}{\partial x} f_i - \frac{\partial f_i}{\partial x} f_j$$

that was also defined in equation (2.5) from Chapter 2, and can be rewritten as an adjoint operator ad_{f_i} to apply on f_j.

There are two trivial involutive distribution cases. One is a simple distribution spanned by only one tangent vector field, say f. Since $[f, f] = 0$ and $[f, 0] = 0$, and the zero vector should always be a member of any distribution, $\Delta = span\{ f \}$ is obviously involutive. Another trivial involutive distribution is the entire vector space $\mathbb{R}^n$.

After the above formal mathematical definition of distribution, we now interpret the concept of *complete integrability*. Let the *complementary distribution* of a distribution Δ, denoted by $\Delta^{\perp}$, be defined by $\Delta^{\perp} =$

$$\left\{ span\{f_{k+1}, \cdots, f_n\} \mid \langle f_j, f_i \rangle = 0, \ k+1 \leq j \leq n, \ 1 \leq i \leq k \text{ and } f_i \in \Delta \right\},$$

where $\langle \cdot, \cdot \rangle$ is the inner product of two vector fields. Clearly, the complementary distribution $\Delta^{\perp}$ collects all the remaining tangent vector fields from the entire space

$\mathbb{R}^n$, excluding the tangent vector fields that have already been defined in Δ. Furthermore, every member in $\Delta^{\perp}$ must be orthogonal to each member of Δ.

The distribution Δ is said to be *completely integrable* if each member $f_j(x)$ of $\Delta^{\perp}$ is a gradient vector of some scalar field $\mu_j(x)$, i.e., $f_j(x) = \partial \mu_j(x)/\partial x$ for $j = k+1, \cdots, n$. This concept implies that a distribution Δ is completely integrable if there exist $n - k$ scalar fields $\mu_{k+1}(x)$ through $\mu_n(x)$, such that

$$\frac{\partial \mu_j(x)}{\partial x} f_i(x) = L_{f_i} \mu_j(x) = 0, \quad \text{for } 1 \le i \le k, \text{ and } k+1 \le j \le n.$$

For example, if we have two vector fields in $\mathbb{R}^3$ to span a distribution Δ:

$$f_1(x) = \begin{pmatrix} x_1 \\ -x_2 \\ 0 \end{pmatrix} \quad \text{and} \quad f_2(x) = \begin{pmatrix} 0 \\ 0 \\ x_3 \end{pmatrix},$$

then with a simple calculation, we obtain

$$[f_1, f_2] = \frac{\partial f_2}{\partial x} f_1 - \frac{\partial f_1}{\partial x} f_2 = 0.$$

Since the above $[f_1, f_2] = 0 \in \Delta$, the distribution Δ is involutive. Furthermore, let a scalar function be $\mu(x) = x_1 x_2$. Its gradient vector becomes

$$\frac{\partial \mu}{\partial x} = (x_2 \ x_1 \ 0).$$

This row vector is obviously orthogonal to either f_1 or f_2, and is a member of the complementary distribution $\Delta^{\perp}$. Based on the above definition, the distribution $\Delta = span\{f_1, f_2\}$ is also completely integrable.

It can be further observed that to determine the involutivity for a distribution Δ, we need only to calculate the Lie bracket of each pair of the vector fields. Whereas to determine the complete integrability, we have to seek all such scalar fields. Thus, between the two properties for a distribution Δ, the involutivity is much easier to be tested than the complete integrability. Nevertheless, it is quite exciting that a direct relationship between the two properties has been found so that the testing difficulty for the complete integrability can be alleviated. This direct relationship is known as the Frobenius Theorem:

Theorem 6.8. (Frobenius) – *A distribution Δ on a smooth manifold M is completely integrable if and only if Δ is involutive.*

Ferdinand Georg Frobenius (1849–1917) was a German mathematician, and his contributions were best known to the theory of elliptic functions, differential equations, number theory, and group theory. Using the Frobenius Theorem, one can readily test whether Δ is completely integrable by repeatedly calculating the Lie brackets to see if Δ is involutive. This will be the key step forward to the success of our linearization theory development.

6.3.1   THE INPUT-STATE LINEARIZATION PROCEDURE

Intuitively, to possibly linearize a nonlinear system, one may find and operate a certain nonlinear coordinate transformation to cancel out the nonlinearity. Here the linearization means to *exactly* linearize a nonlinear system without any residue term to be truncated. To this end, there are two fundamental questions that must be answered:

1. Is a given nonlinear system linearizable? and
2. If linearizable, what is the suitable nonlinear transformation?

In this section, we intend to develop a general theory to answer those two fundamental questions.

Consider a *single-input* nonlinear control system modeled by

$$\dot{x} = f(x) + g(x)u, \quad x \in M \subset \mathbb{R}^n, \text{ and } u \in \mathbb{R}^1, \tag{6.34}$$

and both $f(x)$ and $g(x)$ are n-dimensional smooth vector fields. Our objective is to find a possible coordinate transformation that is a differentiable and one-to-one mapping: $T : M \to Z \subset \mathbb{R}^n$, whose inverse is also differentiable and one-to-one. Mathematically, it is called a *diffeomorphism*.

Let us first study a special distribution that is spanned by $g(x)$ and up to the $(n-2)$nd-order Lie brackets between $f(x)$ and $g(x)$:

$$\Delta_l = span\{g(x), ad_f g(x), \cdots, ad_f^{n-2} g(x)\}. \tag{6.35}$$

Note that in this distribution Δ_l, there are at most $n-1$ independent vectors. Suppose that Δ_l is completely integrable, i.e., there exists a scalar field $\mu(x)$ such that its gradient $\frac{\partial \mu(x)}{\partial x}$ is orthogonal to each of the $n-1$ independent members in Δ_l. Since

$$\frac{\partial \mu(x)}{\partial x} ad_f^i g(x) = L_{ad_f^i g} \mu(x)$$

for each $i = 0, \cdots, n-2$, based on the orthogonality, the following equations should hold:

$$L_g \mu(x) = L_{ad_f g} \mu(x) = \cdots = L_{ad_f^{n-2} g} \mu(x) = 0. \tag{6.36}$$

Furthermore, according to equation (2.6) in the preceding Lie algebra section from Chapter 2, for any scalar field $\mu(x)$ and vector fields $f, g \in \mathbb{R}^n$,

$$L_{[f,g]} \mu(x) \equiv L_{ad_f g} \mu(x) \equiv L_f L_g \mu(x) - L_g L_f \mu(x).$$

Using this identity, we can further prove that equation (6.36) implies that

$$L_g \mu(x) = L_g L_f \mu(x) = \cdots = L_g L_f^{n-2} \mu(x) = 0. \tag{6.37}$$

If the nonlinear single-input system in (6.34) has property (6.37), to find a set of appropriate new coordinates $\{z_1, \cdots, z_n\}$, as the n components of a promising diffeomorphism $z = T(x)$, one can readily start with the definition $z_1 = \mu(x)$ so that

$$\dot{z}_1 = \frac{\partial \mu}{\partial x} \dot{x} = L_f \mu + (L_g \mu)u = L_f z_1.$$

Let the second component of $z = T(x)$ be $z_2 = L_f z_1$. Then, by noticing (6.37) again, we obtain

$$\dot{z}_2 = \frac{\partial L_f \mu}{\partial x} \dot{x} = \frac{\partial L_f \mu}{\partial x}(f + gu) = L_f^2 \mu + (L_g L_f \mu)u = L_f^2 z_1 = L_f z_2.$$

If repeating to define $z_3 = L_f z_2 = L_f^2 z_1$, $\cdots$, until reaching the last component z_n of $z = T(x)$, which yields $z_n = L_f^{n-1} z_1$, then

$$\dot{z}_n = \frac{\partial L_f^{n-1} \mu}{\partial x} \dot{x} = L_f^n \mu + (L_g L_f^{n-1} \mu)u.$$

Therefore, to find a nonlinear transformation candidate $z = T(x)$ for possible exact linearization, one can start with $z_1 = \mu(x)$, and then repeatedly take time-derivatives to achieve a new state-space equation in the z coordinate system:

$$\begin{cases} \dot{z}_1 = z_2 \\ \quad \vdots \\ \dot{z}_{n-1} = z_n \\ \dot{z}_n = L_f^n \mu + (L_g L_f^{n-1} \mu)u. \end{cases}$$

Let a new input v be defined by $v = L_f^n \mu(x) + (L_g L_f^{n-1} \mu(x))u$. We can thus rewrite the above equation into the following matrix form:

$$\dot{z} = Az + Bv, \tag{6.38}$$

where the new state $z = (z_1 \cdots z_n)^T$ and

$$A = \begin{pmatrix} 0 & 1 & 0 & \cdots & 0 \\ \vdots & \vdots & \vdots & \cdots & \vdots \\ 0 & 0 & 0 & \cdots & 1 \\ 0 & 0 & 0 & \cdots & 0 \end{pmatrix}, \quad B = \begin{pmatrix} 0 \\ \vdots \\ 0 \\ 1 \end{pmatrix}.$$

This is known as a **controllable canonical form** seen in the new state space $Z \subset \mathbb{R}^n$ with a new input v that is directly related to the old input u and old state x. It is now clear that if one stands in the new state space Z along with the new input v, then the entire control system can be viewed as a strictly linear time-invariant controllable system given by (6.38). We refer to this procedure as an *exact linearization*. Since the original input u can be resolved algebraically in terms of the original state x and the new input v by

$$u = -\frac{L_f^n \mu(x)}{L_g L_f^{n-1} \mu(x)} + \frac{1}{L_g L_f^{n-1} \mu(x)} v = \alpha(x) + \beta(x)v, \tag{6.39}$$

we also call this procedure an **input-state linearization** [1, 2, 4, 5].

Based on the above development, we can now formally define the **linearizability**.

Definition 6.4. A single-input system (6.34) is said to be *input-state linearizable* if there exists a differentiable scalar function $\mu(x)$ along with its Lie derivatives:

$$z = T(x) = \begin{pmatrix} \mu(x) \\ L_f\mu(x) \\ \vdots \\ L_f^{n-1}\mu(x) \end{pmatrix}$$

such that it can transform (6.34) into the linear controllable canonical system in (6.38) with a new input v, as viewed in the z-space.

According to this definition as well as the above discussions, including the Frobenius Theorem, we now formally state the following theorem to summarize the testing criterion of the input-state linearizability for a *single-input* control system:

Theorem 6.9. *An n-dimensional single-input control system (6.34) is input-state linearizable if*

1. *$rank\{g, ad_fg, \cdots, ad_f^{n-1}g\} = n$, and*
2. *the distribution $\Delta_l = span\{g, ad_fg, \cdots, ad_f^{n-2}g\}$ is involutive.*

In this theorem, the first condition is related to the controllability criterion, while the second one, based on the Frobenius Theorem, guarantees that the distribution Δ_l is completely integrable so that a desired scalar function $\mu(x)$ exists and the linearization procedure can always be initiated and fulfilled.

For example, consider a nonlinear control system with a single input u as follows:

$$\dot{x} = \begin{pmatrix} x_2 \\ (1-\ln x_3)x_2 \\ -2x_1x_3 \end{pmatrix} + \begin{pmatrix} 0 \\ 0 \\ 1 \end{pmatrix} u, \tag{6.40}$$

where $x_1 > 0$, $x_2 > 0$ and $x_3 > 0$. In order to linearize it, we have to first test its linearizability by applying the above Theorem 6.9. Since $g(x) = (0 \; 0 \; 1)^T$ is a constant vector, the first-order Lie bracket between $f(x)$ and $g(x)$ becomes

$$ad_fg = \frac{\partial g}{\partial x}f - \frac{\partial f}{\partial x}g = -\begin{pmatrix} 0 & 1 & 0 \\ 0 & 1-\ln x_3 & -\frac{x_2}{x_3} \\ -2x_3 & 0 & -2x_1 \end{pmatrix}\begin{pmatrix} 0 \\ 0 \\ 1 \end{pmatrix} = \begin{pmatrix} 0 \\ \frac{x_2}{x_3} \\ 2x_1 \end{pmatrix}.$$

To test the first condition of the theorem, we need to further calculate

$$ad_f^2g = \frac{\partial(ad_fg)}{\partial x}f - \frac{\partial f}{\partial x}ad_fg = \begin{pmatrix} -\frac{x_2}{x_3} \\ \frac{4x_1x_2}{x_3} \\ 2x_2 + 4x_1^2 \end{pmatrix}.$$

Thus, the 3 by 3 matrix $(g \; ad_fg \; ad_f^2g)$ has its determinant $-x_2^2/x_3^2 \neq 0$. Therefore, the first condition of Theorem 6.9 holds.

To check the second condition, we have to see if the distribution Δ_l spanned only by g and ad_fg due to $n = 3$ is involutive. To do so, let us compute the Lie bracket $[g, ad_fg]$ to find out if either it is linearly dependent on g and/or ad_fg, or the 3 by 3 matrix $H = (g \ ad_fg \ [g, ad_fg])$ has rank$(H) = 2$, i.e., H is singular. Because

$$[g, ad_fg] = \frac{\partial(ad_fg)}{\partial x}g - \frac{\partial g}{\partial x}ad_fg = \begin{pmatrix} 0 & 0 & 0 \\ 0 & 1 & x_2 \\ 0 & x_3 & x_3^2 \\ 2 & 0 & 0 \end{pmatrix}\begin{pmatrix} 0 \\ 0 \\ 1 \end{pmatrix} = \begin{pmatrix} 0 \\ -\frac{x_2}{x_3^2} \\ 0 \end{pmatrix},$$

the matrix H becomes

$$H = \begin{pmatrix} 0 & 0 & 0 \\ 0 & \frac{x_2}{x_3} & -\frac{x_2}{x_3^2} \\ 1 & 2x_1 & 0 \end{pmatrix},$$

which has a zero first row and is obviously singular. Therefore, the second condition of Theorem 6.9 also holds and the system is input-state linearizable.

In order to find the first new state component z_1 as a starting point, let us examine the two vector fields $g = (0 \ 0 \ 1)^T$ and $ad_fg = (0 \ x_2/x_3 \ 2x_1)^T$. Since they both have the zero first element, one may define a scalar function $z_1 = T_1(x)$ such that its gradient has only the first element being non-zero while the last two are all zeros, i.e., $\partial T_1/\partial x = (\times \ 0 \ 0)$ with some non-zero element "$\times$". Such a gradient vector can always be orthogonal to both g and ad_fg, and can be realized by defining the first new state component z_1 as a function of x_1 only. For the sake of simplicity, let $z_1 = x_1$. Then, according to the controllable canonical form (6.38) and the system equation (6.40), $z_2 = \dot{z}_1 = \dot{x}_1 = x_2$, and $z_3 = \dot{z}_2 = \dot{x}_2 = (1 - \ln x_3)x_2$. Finally,

$$\dot{z}_3 = \dot{x}_2(1 - \ln x_3) - \frac{x_2}{x_3}\dot{x}_3 = (1 - \ln x_3)^2 x_2 + 2x_1x_2 - \frac{x_2}{x_3}u.$$

Let the new input v be

$$v = (1 - \ln x_3)^2 x_2 + 2x_1x_2 - \frac{x_2}{x_3}u. \tag{6.41}$$

Then, under the new state $z = (z_1 \ z_2 \ z_3)^T$ and the new input v, the system is linearized to the controllable canonical system. Therefore, after the original nonlinear system is linearized, the old input u can be resolved in terms of the new input v as well as the old state variables x_1 through x_3, i.e.,

$$u = \alpha(x) + \beta(x)v, \tag{6.42}$$

where, for this particular example,

$$\alpha(x) = (1 - \ln x_3)^2 x_3 + 2x_1x_3 \quad \text{and} \quad \beta(x) = -\frac{x_3}{x_2}.$$

Because equation (6.42) provides an algebraic relation between u and v and it does not contain any term with time-derivatives of u or v, equation (6.42) is further called a *static state-feedback control*.

Based on equation (6.39), a more general form for the new input, i.e., the static state-feedback control is given by

$$\alpha(x) = -\frac{L_f^n \mu(x)}{L_g L_f^{n-1} \mu(x)}, \quad \text{and} \quad \beta(x) = \frac{1}{L_g L_f^{n-1} \mu(x)}. \tag{6.43}$$

Therefore, to determine a static state-feedback control for a single-input nonlinear system, it is required to find a certain scalar function $\mu(x)$. Moreover, it can also be observed from the above example that $z_1 = \mu(x)$ is not unique in general. However, one can always choose a desired state such that its gradient vector is orthogonal to all the vector fields in the distribution Δ_l, as illustrated in the above example.

The second example for a single-input nonlinear control system is given by

$$\begin{cases} \dot{x}_1 = x_2^2 + x_2 u \\ \dot{x}_2 = u \\ \dot{x}_3 = x_1. \end{cases}$$

This means that the two vector fields $f(x)$ and $g(x)$ of the system are as follows:

$$f(x) = \begin{pmatrix} x_2^2 \\ 0 \\ x_1 \end{pmatrix}, \quad \text{and} \quad g(x) = \begin{pmatrix} x_2 \\ 1 \\ 0 \end{pmatrix}.$$

By following the criteria in Theorem 6.9, let us first calculate

$$ad_f g = [f, g] = \frac{\partial g}{\partial x} f(x) - \frac{\partial f}{\partial x} g(x) = \begin{pmatrix} -2x_2 \\ 0 \\ -x_2 \end{pmatrix}.$$

Continuing the same procedure, we have

$$ad_f^2 g = [f, [f, g]] = \frac{\partial [f, g]}{\partial x} f(x) - \frac{\partial f}{\partial x} [f, g] = \begin{pmatrix} 0 \\ 0 \\ 2x_2 \end{pmatrix}.$$

By testing the matrix

$$(g \ ad_f g \ ad_f^2 g) = \begin{pmatrix} x_2 & -2x_2 & 0 \\ 1 & 0 & 0 \\ 0 & -x_2 & 2x_2 \end{pmatrix},$$

its determinant is $4x_2^2$, which should not be zero. Therefore, the first criterion is passed as a controllable system.

Since the distribution $\Delta_l = span\{g, ad_f g\}$ needs to be tested to see if it is involutive or not, let us calculate

$$[g, ad_f g] = [g, [f, g]] = \frac{\partial [f, g]}{\partial x} g(x) - \frac{\partial g}{\partial x} [f, g] = \begin{pmatrix} -2 \\ 0 \\ -1 \end{pmatrix}.$$

Thus, the matrix

$$H = (g \ [f,g] \ [g,[f,g]]) = \begin{pmatrix} x_2 & -2x_2 & -2 \\ 1 & 0 & 0 \\ 0 & -x_2 & -1 \end{pmatrix}$$

has a determinant $\det(H) = -(2x_2 - 2x_2) = 0$ so that H is singular and the distribution Δ_l is involutive. Based on the Frobenius Theorem, it is also completely integrable. After the second criterion is passed, let us now find a scalar function $\mu(x)$ to be orthogonal to both g and $ad_f g$, which are the first two columns of the matrix H. It can be seen without difficulty that one of the orthogonal row vectors is $(1 \ -x_2 \ -2)$, which is exactly the gradient vector of the following scalar function:

$$\mu(x) = x_1 - \frac{1}{2}x_2^2 - 2x_3.$$

Now, since

$$L_f\mu(x) = (1 \ -x_2 \ -2)f(x) = x_2^2 - 2x_1, \quad \text{and} \quad L_f^2\mu(x) = (-2 \ 2x_2 \ 0)f(x) = -2x_2^2,$$

$L_f^3\mu(x) = 0$, while $L_gL_f^2\mu(x) = -4x_2$. Thus, let $z_1 = \mu(x) = x_1 - \frac{1}{2}x_2^2 - 2x_3$. Then, the input-state linearized system turns out to be

$$\begin{cases} \dot{z}_1 = z_2 \\ \dot{z}_2 = z_3 \\ \dot{z}_3 = L_f^3\mu(x) + L_gL_f^2\mu(x)u = -4x_2u. \end{cases}$$

In this case, the exactly linearized system has a new input $v = -4x_2u$, which indicates that $\alpha(x) = 0$ and $\beta(x) = -\frac{1}{4x_2}$. If $x_2 \neq 0$, the original input u can be resolved by $u = \alpha(x) + \beta(x)v = -\frac{v}{4x_2}$.

In order to extend the input-state linearization from a single-input case to a multi-input case, where the input has more than one coefficient vectors, i.e., $g_1(x),\cdots,g_r(x)$, one has to examine the involutivity for every sub-distribution $\Delta_{l0} = span\{g_1,\cdots,g_r\}$, $\Delta_{l1} = span\{g_1,\cdots,g_r,ad_fg_1,\cdots,ad_fg_r\}$, $\cdots$, $\Delta_{li} = span\{ad_f^k g_j : 0 \leq k \leq i, 1 \leq j \leq r\}$ [4, 5]. This becomes a very lengthy and tedious task, and could even be impractical. Therefore, beyond the single-input case, the input-state linearization approach is often replaced by a more practical input-output linearization procedure.

6.3.2 INPUT-OUTPUT MAPPING, RELATIVE DEGREES AND SYSTEMS INVERTIBILITY

Starting with this section, we will model and study a *complete nonlinear control system* that contains both input (control) and output vectors in a more global fashion.

The general model for a complete nonlinear autonomous (time-invariant) control system that is affine of the inputs can be written as

$$\begin{cases} \dot{x} = f(x) + \sum_{i=1}^{r} g_i(x)u_i = f(x) + G(x)u \\ y = h(x), \end{cases} \tag{6.44}$$

where $G(x) = (g_1(x) \; \cdots \; g_r(x))$ is an n by r coefficient matrix of the input $u = (u_1 \cdots u_r)^T$. For instance, a complete linear time-invariant control system is a special case of (6.44), where $f(x) = Ax$, $G(x) = B$ and $h(x) = Cx$, and A, B and C are n by n, n by r and m by n constant matrices, respectively. Thus, the complete linear time-invariant control system equation is well-known as

$$\begin{cases} \dot{x} = Ax + Bu \\ y = Cx. \end{cases} \tag{6.45}$$

A complete control system model either in (6.44) or in (6.45) represents a certain but indirect relationship between the input (control) u and the output y through the state vector x, unless $y = h(x, u)$ or $y = Cx + Du$ in linear cases is defined to contain a feedforward portion. However, we do not consider any feedforward part in our discussion here and only treat the input and output relationship to be purely indirect through the state. Such an indirect relationship between u and y is often referred to as an *input-output mapping* [5]. For the linear time-invariant system (6.45), we can take Laplace transformations for both sides of the equations, and obtain the following simultaneous algebraic equations in s-domain:

$$\begin{cases} sX(s) - x(0) = AX(s) + BU(s) \\ Y(s) = CX(s), \end{cases} \tag{6.46}$$

where $X(s)$, $U(s)$ and $Y(s)$ are the Laplace transforms of $x(t)$, $u(t)$ and $y(t)$, respectively. Solving the two s-domain algebraic equations in (6.46) yields

$$Y(s) = C(sI - A)^{-1}x(0) + C(sI - A)^{-1}BU(s). \tag{6.47}$$

This general solution reveals that a system output response always consists of two terms: the first term is a state *self-transition*, or called a *zero-input response* (ZIR) part, and the second term is the *input-output mapping*, or called a *zero-state response* (ZSR) part. If the initial state $x(0) = 0$, the input-output mapping of the complete linear time-invariant system is given by an m by r *transfer matrix* $T(s) = C(sI - A)^{-1}B$ in s-domain. In general, each element of the transfer matrix $T(s)$, called a transfer function, is an s-domain polynomial fraction of the form:

$$\frac{b_0 s^q + b_1 s^{q-1} + \cdots + b_q}{s^p + a_1 s^{p-1} + \cdots + a_p}, \tag{6.48}$$

where the highest power of the polynomial in the denominator, p, can be up to n, the dimension of the state space, while the highest order of the numerator is $q < p$ in general as a *proper transfer function* case, and all the coefficients a_i's and b_j's are real [7, 8]. According to the linear systems theory, the highest-order difference between the denominator and numerator of a transfer function is equal to the difference between the number of poles and number of zeros. We call the difference integer $r_d = p - q$ the **relative degree** (or the *relative order*) of the system.

In general control design cases, we often demand to find a control law, i.e., an input function such that the system output can satisfy a desired performance objective. Intuitively, the control problem can also be interpreted as *inverting* the input-output system. In other words, given a desired output function, the control problem is to determine an input function in terms of the desired output under the system equation as a constraint. Therefore, whether an input-output system is invertible and what is the bottom line of the invertibility become two fundamental questions requiring for a further investigation.

At a qualitative standpoint, for a single-input/single-output (SISO) linear system, its transfer function is often defined as a ratio of the output Laplace transform $Y(s)$ to the input Laplace transform $U(s)$ and is usually in the form given by (6.48). The number of poles p is greater than the number of zeros q, $p > q$, in a *proper system* case. Now, if we want to invert the system, i.e., to find $U(s)$ in terms of $Y(s)$, the new transfer function after the inversion becomes the reciprocal of (6.48). In this case, since the zeros and poles are swapped over, the number of zeros in the inverted system is now greater than the number of poles so that the inverted system becomes improper.

It is also well known from the linear control theory [7, 8] that each pole (if it is a real number) contributes a slope of -20 db/decade to the frequency response of the transfer function magnitude, while each zero contributes $+20$ db/decade to it in a steady-state analysis of the linear system, i.e., $s = j\omega$. An improper transfer function with more zeros than poles will cause energy divergence as frequency goes to infinity in the system frequency-spectrum, such as the Bode plot. Therefore, we need to add more zeros to the original transfer function before inverting the system. What would be the least number of zeros to add on to ensure the inverted system to be critically proper or proper? The answer is now quite obvious that it at least needs $r_d = p - q$ more zeros.

In order to add r_d more zeros into the original system transfer function, the simplest way is to multiply the output function $Y(s)$ by s^{r_d}, which will result in the same numbers for both zeros and poles in the inverted system transfer function $U(s)/[s^{r_d}Y(s)]$. Because the inverse Laplace transform of $s^{r_d}Y(s)$ is equivalent to differentiating $y(t)$ by r_d times if zero initial conditions of the output are assumed, we conclude that to invert an input-output system with a relative degree r_d, the output function must be differentiated up to the r_d-th order.

For a more rigorous explanation, let us visit the SISO linear system given by (6.45) with scalar u and y in time-domain. First, the output equation $y = Cx$ does not explicitly contain the input u so that the current output equation cannot immediately be used for the system inversion. We now take time-derivatives for both sides of the output equation, and then substitute the state equation into it to obtain

$$\dot{y} = C\dot{x} = CAx + CBu.$$

If $CB \neq 0$, then the input u appears in the above equation so that we can invert the system by solving u in terms of y and x. However, if $CB = 0$, we have to continue taking the second-order time-derivative for y, i.e.,

$$\ddot{y} = CA\dot{x} = CA^2x + CABu.$$

If $CAB \neq 0$, we say the system is invertible with $r_d = 2$. Otherwise, we have to keep taking the higher order of time-derivatives until the input u shows up in the equation.

After we reviewed the input-output mapping for linear control systems, we now turn our attention to the SISO complete nonlinear control system given by (6.44) with $r = m = 1$, i.e.,

$$\begin{cases} \dot{x} = f(x) + g(x)u \\ y = h(x). \end{cases} \tag{6.49}$$

Since there are no constant coefficient matrices A, B and C in (6.49), the Laplace transformation is no longer effective. Nevertheless, we can follow the same procedure as presented for the linear system cases in time-domain, and adopt the Lie derivative concept that is defined by

$$L_f h(x) = \frac{\partial h(x)}{\partial x} f(x)$$

as a generalized directional derivative to reach a higher-order differentiation of the output function. For the first-order time-derivative,

$$\dot{y} = \frac{\partial h(x)}{\partial x}\dot{x} = \frac{\partial h(x)}{\partial x}f(x) + \frac{\partial h(x)}{\partial x}g(x)u = L_f h(x) + (L_g h(x))u.$$

Likewise, if $L_g h(x) = 0$, the second-order time-derivative is needed for y and results in

$$\ddot{y} = \frac{\partial \dot{y}}{\partial x}\dot{x} = L_f^2 h(x) + (L_g L_f h(x))u.$$

The differentiation keeps going on until the input u shows up in the equation. Therefore, the relative degree r_d for an SISO nonlinear system can be defined formally as follows:

Definition 6.5. The system (6.49) is said to have a relative degree r_d in some $M \subset \mathbb{R}^n$ if for any $x \in M$,

$$L_g L_f^k h(x) = 0, \quad 0 \leq k < r_d - 1, \quad \text{and}$$

$$L_g L_f^{r_d - 1} h(x) \neq 0.$$

This definition of relative degrees can be extended to an MIMO complete nonlinear system given in (6.44) [5,6]:

Definition 6.6. The system (6.44) is said to have relative degrees (vector) $\{r_1, \cdots, r_m\}$ at a point $x = x_0 \in M$ if

1. $L_{g_j} L_f^k h_i(x) = 0$, for $1 \leq j \leq r$, $1 \leq i \leq m$, $0 \leq k < r_i - 1$ and $x \in U(x_0) \subset M$, and

2. the matrix

$$D(x) = \begin{pmatrix} L_{g_1} L_f^{r_1-1} h_1(x) & \cdots & L_{g_r} L_f^{r_1-1} h_1(x) \\ \vdots & \cdots & \vdots \\ L_{g_1} L_f^{r_m-1} h_m(x) & \cdots & L_{g_r} L_f^{r_m-1} h_m(x) \end{pmatrix}$$

has m independent rows at $x = x_0$.

The matrix $D(x)$ is called a **Decoupling Matrix** [2,5]. If $r = m$, i.e., the input and output vectors have the same dimension, we call such a system a *square input-output system*, or just a *square system*, and the second condition of the above definition can be re-stated that the matrix $D(x)$ is nonsingular at $x = x_0$. Moreover, in the MIMO case, where the system has m output channels y_1 through y_m, and each channel has its individual relative degree r_i for $i = 1, \cdots, m$, the *total relative degree* r_d of the MIMO system is defined by the sum of all the individual r_i's, i.e.,

$$r_d = r_1 + \cdots + r_m.$$

Let us take the following control system as an example to illustrate the determination of relative degree for an SISO system:

$$\begin{cases} \dot{x}_1 = x_2 \\ \dot{x}_2 = -\sin x_2 + u. \end{cases} \tag{6.50}$$

In the first case, let the output function be $y = h(x) = x_1$. Based on Definition 6.5, we calculate

$$L_g h(x) = \frac{\partial h}{\partial x} g(x) = (1 \ 0) \begin{pmatrix} 0 \\ 1 \end{pmatrix} = 0.$$

Then, we need to calculate the second-order Lie derivative $L_g L_f h(x)$. Since

$$L_f h(x) = (1 \ 0) \begin{pmatrix} x_2 \\ -\sin x_2 \end{pmatrix} = x_2,$$

and

$$L_g L_f h(x) = L_g(x_2) = (0 \ 1) \begin{pmatrix} 0 \\ 1 \end{pmatrix} = 1 \neq 0,$$

the relative degree $r_d = 2$ if the output is defined by $y = x_1$. In fact, according to (6.50), $\ddot{x}_1 = \dot{x}_2 = -\sin x_2 + u$. This shows that by taking up to the second time-derivative for the output $y = x_1$, the input u shows up.

In the second case, let the output be changed to $y = h(x) = x_2$. Using the same procedure,

$$L_g h(x) = (0\ \ 1)\begin{pmatrix} 0 \\ 1 \end{pmatrix} = 1 \neq 0,$$

thus, $r_d = 1$. It can also be observed from the second equation of (6.50) that taking only the first time-derivative for $y = x_2$ is sufficient enough to make the input u show up, and it confirms that the relative degree $r_d = 1$. This example tells us an interesting fact that the relative degree for a system also depends on the definition of the output function.

We now look at a following fifth-order MIMO system:

$$\dot{x} = \begin{pmatrix} -x_1^3 + x_2 \\ x_1 x_3 \\ -x_1 + x_3 \\ x_2 \\ x_5 + x_3^2 \end{pmatrix} + \begin{pmatrix} 0 \\ 1 \\ 1 \\ 0 \\ x_2 \end{pmatrix} u_1 + \begin{pmatrix} 0 \\ 1 \\ 0 \\ 0 \\ 0 \end{pmatrix} u_2, \quad \text{and} \quad y = \begin{pmatrix} x_3 \\ x_4 \end{pmatrix}. \tag{6.51}$$

According to Definition 6.6 for an MIMO system, we need to compute the decoupling matrix $D(x)$ in order to determine r_d for (6.51). Since

$$L_{g_1} h_1(x) = \frac{\partial x_3}{\partial x} g_1 = (0\ 0\ 1\ 0\ 0)\begin{pmatrix} 0 \\ 1 \\ 1 \\ 0 \\ x_2 \end{pmatrix} = 1,$$

and

$$L_{g_2} h_1(x) = (0\ 0\ 1\ 0\ 0)\begin{pmatrix} 0 \\ 1 \\ 0 \\ 0 \\ 0 \end{pmatrix} = 0,$$

we complete the first row of $D(x)$, because one of the above two components has been non-zero.

Next, let us calculate

$$L_{g_1} h_2(x) = (0\ 0\ 0\ 1\ 0)\begin{pmatrix} 0 \\ 1 \\ 1 \\ 0 \\ x_2 \end{pmatrix} = 0,$$

and

$$L_{g_2}h_2(x) = (0\ 0\ 0\ 1\ 0)\begin{pmatrix} 0 \\ 1 \\ 0 \\ 0 \\ 0 \end{pmatrix} = 0.$$

Since both are zeros, we have to continue calculating

$$L_f h_2(x) = (0\ 0\ 0\ 1\ 0)f = x_2,$$

$$L_{g_1}L_f h_2(x) = (0\ 1\ 0\ 0\ 0)\begin{pmatrix} 0 \\ 1 \\ 1 \\ 0 \\ x_2 \end{pmatrix} = 1,$$

and

$$L_{g_2}L_f h_2(x) = (0\ 1\ 0\ 0\ 0)\begin{pmatrix} 0 \\ 1 \\ 0 \\ 0 \\ 0 \end{pmatrix} = 1.$$

Thus, the decoupling matrix turns out to be

$$D(x) = \begin{pmatrix} 1 & 0 \\ 1 & 1 \end{pmatrix}$$

which is always non-singular. Therefore, we conclude that $r_1 = 1$ for the first channel of the output $y = h(x)$, and $r_2 = 2$ for the second channel of the output. The total relative degree of the fifth-order MIMO system (6.51) becomes $r_d = r_1 + r_2 = 3$, which is less than the system dimension $n = 5$.

6.3.3 SYSTEMS INVERTIBILITY AND APPLICATIONS

With the definitions of relative degree, we now formally state that an input-output system is *invertible* if its relative degree $r_d \leq n$, the dimension of the state space. Note that in this *invertibility* definition, if the system is MIMO, then r_d should be the total relative degree, i.e., $r_d = r_1 + \cdots + r_m \leq n$.

In contrast, if $r_d > n$, the input-output system is *non-invertible* [4, 5]. Therefore, to determine the invertibility for a complete system, one needs only to calculate the (total) relative degree r_d and compare it with n, the dimension of the state space.

The physical meaning of the system invertibility can be viewed as a *sufficiency of output channels*. In other words, if a system has sufficient output channels, through which all the input variables can be seen through the internal state variables, then this system must be invertible, or say, the inputs can all be determined in terms of the outputs and states.

The previous two examples in the last subsection are both invertible. More specifically, the first example includes two cases of the output definitions: one has $r_d = 2 = n$, and the other one has $r_d = 1 < n = 2$. The second example is an MIMO system with $n = 5$. Since the total relative degree $r_d = r_1 + r_2 = 1 + 2 = 3 < n = 5$, it is also invertible.

In order to gain more physical insight into the relative degree and invertibility, let us take a 6-joint fully-actuated open serial-chain industrial robot as a realistic nonlinear MIMO system to test its invertibility. Note that the adjective "fully-actuated" means that each joint of the robot is driven by a motor. In other words, the joint torque vector τ of such a fully-actuated robot has the same dimension as the joint position vector q. Let us now derive the decoupling matrix $D(x)$ and determine its total relative degree r_d.

For the 6-joint robot dynamic system, $x = \begin{pmatrix} q \\ \dot{q} \end{pmatrix} \in \mathbb{R}^{12}$ is the state vector that augments the joint position vector q and joint velocity vector $\dot{q}$ together. According to the robot dynamic formulation studied in Chapter 3, a general state-space equation (6.44) can be expressed as

$$f(x) = \begin{pmatrix} \dot{q} \\ W^{-1}((\frac{1}{2}W_d - W_d^T)\dot{q} + \tau_g) \end{pmatrix} \quad \text{and} \quad G(x) = \begin{pmatrix} O \\ W^{-1} \end{pmatrix}, \tag{6.52}$$

where O is the 6 by 6 zero matrix.

Let the 6 d.o.f. robot end-effector motion vector (three for position and three for orientation) in a 3D Cartesian space be defined as an output function $y = h(x) \in \mathbb{R}^6$ if it could be available. Such an output definition $y = h(x)$, in most robotic application cases, is a function of the joint position q only, and is independent of $\dot{q}$ so that $y = h(x) = h(q)$ for most open serial-chain robotic systems.

It is also presumable that the six output channels, $h_1(q)$ through $h_6(q)$, represent all the six degrees of freedom motion, and each channel should have an "equal opportunity" in motion that is controlled by all the six joint torques. Therefore, it is reasonable to assume that each output channel for the 6-joint fully-actuated robot manipulator has a common relative degree, i.e., $r_1 = \cdots = r_6 = r_{cd}$. This is referred to as an *equal-RD (relative-degree) hypothesis*.

Under the equal-RD hypothesis, every row in the targeting decoupling matrix $D(x)$ will arrive at the non-zero status concurrently. Therefore, we can compute the common relative degree r_{cd} through a more compact and global derivation, instead of repeating six times to compute each individual relative degree r_i for each output channel.

To do so, let us first compute $L_{g_1}h(q)$ through $L_{g_6}h(q)$, which can actually be augmented to form a 6 by 6 matrix and denoted by $L_Gh(q)$ for a notation convenience, where $G(x) = (g_1(x) \cdots g_6(x))$ is the 12 by 6 coefficient matrix of the input u in equations (6.44) and (6.52). In fact,

$$L_Gh(q) = \frac{\partial h(q)}{\partial x}G(x) = \left(\frac{\partial h(q)}{\partial q} \quad \frac{\partial h(q)}{\partial \dot{q}}\right)G(x) = (J \ O)\begin{pmatrix} O \\ W^{-1} \end{pmatrix} = O,$$

where we have defined $\partial h(q)/\partial q = J$, the 6 by 6 kinematic Jacobian matrix of the robot, and $\partial h(q)/\partial \dot{q} = O$, the 6 by 6 zero matrix. This zero result suggests that the common relative degree r_{cd} for each output channel be greater than one according to Definition 6.6 for MIMO systems in the last subsection.

We now proceed to the second-order differentiation, and further calculate that $L_f h(q) = (J\ O)f(x) = J\dot{q}$. Thus,

$$L_G L_f h(q) = \left(\frac{\partial J\dot{q}}{\partial q} \; \frac{\partial J\dot{q}}{\partial \dot{q}} \right) \begin{pmatrix} O \\ W^{-1} \end{pmatrix} = JW^{-1},$$

because $\partial J\dot{q}/\partial \dot{q} = J$. Since the inertial matrix W of the robot is always non-singular, the above result clearly exhibits that if the Jacobian matrix J is also non-singular, then the decoupling matrix

$$D(x) = L_G L_f h(q) = JW^{-1}$$

can be non-singular, too, so that the common relative degree $r_{cd} = 2$. In this case, the total relative degree $r_d = 6 r_{cd} = 12$ that is exactly equal to the dimension of the state space.

Therefore, we conclude that a 6-joint fully-actuated industrial robot with the six-dimensional Cartesian position plus orientation to form an output vector $h(q)$ is invertible if the 6 by 6 Jacobian matrix $J = \partial h(q)/\partial q$ does not fall into singularity. In other words, only at a singular point of J, the fully-actuated robotic system will destroy its invertibility.

It should be pointed out that the above justification is under the assumption of forming all 6 d.o.f. with a single 6 by 1 Cartesian position/orientation vector to be ready to take time-derivatives. As we have already investigated in Chapter 2, to find an unified 6 by 1 vector and its time-derivative to uniquely represent the 6 d.o.f. motion: three for position/translation and three for orientation/rotation is almost impossible. Therefore, unless we use the isometric embedding dynamics modeling approach developed in Chapter 4, defining a 6-dimensional output function $y = h(q)$ in the above 6-joint industrial robot example may just be a conceptual model.

The above real robotic system example also reveals a fact that to invert a robot dynamic system, or equivalently, to design a control law for the robotic system to meet a desired output objective, it is required to compute up to the second-order time-derivative for an output function. In other words, we need *acceleration* information on the output in order to resolve the global control design problem for a robot dynamic system.

6.3.4   THE INPUT-OUTPUT LINEARIZATION PROCEDURE

The main motivation for developing this **input-output linearization** procedure is to take advantage of an existing output function that is often specified by the physical requirements of a given nonlinear system to possibly generate a nonlinear transformation for the linearization. Clearly, this provides us with the following two major advantages: (1) to avoid the most difficult steps of seeking some scalar functions as required by the input-state linearization procedure, especially for a multi-input nonlinear system; and (2) the new state of the linearized system can immediately be employed to represent any physical task required by the original system.

As was discussed in the last section, an MIMO complete nonlinear autonomous (time-invariant) control system can be generally expressed by

$$\begin{cases} \dot{x} = f(x) + \sum_{i=1}^{r} g_i(x) u_i \\ y = h(x). \end{cases} \tag{6.53}$$

If the dimension m of the output function $h(x)$ is equal to the input dimension r, the system in (6.53) is called a *square nonlinear system*. For such a square system, the decoupling matrix $D(x)$, as formulated in Definition 6.6, is an m by m square matrix ($m = r$). If $D(x)$ at a point of interest is non-singular, then the total relative degree r_d of the square system should be either less than or equal to n, the dimension of the state vector x.

Definition 6.7. A complete nonlinear control system (6.53) is said to be input-output linearizable if its input-output mapping is invertible and the total relative degree $r_d = n$, the dimension of the state space.

Under this definition, let us derive an input-output relation that is required for the linearization by taking the time-derivative for each component of the output $y = h(x)$. Since for the j-th component $y_j = h_j(x)$, $\dot{y}_j = (\partial h_j(x)/\partial x)\dot{x}$, substituting the first equation of (6.53) yields

$$\dot{y}_j = \frac{\partial h_j(x)}{\partial x}\left[f(x) + \sum_{i=1}^{m} g_i(x) u_i \right] = L_f h_j(x) + \sum_{i=1}^{m}(L_{g_i} h_j(x)) u_i.$$

Continuing to take higher-order time-derivatives until reaching the r_j-th order, where r_j is the relative degree of the j-th output channel $h_j(x)$, we obtain

$$y_j^{(r_j)} = L_f^{r_j} h_j(x) + \sum_{i=1}^{m}(L_{g_i} L_f^{r_j-1} h_j(x)) u_i.$$

Augmenting all the output components together, one can finally express

$$\begin{pmatrix} y_1^{(r_1)} \\ \vdots \\ y_m^{(r_m)} \end{pmatrix} = \begin{pmatrix} L_f^{r_1} h_1(x) \\ \vdots \\ L_f^{r_m} h_m(x) \end{pmatrix} + \begin{pmatrix} L_{g_1} L_f^{r_1-1} h_1(x) & \cdots & L_{g_r} L_f^{r_1-1} h_1(x) \\ \vdots & \cdots & \vdots \\ L_{g_1} L_f^{r_m-1} h_m(x) & \cdots & L_{g_r} L_f^{r_m-1} h_m(x) \end{pmatrix} \begin{pmatrix} u_1 \\ \vdots \\ u_m \end{pmatrix}$$

$$= b(x) + D(x)u,$$

where $b(x) = (L_f^{r_1} h_1(x) \cdots L_f^{r_m} h_m(x))^T$ is an m by 1 vector, $D(x)$ is just the m by m decoupling matrix ($r = m$), and $u = (u_1 \cdots u_m)^T$ is the input vector of the square system. By defining a new input $v = (y_1^{(r_1)} \cdots y_m^{(r_m)})^T$, we arrive at the desired destination:

$$v = b(x) + D(x)u, \tag{6.54}$$

where the old input u can be resolved in terms of the new one v through the internal states by

$$u = \alpha(x) + \beta(x)v = -D^{-1}(x)b(x) + D^{-1}(x)v. \tag{6.55}$$

This form is identical to the static state-feedback control law given in (6.42) except that both u and v are now two vectors. Because the above control law requires inverting $D(x)$, the nonlinear system should be at least invertible before it is possibly input-output linearizable.

With the new input v defined, we now try to find a new state vector. Since the input or output dimension m is, in general, less than (at most equal to) the state dimension n, and the input-output linearization also requires the total relative degree $r_d = r_1 + \cdots + r_m = n$, we may divide the n promising new state variables z_1 through z_n into m groups, with the i-th group having r_i members ($i = 1, \ldots, m$). Each member of the i-th group is defined by the i-th output component $y_i = h_i(x)$ and its time-derivatives.

In order to more explicitly illustrate the definition of z, without loss of generality, let us assume that $m = 2$, $n = 5$, $r_1 = 2$ and $r_2 = 3$. Then, defining $z_1 = y_1$, $z_2 = \dot{y}_1$, $z_3 = y_2$, $z_4 = \dot{y}_2$ and $z_5 = \ddot{y}_2$, and noticing that the proper definition of the new input v in this case is $v = (y_1^{(r_1)} \; y_2^{(r_2)})^T = (y_1^{(2)} \; y_2^{(3)})^T$, the linearized system becomes

$$\dot{z} = Az + Bv, \tag{6.56}$$

where $z = (z_1 \; \cdots \; z_5)^T$ and

$$A = \begin{pmatrix} 0 & 1 & 0 & 0 & 0 \\ 0 & 0 & 0 & 0 & 0 \\ 0 & 0 & 0 & 1 & 0 \\ 0 & 0 & 0 & 0 & 1 \\ 0 & 0 & 0 & 0 & 0 \end{pmatrix}, \quad \text{and} \quad B = \begin{pmatrix} 0 & 0 \\ 1 & 0 \\ 0 & 0 \\ 0 & 0 \\ 0 & 1 \end{pmatrix}.$$

It can be observed from the above definition of z that the total number of the new state variables is equal to r_d, which is equal to the old state space dimension. This is the main reason why the input-output linearization should also require $r_d = n$.

Because an output function and its time-derivatives can be chosen to form a set of new state variables in an input-output linearizable system, any physical task-planning for a real system can directly be represented in terms of the new state variables. A typical application is to carry out a *trajectory-tracking* task for a dynamic system, such as for an industrial robot manipulator to track a desired path. For instance, the above input-output linearizable example was assumed to have $m = 2$, $n = 5$, $r_1 = 2$ and $r_2 = 3$. If a desired output trajectory $y_d(t) = (y_{1d}(t) \; y_{2d}(t))^T$ is specified for the actual system output to follow, then the following PD control law for the new input v can guarantee the trajectory-tracking convergence:

$$v = \begin{pmatrix} \ddot{y}_{1d} + k_{1v}\Delta\dot{y}_1 + k_{1p}\Delta y_1 \\ y_{2d}^{(3)} + k_{2a}\Delta\ddot{y}_2 + k_{2v}\Delta\dot{y}_2 + k_{2p}\Delta y_2 \end{pmatrix}, \tag{6.57}$$

where, and hereafter $y^{(3)}$ is the third time-derivative of $y(t)$, $\Delta y_1 = y_{1d} - y_1$ and $\Delta y_2 = y_{2d} - y_2$ are the two output error components between the desired and actual output trajectories, and k_{1v}, k_{1p}, k_{2a}, k_{2v} and k_{2p} are positive gain constants. In fact, since $v = (\ddot{y}_1 \ y_2^{(3)})^T$, substituting it into equation (6.57), with some manipulation, we obtain

$$\begin{pmatrix} \Delta \ddot{y}_1 + k_{1v}\Delta \dot{y}_1 + k_{1p}\Delta y_1 \\ \Delta y_2^{(3)} + k_{2a}\Delta \ddot{y}_2 + k_{2v}\Delta \dot{y}_2 + k_{2p}\Delta y_2 \end{pmatrix} = 0. \tag{6.58}$$

This clearly demonstrates that the output error vector Δy can asymptotically converge to zero if the gain constants for each component of the vector in (6.58) can make each equation be Hurwitz, i.e., all the roots of their characteristic equations have strictly negative real parts.

To illustrate the input-output linearization procedure, let us look at the following 3-dimensional nonlinear control system:

$$\begin{cases} \dot{x}_1 = x_2 \\ \dot{x}_2 = -x_1 + x_3^2 + u_1 + 3u_2 \\ \dot{x}_3 = -x_1^3 + 2u_1 + 4u_2. \end{cases} \tag{6.59}$$

If an output vector is defined by $y = h(x) = \begin{pmatrix} x_1 \\ x_2 \end{pmatrix}$, then we follow the procedure to find its relative degree. In this case, the input coefficients are $g_1 = (0 \ 1 \ 2)^T$ and $g_2 = (0 \ 3 \ 4)^T$, and the first-order Lie derivatives become

$$L_{g_1}h_1(x) = \frac{\partial h_1}{\partial x}g_1(x) = (1 \ 0 \ 0)g_1(x) = 0,$$

$$L_{g_2}h_1(x) = \frac{\partial h_1}{\partial x}g_2(x) = (1 \ 0 \ 0)g_2(x) = 0,$$

$$L_{g_1}h_2(x) = \frac{\partial h_2}{\partial x}g_1(x) = (0 \ 1 \ 0)g_1(x) = 1,$$

$$L_{g_2}h_2(x) = \frac{\partial h_2}{\partial x}g_2(x) = (0 \ 1 \ 0)g_2(x) = 3.$$

Because the first two Lie derivatives for $h_1(x)$ are all zeros and the last two for $h_2(x)$ are non-zero, we should stop differentiating $h_2(x)$ while continue to take higher-order Lie derivatives for $h_1(x)$. Since

$$L_f h_1(x) = \frac{\partial h_1}{\partial x}f(x) = (1 \ 0 \ 0)f(x) = x_2,$$

$$L_{g_1}L_f h_1(x) = \frac{\partial L_f h_1}{\partial x}g_1(x) = (0 \ 1 \ 0)g_1(x) = 1,$$

$$L_{g_2}L_f h_1(x) = \frac{\partial L_f h_1}{\partial x}g_2(x) = (0 \ 1 \ 0)g_2(x) = 3,$$

both of which are non-zero, but the decoupling matrix

$$D(x) = \begin{pmatrix} 1 & 3 \\ 1 & 3 \end{pmatrix}$$

is singular. Therefore, the system with $y = \begin{pmatrix} x_1 \\ x_2 \end{pmatrix}$ is not invertible, nor is it input-output linearizable.

If the output vector is now redefined to be $y = h(x) = \begin{pmatrix} x_1 \\ x_3 \end{pmatrix}$, then we can repeat the above derivations again and show that the relative degrees $r_1 = 2$ and $r_2 = 1$, and the decoupling matrix

$$D(x) = \begin{pmatrix} 1 & 3 \\ 2 & 4 \end{pmatrix}$$

is obviously non-singular. Therefore, with the output redefined, due to $r_d = r_1 + r_2 = 3 = n$, the system is input-output linearizable.

It is time now to define the new state variables $z_1 = h_1(x) = x_1$, $z_2 = \dot{x}_1$ and $z_3 = h_2(x) = x_3$, and define a new input $v = (\ddot{y}_1 \; \dot{y}_2)^T$. Based on equation (6.55), the old input u can be resolved by calculating the vector $b(x)$ and the decoupling matrix inverse $D^{-1}(x)$. Since $b(x) = (L_f^2 h_1(x) \; L_f h_2(x))^T = (-x_1 + x_3^2 \; -x_1^3)^T$, we obtain

$$\alpha(x) = -D^{-1}(x)b(x) = \begin{pmatrix} -2x_1 + 2x_3^2 + \frac{3}{2}x_1^3 \\ x_1 - x_3^2 - \frac{1}{2}x_1^3 \end{pmatrix}$$

and

$$\beta(x) = D^{-1}(x) = \begin{pmatrix} -2 & \frac{3}{2} \\ 1 & -\frac{1}{2} \end{pmatrix}.$$

Finally, the static state-feedback control for the system given in (6.59) with $y = \begin{pmatrix} x_1 \\ x_3 \end{pmatrix}$ is determined by $u = \alpha(x) + \beta(x)v$.

As a second realistic example, let us revisit the fully-actuated 6-joint robot arm. It has been shown in the previous section that the robotic system along with the end-effector Cartesian position and orientation as a 6-dimensional output function is invertible with the total relative degree $r_d = 12 = n$ if the Jacobian matrix J is non-singular, and thus, it is input-output linearizable. Although we have also derived its decoupling matrix $D(x) = JW^{-1}$ under the equal-RD assumption, the vector $b(x) = L_f^2 h(q)$ is quite difficult to calculate symbolically. Because a fully-actuated open serial-chain robot arm is always a typical mechanical system with the relative degree $r_{cd} = 2$ for each output channel, we may alternatively develop its static state-feedback control law from its dynamic equation in a more compact manner.

Since $y = h(q)$ and $\dot{y} = J\dot{q}$, one can find $\ddot{y} = J\ddot{q} + \dot{J}\dot{q}$ so that $\ddot{q} = J^{-1}\ddot{y} - J^{-1}\dot{J}\dot{q}$. On the other hand, the dynamic equation of the robotic system can be expressed as

$$W\ddot{q} + c(q,\dot{q}) - \tau_g = \tau.$$

Substituting the above $\ddot{q}$ into the robot dynamic equation, we obtain

$$WJ^{-1}\ddot{y} - WJ^{-1}\dot{J}\dot{q} + c(q,\dot{q}) - \tau_g = \tau.$$

If we define the old input $u = \tau$ and the new input $v = \ddot{y}$, then the above equation becomes

$$u = WJ^{-1}v - WJ^{-1}\dot{J}\dot{q} + c(q,\dot{q}) - \tau_g.$$

Comparing this with equation (6.55), we conclude that

$$\alpha(x) = -WJ^{-1}\dot{J}\dot{q} + c(q,\dot{q}) - \tau_g, \quad \text{and} \quad \beta(x) = WJ^{-1}. \tag{6.60}$$

This straightforward and compact derivation showcases that instead of the required computation for both $D(x)$ and $b(x)$, we can determine a static state-feedback control law directly from the robot dynamic equation.

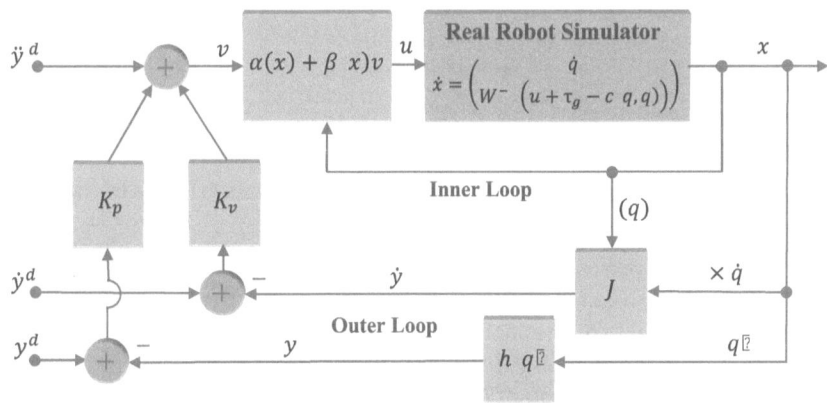

Figure 6.3 A block diagram for an input-output linearized trajectory-tracking system

It can also be seen that the new state vector, after the input-output linearization is applied to the robotic system, is $z = \begin{pmatrix} y \\ \dot{y} \end{pmatrix}$. If a trajectory-tracking task is planned for the robot, then we can define the following PD control law for the linearized system:

$$v = \ddot{y}^d + K_v\Delta\dot{y} + K_p\Delta y, \tag{6.61}$$

where $y^d = y^d(t)$ represents a desired output trajectory, $\Delta y = y^d - y$ is the output error vector between the desired and actual output vectors, and K_v and K_p are two positive-definite diagonal constant gain matrices. It is quite interesting that the entire robotic control system, as shown in Figure 6.3, contains an *inner nonlinear feedback loop* representing the static state-feedback control law (6.55) and an *outer linear feedback loop* that represents the PD control law (6.61).

6.4 ISOMETRIC EMBEDDING DYNAMIC MODEL AND CONTROL

If an isometric embedding dynamic model is adopted for a robotic system, the control law design will almost be a linearization-free process. As we developed in Chapter 4, the isometric embedding model for either an open serial-chain or a closed hybrid chain robotic system looks in appearance like the Newton's Second Law, which has already been a controllable canonical form.

If the robot is of open serial-chain, like the 6-joint industrial robot example in Figure 4.3, the dynamic equation is

$$u = \tau = J^T \ddot{z} - \tau_g, \tag{6.62}$$

where $z = \zeta(q)$ is an isometric embedding as a function of the joint vector q, and J is a genuine Jacobian matrix of the embedding, $J = \frac{\partial \zeta}{\partial q}$. If we define two state vectors $x_1 = z$ and $x_2 = \dot{z}$, then,

$$\begin{cases} \dot{x}_1 = x_2 \\ \dot{x}_2 = J^{T+} \tau_g + J^{T+} u, \end{cases}$$

where $J^{T+} = J(J^T J)^{-1}$ is the pseudo-inverse of the Jacobian matrix J^T. The real input obviously follows equation (6.62) with $\alpha(x) = -\tau_g$ and $\beta(x) = J^T$. The new input $v = \ddot{z}$ can be directly specified as a PD control for the trajectory-tracking task planning:

$$v = \ddot{z}^d + K_v(\dot{z}^d - \dot{z}) + K_p(z^d - z) \tag{6.63}$$

to track a desired Cartesian path specified by $z^d(t)$.

As described in Table 5.2, if the robot is a closed parallel-chain mechanism, like the Stewart platform in Figures 5.1 and 5.2 from Chapter 5, the dynamic equation should obey the dual counterpart of (6.62). Namely,

$$\ddot{z} = J(\tau + \tau_g) = Ju + J\tau_g \quad \text{so that} \quad u = J^+ \ddot{z} - \tau_g, \tag{6.64}$$

where $J^+ = (J^T J)^{-1} J^T$ is the pseudo-inverse of the Jacobian matrix J. In this case, $\alpha(x) = -\tau_g$ and $\beta(x) = J^+$, and the new input $v = \ddot{z}$ can also be defined as (6.63) for the trajectory-tracking control purpose. The state-space equation becomes

$$\begin{cases} \dot{x}_1 = x_2 \\ \dot{x}_2 = J\tau_g + Ju, \end{cases} \tag{6.65}$$

which has already been in a linearized controllable canonical form.

A simulation study with 3D animation for the Stewart platform has been performed, as shown in Figure 6.4. The detailed procedure to implement the dynamic equation and control scheme is outlined in the following algorithm:

Algorithm 6.1. – Simulation Study on Dynamic Control of a Stewart Platform

1. Given an initial position $p_0^c(0)$ and an initial orientation $R_0^p(0)$ of the mass center of the top dominant panel for the Stewart platform;
2. Given an initial linear velocity $v_0^c(0)$ and an initial angular velocity $\omega_0(0)$ of the top dominant panel;
3. Specify a desired trajectory $p_0^c(t)$ and $R_0^p(t)$ $(t > 0)$ for the top dominant panel;
4. Provide every required parameter.

Then,

Figure 6.4 Simulation study on dynamic control of Stewart Platform: the initial posture on left and the final posture on right

1. Construct an isometric embedding z and its time-derivative $\dot{z}$ based on equations (5.5) and (5.6), where r_x, r_y and r_z are the three columns of $R_0^p = (r_x\ r_y\ r_z)$, m is the top panel mass, and the coefficients β_1, β_2 and β_3 are related to the moments of inertial for the top panel by equations (4.6) and (4.7). The linear velocity follows $v_0^c = \dot{p}_0^c$, while the time-derivation of R_0^p follows equation (2.52) so that $\dot{R}_0^p = \Omega_0 R_0^p$, where Ω_0 is a skew-symmetric matrix representing the cross-product operator of the angular velocity ω_0, i.e., $\Omega_0 = \omega_0\times$.
2. Construct the Jacobian matrix J by exactly following equation (5.7), where the unit vector l_i^o of each prismatic piston leg for $i = 1,\cdots,6$ is determined by the inverse kinematics (IK) of the Stewart platform in equation (5.1).
3. Following the same formation of the above isometric embedding, build a desired vector $z^d(t)$ and its time-derivative vector $\dot{z}^d$ as well as the second derivative $\ddot{z}^d$ in terms of the given desired trajectory.
4. The new input $v = \ddot{z}$ is defined based on the PD control equation (6.63), while the original control input u is determined by equation (6.64).
5. The real-plant simulator relies on the state-space equation given in (6.65) with the input u determined in Step 4.
6. The joint torque τ_g due to gravity can be determined using equation (5.14). Since the z component of the mass center position vector p_0^c is the height h with respect to the base, the gravitational potential energy P_g can be found as

$$P_g = mgh = \frac{mg}{6}\sum_{i=1}^{6} d_i l_i^o(3) \quad \text{so that} \quad \tau_g = -\frac{\partial P_g}{\partial q} = -\frac{mg}{6}\begin{pmatrix} l_1^o(3) \\ \vdots \\ l_6^o(3) \end{pmatrix},$$

where each element of the last column vector is the z component of the unit directional vector l_i^o for piston leg i. Although each l_i^o may also be a function of every piston displacement d_j for $j = 1,\cdots,6$, the above equation for τ_g is just an estimation.

The simulation study with 3D graphical animation has demonstrated an expected good result. The left picture of Figure 6.4 is an initial posture of the system and the right one exhibits reaching the final destination. The top panel as a dominant body has followed the desired trajectory specified in terms of both position of the mass center and orientation of the panel. However, the tracking errors for both the position and orientation are not converging to zero exactly. The major reason is due to the pseudo-inverse $J^+ = (J^T J)^{-1} J^T$ of the 12 by 6 tall Jacobian matrix J in the control law equation (6.64). In this case, when the control input u is fed back to the system equation (6.65), $J(u + \tau_g) = J J^+ v = J(J^T J)^{-1} J^T v \neq v$. If the Cartesian position and orientation could be unified to a 6 by 1 vector for the isometric embedding z, the error would automatically vanish. However in the reality, it would never happen, and we have to pay cost to define a longer embedding vector z to represent the 6 d.o.f. Cartesian motion.

While the isometric embedding model offers an exclusive linearization-free advantage for the trajectory-tracking control applications, it must maintain r_x, r_y and r_z as the last nine components of the isometric embedding z to be the three columns of a rotation matrix during the updating process at each sampling point. Violating this $SO(3)$ group membership could cause an additional error, divergence, or possible distortion of the graphical animation.

Recalling the simulation study for the industrial robot carrying a heavy load in Figures 4.3 and 4.5. Since the 6-joint industrial robot is an open serial-chain system, we can directly use the compact dynamic model $\tau = J^T \ddot{z} - \tau_g$ with the isometric embedding $z = \zeta(q)$ as an explicit function of the joint variables in q. Although inside the embedding $z = \zeta(q)$, there are also three columns augmented from a rotation matrix, they are all the functions of q that can always keep forming a special orthogonal rotation matrix. In contrast, for the Stewart platform in such a closed parallel-chain mechanism case, its dynamics must be formulated by (6.64) as a dual counterpart of the dynamic model for the 6-joint industrial robot. Thus, the function arguments for the isometric embedding $z = \zeta(\cdot)$ are now the Cartesian position p_0^c and orientation R_0^p, instead of the joint variables q. Therefore, keeping the integrity of rotation matrix is no longer automatic.

6.5 LINEARIZABLE SUBSYSTEMS AND INTERNAL DYNAMICS

As we have just discussed, for a complete nonlinear control system, the invertibility and the total relative degree $r_d = n$ imply that the system is input-output linearizable. However, in many practical cases, a system may be just invertible, but $r_d < n$. This situation, though excluded by the definition of input-output linearizability, can still be linearized *partially*. In other words, an invertible system with $r_d < n$ can be decomposed into two portions: one is an input-output linearizable subsystem of dimension r_d, and the other one is an unlinearizable subsystem of dimension $n - r_d$, which is conventionally called an **internal dynamics** [2, 5, 6].

Recalling the fifth-order nonlinear control system given in (6.51), let us repeat the equation here:

$$\dot{x} = \begin{pmatrix} -x_1^3 + x_2 \\ x_1 x_3 \\ -x_1 + x_3 \\ x_2 \\ x_5 + x_3^2 \end{pmatrix} + \begin{pmatrix} 0 \\ 1 \\ 1 \\ 0 \\ x_2 \end{pmatrix} u_1 + \begin{pmatrix} 0 \\ 1 \\ 0 \\ 0 \\ 0 \end{pmatrix} u_2, \quad \text{and} \quad y = \begin{pmatrix} x_3 \\ x_4 \end{pmatrix}. \tag{6.66}$$

As we have calculated earlier, the relative degrees $r_1 = 1$, $r_2 = 2$ and the total relative degree $r_d = 3 < n$. Therefore, the entire system can be decomposed into a 3-dimensional linearizable subsystem and a 2-dimensional internal dynamics. For the linearizable one, the new state variables are $z_1 = y_1 = x_3$, $z_2 = y_2 = x_4$, and $z_3 = \dot{y}_2 = \dot{x}_4 = x_2$ that is based on the fourth equation of (6.66). Therefore, with $z = (z_1 \; z_2 \; z_3)^T = (x_3 \; x_4 \; x_2)^T$, the 3-dimensional linearized subsystem can be described by $\dot{z}_1 = v_1$, $\dot{z}_2 = z_3$ and $\dot{z}_3 = v_2$, or

$$\dot{z} = \begin{pmatrix} 0 & 0 & 0 \\ 0 & 0 & 1 \\ 0 & 0 & 0 \end{pmatrix} z + \begin{pmatrix} 1 & 0 \\ 0 & 0 \\ 0 & 1 \end{pmatrix} \begin{pmatrix} v_1 \\ v_2 \end{pmatrix}, \tag{6.67}$$

where the new input

$$
\begin{aligned}
v = \begin{pmatrix} v_1 \\ v_2 \end{pmatrix} = b(x) + D(x)u &= \begin{pmatrix} L_f h_1(x) \\ L_f^2 h_2(x) \end{pmatrix} + D(x)u \\
&= \begin{pmatrix} -x_1 + x_3 \\ x_1 x_3 \end{pmatrix} + \begin{pmatrix} 1 & 0 \\ 1 & 1 \end{pmatrix} \begin{pmatrix} u_1 \\ u_2 \end{pmatrix}.
\end{aligned}
\tag{6.68}
$$

By taking aside the second, third and fourth rows of equation (6.66), which correspond to the new states of the linearized subsystem, the remaining two rows are supposed to represent the internal dynamics. If we denote $\zeta_1 = x_1$ and $\zeta_2 = x_5$, then the first and the fifth rows in (6.66) can be rewritten by

$$\begin{cases} \dot{\zeta}_1 = -\zeta_1^3 + z_3 \\ \dot{\zeta}_2 = \zeta_2 + z_1^2 + z_3 u_1, \end{cases} \tag{6.69}$$

where u_1 can be solved using the first row of (6.68), i.e., $u_1 = \zeta_1 - z_1 + v_1$.

In general, for an n-dimensional complete nonlinear control system with m-dimensional input and output, if it is invertible and the total relative degree $r_d < n$, then the new state vector z for the input-output linearizable subsystem is r_d-dimensional, while the new state vector ζ for the internal dynamics is $(n - r_d)$-dimensional. The entire control system sitting in the new coordinate system can be expressed as

$$\begin{cases} \dot{z} = Az + Bv \\ \dot{\zeta} = \phi(\zeta, z) + \psi(\zeta, z)v, \end{cases} \tag{6.70}$$

where A and B are r_d by r_d and r_d by m constant matrices, respectively, in a controllable canonical form, and $\phi(\cdot,\cdot)$ and $\psi(\cdot,\cdot)$ are $n-r_d$ by 1 and $n-r_d$ by m vector fields representing the unlinearizable subsystem, i.e., the internal dynamics. Equation (6.70) is called a **normal form** of the control system [2, 5, 6]. The m-dimensional original input u and the new input v are related by either $v = b(x) + D(x)u$, or $u = \alpha(x) + \beta(x)v$ that is still called a *static state-feedback control*. The block diagram that represents this decomposition and the state-feedback control is shown in Figure 6.5.

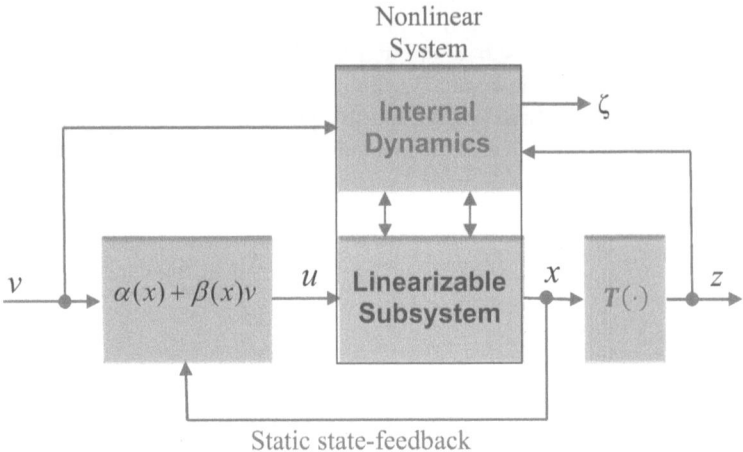

Figure 6.5 A block diagram for a partially input-output linearizable system

It should be noted that the above discussion presumes *square* nonlinear systems, i.e., the input u and output y have the same dimension so that the new input v is also of the same dimension. The main reason for the equal dimensions between u and v is that the decoupling matrix is a square matrix and can be uniquely inverted to resolve the original input u. If a practical system does not meet the same dimensional condition between its input and output, one may increase or decrease the number of output variables to match the number of inputs. This is a reasonable method in most applications, because a certain number of independent inputs, intuitively, are supposed to produce the same number of output responses.

In fact, the category of under-actuated robotic systems, such as a space robot or a floating base robot, inherently has a non-trivial internal dynamics. For example, if a 3-joint robot arm is mounted on a floating base that can free move in 3 d.o.f.: two translational axes and one rotational axis. Then, such a floating base robotic system can be modeled as a 6-joint robot with respect to the fixed base on the earth. The total dimension of the state space is $2 \times 6 = 12$ due to $x = \begin{pmatrix} q \\ \dot{q} \end{pmatrix}$. In this case, only three joints of the robot arm sitting on the 3-axis floating base can be driven by their motors so that the dimension of the input is three, i.e., $u = (\tau_4 \ \tau_5 \ \tau_6)^T$.

If an output is defined to represent the robot hand 3 d.o.f. motion, the dimension of the output $y = h(q)$ should also be three. Its Jacobian matrix can be determined by $J = \frac{\partial h(q)}{\partial q}$, which is a 3 by 6 short matrix because the joint position vector q of this floating base robot model is 6-dimensional. Now, according to the equal-RD condition, each of the three output channels has a relative degree $r_i = 2$ so that the total relative degree is $r_d = 3r_i = 6 < n = 12$. Therefore, we can predict that the linearizable subsystem for this floating base robotic system referred to the fixed base on the earth is $r_d = 6$-dimensional and the dimension of the internal dynamics becomes $n - r_d = 6$. For this non-fully linearizable system, we cannot employ the same formula as derived in equation (6.60) to determine its state-feedback control law due to the non-square Jacobian matrix.

Although we can directly apply the general theory and procedures of the exact linearization to find a nonlinear state-feedback control, as expressed in equation (6.55) through a determination of the decoupling matrix $D(x)$ and $b(x)$, the derivation for such an under-actuated robotic system with a non-trivial internal dynamics will be very tedious. In order to seek a better and more compact symbolical derivation, let us start from the robot dynamic equation:

$$W\ddot{q} + c(q,\dot{q}) - \tau_g = \tau. \tag{6.71}$$

Since the inertial matrix W is always non-singular, pre-multiplying the W inverse on both sides of (6.71) yields

$$\ddot{q} + W^{-1}[c(q,\dot{q}) - \tau_g] = W^{-1}\tau.$$

For a certain output function $y = h(q)$, $\dot{y} = J\dot{q}$ so that $\ddot{y} = J\ddot{q} + \dot{J}\dot{q}$. By further pre-multiplying the Jacobian matrix J, even though it is not a square matrix, to the above dynamic equation, we obtain

$$J\ddot{q} + JW^{-1}[c(q,\dot{q}) - \tau_g] = \ddot{y} - \dot{J}\dot{q} + JW^{-1}[c(q,\dot{q}) - \tau_g] = JW^{-1}\tau. \tag{6.72}$$

Since τ is a nominal n-dimensional external torque vector, in an under-actuated robot case, such as in the above floating base robotic system, only the last three components of τ are non-zero to form a non-zero input $u = (\tau_4 \ \tau_5 \ \tau_6)^T$, and the first three components of τ that belong to the floating base model are all zero. Therefore, without loss of generality, staying in the floating base robot case as a typical under-actuated robotic system, the 6 by 1 torque vector τ in the above dynamic equation (6.72) can be replaced by

$$\tau = \begin{pmatrix} 0_{3\times1} \\ u \end{pmatrix}.$$

Let the last three columns of W^{-1} form a new 6 by 3 matrix B, i.e.,

$$B = W^{-1}(:,4:6).$$

Then, equation (6.72) becomes

$$\ddot{y} - \dot{J}\dot{q} + JW^{-1}[c(q,\dot{q}) - \tau_g] = JBu.$$

Clearly, the 3 by 3 square coefficient matrix $JB = D$ of the input u is just the decoupling matrix. If $D = JB$ is non-singular, then, the static state-feedback control of the floating base robotic system can be resolved by

$$u = D^{-1}\ddot{y} - D^{-1}\dot{J}\dot{q} + D^{-1}JW^{-1}[c(q,\dot{q}) - \tau_g] = \alpha(x) + \beta(x)v, \qquad (6.73)$$

where the new input $v = \ddot{y}$ and

$$\alpha(x) = -D^{-1}\dot{J}\dot{q} + D^{-1}JW^{-1}[c(q,\dot{q}) - \tau_g], \quad \text{and} \quad \beta(x) = D^{-1} = (JB)^{-1}.$$

In fact, the robot dynamic equation in (6.71) has 6 elements for each term. Let $W(1:3,:)$ be the top 3 by 6 block of the inertial matrix W, $c(1:3)$ be the top 3 elements of the centrifugal and Coriolis term $c(q,\dot{q})$, and $\tau_g(1:3)$ be the top 3 elements of the joint torque due to gravity τ_g. Then, because the top 3 elements of τ are all zero, we obtain

$$W(1:3,:)\ddot{q} + c(1:3) - \tau_g(1:3) = 0_{3\times 1}.$$

This homogeneous equation exactly represents an internal dynamic constraint as an internal dynamics of the under-actuated robotic system. Clearly, this internal dynamics cannot be controlled by the state-feedback control law in (6.73), but it is coupled with the linearizable subsystem.

While it is a success to achieve the above compact nonlinear feedback control law in (6.73) for an under-actuated robotic system, to find an explicit form of its internal dynamics may still be a tough job. In particular, the two subsystems: the linearizable one and the internal dynamics are often coupled to each other. Nevertheless, as long as the stability of the internal dynamics can be justified, finding its explicit form becomes unimportant.

6.6   CONTROL OF A MINIMUM-PHASE SYSTEM

Having introduced the special decomposition for a partially input-output linearizable nonlinear control system, we now turn our attention to discussing the stability issue. It is now very clear that a linearizable subsystem can always be stabilized by designing a certain linear control law, for instance, using a PD controller to asymptotically stabilize the trajectory-tracking system. However, what is the stability of the internal dynamics? Does it interfere the stabilized linearizable subsystem if it is unstable?

To answer those fundamental questions, let us first consider the previous example in (6.66). Its 3-dimensional linearized subsystem was given by (6.67) and the 2-dimensional internal dynamics was described in (6.69). If the system is to be applied for executing a trajectory-tracking task, under a given desired trajectory $y_d(t) = (y_{1d}(t) \ y_{2d}(t))^T$, we can always define a PD control law for the new input:

$$v = \begin{pmatrix} \dot{y}_{1d} + k_{1p}\Delta y_1 \\ \ddot{y}_{2d} + k_{2v}\Delta\dot{y}_2 + k_{2p}\Delta y_2 \end{pmatrix}. \qquad (6.74)$$

This ensures that the actual output vector $y(t)$ of the linearized subsystem will converge to the desired trajectory $y_d(t)$, as long as the constant gains k_{1p}, k_{2v}, and k_{2p} can make sure each component of (6.74) be Hurwitz.

However, the non-trivial internal dynamics (6.69) may interfere and destroy the entire system's stability even if it has a small unstable region. In general, according to equation (6.69), an internal dynamics may still be coupled with the linearized subsystem through the new state z. Although the linearizable subsystem and the internal dynamics can be *decomposed* in concept, they may not be thoroughly *decoupled* in reality for many nonlinear systems. This causes a difficulty in testing whether an internal dynamics is stable [9–11].

To answer the fundamental questions, we need more insight into the concept of decomposition and the related stability. Let us go backwards to revisit the typical linear time-invariant SISO system, as given in equation (6.45). After taking the Laplace transformation under the zero initial condition, the entire system is converted to (6.46) so that the transfer function can be deduced by

$$G(s) = \frac{Y(s)}{U(s)} = C(sI - A)^{-1}B = \frac{b_0 s^k + b_1 s^{k-1} + \cdots + b_k}{s^n + a_1 s^{n-1} + \cdots + a_n} = \frac{N(s)}{D(s)},$$

where we assume that the system dimension is n. Clearly, the denominator $D(s)$ is an n-th order s-domain polynomial and the numerator $N(s)$ is defined as a k-th order polynomial with $k < n$ for a proper transfer function. In other words, the transfer function $G(s)$ has n poles and k zeros and the difference $n - k = r_d$ is just the relative degree of this linear system.

Let $D(s)$ be divided by $N(s)$ through a polynomial long division procedure. We can always expect to have the following general result:

$$D(s) = Q(s)N(s) + R(s),$$

where $Q(s)$ is the r_d-th order *quotient* polynomial and $R(s)$ is up to $k - 1$st order *remainder* polynomial. For example, suppose that

$$G(s) = \frac{2s^2 + 2s + 1}{s^4 + 3s^3 + 4s^2 + s + 1}.$$

After a long division of the 4th order $D(s)$ by the 2nd order $N(s)$, i.e., $n = 4$ and $k = 2$, we obtain a quotient $Q(s) = 0.5s^2 + s + 0.75$ that is the 2nd order due to $r_d = n - k = 2$ and a remainder $R(s) = -1.5s + 0.25$. With such a long division, in general, we can rewrite

$$G(s) = \frac{N(s)}{D(s)} = \frac{N(s)}{Q(s)N(s) + R(s)} = \frac{\frac{1}{Q(s)}}{1 + \frac{1}{Q(s)} \cdot \frac{R(s)}{N(s)}}. \tag{6.75}$$

In comparison to the well-known negative feedback equation for finding the closed-loop transfer function:

$$G(s) = \frac{H(s)}{1 + H(s)F(s)}$$

with an open-loop gain $H(s)$ and a feedback gain $F(s)$, we conclude that the transfer function $G(s)$ in (6.75) is virtually a negative feedback closed-loop gain with its open-loop gain $H(s) = \frac{1}{Q(s)}$ and the feedback gain $F(s) = \frac{R(s)}{N(s)}$.

Since the open-loop gain $\frac{1}{Q(s)}$ has r_d poles and no zeros, this block is analog to the linearizable subsystem. Whereas the feedback gain $\frac{R(s)}{N(s)}$ has $k = n - r_d$ poles, the same number as the zeros of the closed-loop transfer function $G(s)$, because the numerator of $G(s)$ appears now in the denominator of the feedback $\frac{R(s)}{N(s)}$. Thus, this feedback block is analog to the internal dynamics. Conventionally, we say a linear system is *minimum phase* if both its zeros and poles are located on the left half of s-plane (LHP). Therefore, by borrowing the concept from the linear systems theory, we can state that *if the internal dynamics is asymptotically stable, then the entire nonlinear system should be asymptotically minimum-phase.*

A further study shows that we can determine the stability of an internal dynamics by "turning off" the linearized subsystem to ease the testing difficulty. Namely, first set all the output variables and their time-derivatives to be zero and then test the internal dynamics stability under such a zero-output condition. We refer to the internal dynamics under the zero-output condition as the **zero dynamics** [2,5,6].

According to the normal form (6.70) in the sense of input-output linearization, a general internal dynamics can be written by

$$\dot{\zeta} = \phi(\zeta,z) + \psi(\zeta,z)v.$$

Since the zero-output condition implies that $z = 0$ and also $v = 0$, the corresponding zero dynamics becomes

$$\dot{\zeta} = \phi(\zeta,\, 0). \tag{6.76}$$

Therefore, with the zero dynamics as a reduced form of the original internal dynamics, we may have more successful chances to test its stability.

It can be further shown that a complete nonlinear control system (6.53) is locally (globally) asymptotically *minimum phase* if its zero dynamics is locally (globally) asymptotically stable [1,2,5,6]. In addition, if (6.53) is locally (globally) asymptotically minimum phase, and the new input v of the linearized subsystem is defined by a PD trajectory-tracking control law, like equation (6.74) with each component being Hurwitz, then, the entire trajectory-tracking control system is at least locally asymptotically stable. This conclusion will be very useful in most practical cases for nonlinear trajectory-tracking control systems.

Let us now revisit the last example given in (6.66). We have found its internal dynamics described in (6.69) that is 2-dimensional. Under the zero-output condition, $y_1 = 0$ and $\dot{y}_1 = 0$ due to $r_1 = 1$, and $y_2 = 0$, $\dot{y}_2 = 0$ and $\ddot{y}_2 = 0$ due to $r_2 = 2$, so that $z_1 = z_2 = z_3 = 0$ and $v = (\dot{y}_1\ \ddot{y}_2)^T = 0$ as well. Thus, the zero dynamics becomes

$$\begin{cases} \dot{\zeta}_1 = -\zeta_1^3 \\ \dot{\zeta}_2 = \zeta_2. \end{cases} \tag{6.77}$$

Obviously, the two equations in (6.77) are decoupled in this particular case. We can immediately show that the first equation is globally asymptotically stable at its equilibrium point $\zeta_1 = 0$, while the second one is unstable due to its solution $\zeta_2(t) = \zeta_2(0)e^t$ that grows exponentially to ∞ as $t \to \infty$. Therefore, the system (6.66) is non-minimum phase, nor is it stable.

However, if we replace x_5 in the last row of (6.66) by $-x_5$, then everything will remain the same except that the second equation of the zero dynamics (6.77) will be changed to $\dot{\zeta}_2 = -\zeta_2$, which is obviously globally asymptotically stable. Therefore, under such a relatively simple modification in the model, the entire control system becomes globally asymptotically minimum phase, and with the PD trajectory-tracking control law (6.74) being Hurwitz, the remodeled system can be globally asymptotically stable.

6.7 EXAMPLES OF PARTIALLY LINEARIZABLE SYSTEMS WITH INTERNAL DYNAMICS

Example 6.1. Let us now study another 5th-order control system given as follows:

$$
\begin{cases}
\dot{x}_1 = x_2 - x_1 x_5 \\
\dot{x}_2 = -x_1 + x_3 + x_1 u_2 \\
\dot{x}_3 = x_2 - x_4 - u_1 \qquad \text{with an output} \quad y = h(x) = \begin{pmatrix} x_1 \\ x_2 \end{pmatrix}. \qquad (6.78) \\
\dot{x}_4 = -x_1 x_3 \\
\dot{x}_5 = x_2 - x_5 + u_2
\end{cases}
$$

It has two inputs and the vector fields are given by

$$
f(x) = \begin{pmatrix} x_2 - x_1 x_5 \\ -x_1 + x_3 \\ x_2 - x_4 \\ -x_1 x_3 \\ x_2 - x_5 \end{pmatrix}, \quad G(x) = (g_1(x) \ g_2(x)) = \begin{pmatrix} 0 & 0 \\ 0 & x_1 \\ -1 & 0 \\ 0 & 0 \\ 0 & 1 \end{pmatrix}.
$$

To find the relative degrees for both channels of the output $y = h(x)$, let us calculate

$$
L_G h(x) = \frac{\partial h(x)}{\partial x} G(x) = \begin{pmatrix} 1 & 0 & 0 & 0 & 0 \\ 0 & 1 & 0 & 0 & 0 \end{pmatrix} G(x) = \begin{pmatrix} 0 & 0 \\ 0 & x_1 \end{pmatrix}.
$$

It can be clearly seen that the second output channel has a nonzero coefficient of the input u so that the relative degree $r_2 = 1$ for $y_2 = h_2(x) = x_2$. We have to continue finding the relative degree r_1 for the first output channel $y_1 = h_1(x) = x_1$.

Since

$$
L_f h_1(x) = (1 \ 0 \ 0 \ 0 \ 0) f(x) = x_2 - x_1 x_5,
$$

$$L_G L_f h_1(x) = (-x_5 \ 1 \ 0 \ 0 \ -x_1) \begin{pmatrix} 0 & 0 \\ 0 & x_1 \\ -1 & 0 \\ 0 & 0 \\ 0 & 1 \end{pmatrix} = (0 \ 0),$$

which is still a zero row vector. Keeping the same procedure,

$$L_f^2 h_1(x) = (-x_5 \ 1 \ 0 \ 0 \ -x_1) f(x) = -x_2 x_5 + x_1 x_5^2 - x_1 + x_3 - x_1 x_2 + x_1 x_5,$$

then,

$$L_G L_f^2 h_1(x) = (-1 - x_2 + x_5 + x_5^2 \ -x_1 - x_5 \ 1 \ 0 \ 2x_1 x_5 + x_1 - x_2) G(x)$$

$$= (-1 \ x_1 - x_2 - x_1^2 + x_1 x_5).$$

Thus, the decoupling matrix becomes

$$D(x) = \begin{pmatrix} -1 & x_1 - x_2 - x_1^2 + x_1 x_5 \\ 0 & x_1 \end{pmatrix},$$

which is non-singular as long as $x_1 \neq 0$. Therefore, the relative degree $r_2 = 3$, and total relative degree $r_d = r_1 + r_2 = 4$. This implies that a 4-dimensional subsystem is state-feedback linearizable and an internal dynamics is predicted as a 1-dimensional subsystem.

From the above derivation, we can see that the 4 new state-space variables should be defined by

$$z = \begin{pmatrix} z_1 \\ z_2 \\ z_3 \\ z_4 \end{pmatrix} = \begin{pmatrix} x_1 \\ \dot{x}_1 \\ \ddot{x}_1 \\ x_2 \end{pmatrix} = \begin{pmatrix} x_1 \\ x_2 - x_1 x_5 \\ z_3 \\ x_2 \end{pmatrix},$$

where $z_3 = \ddot{x}_1 = -x_1 + x_3 - x_1 x_2 + x_1 x_5 - x_2 x_5 + x_1 x_5^2$. Actually, the above equation form a nonlinear transformation $z = T_z(x)$ to linearize the 5th-order nonlinear system in (6.78). If a new state variable ζ can be found to represent the internal dynamics, then the complete nonlinear transformation:

$$\begin{pmatrix} z \\ \zeta \end{pmatrix} = T(x)$$

should be interpreted as an $n = 5$-dimensional manifold M^5 for this particular system, which must be non-singular, i.e., its Jacobian matrix must be non-singular. We can now use this criterion to find the new state variable ζ for the 1-dimensional unlinearizable internal dynamics.

First, the Jacobian matrix J_z for the linearizable subsystem $z = T_z(x)$ is

$$J_z = \frac{\partial T_z(x)}{\partial x} = \begin{pmatrix} 1 & 0 & 0 & 0 & 0 \\ -x_5 & 1 & 0 & 0 & -x_1 \\ x_5 + x_5^2 - x_2 - 1 & -x - 1 - x_5 & 1 & 0 & x_1 - x_2 + 2x_1 x_5 \\ 0 & 1 & 0 & 0 & 0 \end{pmatrix}.$$

It can be seen that the 4th column of J_z is zero. Thus, to add the 5th row, the new internal dynamics state variable should at least be $\zeta = x_4$. Therefore, the Jacobian matrix of the nonlinear transformation $T(x)$ turns out to be

$$
J = \frac{\partial T(x)}{\partial x} = \begin{pmatrix}
1 & 0 & 0 & 0 & 0 \\
-x_5 & 1 & 0 & 0 & -x_1 \\
x_5 + x_5^2 - x_2 - 1 & -x - 1 - x_5 & 1 & 0 & x_1 - x_2 + 2x_1x_5 \\
0 & 1 & 0 & 0 & 0 \\
0 & 0 & 0 & 1 & 0
\end{pmatrix}.
$$

For nonzero values of the state x, the Jacobian matrix J of the manifold $z = T(x)$ is non-singular.

With the internal dynamics $\zeta = x_4$, $\dot{\zeta} = \dot{x}_4 = -x_1x_3 = -z_1x_3$, and x_3 can be solved in terms of z_1 through z_4 from $z = T_z(x)$. Since the linearized subsystem is given by

$$
\dot{z} = \begin{pmatrix} \dot{z}_1 \\ \dot{z}_2 \\ \dot{z}_3 \\ \dot{z}_4 \end{pmatrix} = \begin{pmatrix} 0 & 1 & 0 & 0 \\ 0 & 0 & 1 & 0 \\ 0 & 0 & 0 & 0 \\ 0 & 0 & 0 & 0 \end{pmatrix} \begin{pmatrix} z_1 \\ z_2 \\ z_3 \\ z_4 \end{pmatrix} + \begin{pmatrix} 0 & 0 \\ 0 & 0 \\ 1 & 0 \\ 0 & 1 \end{pmatrix} \begin{pmatrix} v_1 \\ v_2 \end{pmatrix}
$$

with the new input $v = \begin{pmatrix} v_1 \\ v_2 \end{pmatrix}$, under the non-singular decoupling matrix $D(x)$, the control input u can be resolved by $u = \alpha(x) + \beta(x)v$ according to equation (6.55).

To assess the stability of the internal dynamics, let us test its **zero dynamics** by setting the system outputs and its time-derivatives to be zero, i.e., $y = 0$, $v = 0$ and each $z_i = 0$ for $i = 1, \cdots, 4$. In this case, the internal dynamics $\dot{\zeta} = -z_1x_3 = 0$, which is critically stable. Therefore, the 5th-order nonlinear control system (6.78) is minimum-phase, but is not asymptotically minimum-phase.

Finally, let us test the observability for this 5th-order nonlinear system in (6.78). Because

$$
L_f h(x) = \frac{\partial h(x)}{\partial x} f(x) = \begin{pmatrix} 1 & 0 & 0 & 0 & 0 \\ 0 & 1 & 0 & 0 & 0 \end{pmatrix} f(x) = \begin{pmatrix} x_2 - x_1x_5 \\ -x_1 + x_3 \end{pmatrix},
$$

and

$$
L_f^2 h(x) = \frac{\partial L_f h(x)}{\partial x} f(x) = \begin{pmatrix} -x_5 & 1 & 0 & 0 & -x_1 \\ -1 & 0 & 1 & 0 & 0 \end{pmatrix} f(x)
$$

$$
= \begin{pmatrix} x_1x_5 + x_1x_5^2 - x_1 + x_3 - x_1x_2 - x_2x_5 \\ x_1x_5 - x_4 \end{pmatrix},
$$

the observation space formed by $\mathcal{O} = span\{h(x), L_f h(x), L_f^2 h(x), \cdots\}$. By taking derivatives on $\mathcal{O}$, the observability distribution for the first six rows can be further determined as

$$
d\mathcal{O} = \begin{pmatrix}
1 & 0 & 0 & 0 & 0 \\
0 & 1 & 0 & 0 & 0 \\
-x_5 & 1 & 0 & 0 & -x_1 \\
-1 & 0 & 1 & 0 & 0 \\
x_5^2 - 1 - x_2 + x_5 & -x_1 - x_5 & 1 & 0 & x_1 + 2x_1x_5 - x_2 \\
x_5 & 0 & 0 & -1 & x_1
\end{pmatrix},
$$

which is full-ranked if each $x_i \neq 0$. Thus, the system with the output is observable.

Therefore, a non-trivial internal dynamics does not imply that the system is unobservable. Actually, when we determine the relative degree for each output channel $h_i(x)$, we want to know taking how high-order time-derivatives on the output channel until the input u will show up. Thus, the linearizability and internal dynamics are related to the input-output connection, instead of just the output-state observation. This may also related to the system controllability. Nevertheless, we can directly follow the procedure to determine the relative degree for each output channel in finding the input-output connection, and any non-trivial internal dynamics is just a disconnecting subsystem with respect to the certain output and input. If one remodels the system by changing the output function, an original internal dynamics may turn to be part of the linearizable subsystem. Therefore, it is a relative sense.

Example 6.2. The second example is to revisit the log-arm robotic system, as studied in Chapter 3 and shown in Figure 3.3. This is a 3-body planar system: a cylindrical log rolling on the floor without actuation plus a revolute-prismatic type 2-joint arm sitting on the log with drives. Intuitively, for such an under-actuated robotic system, the 2-joint arm will establish a linearizable subsystem, while the unactuated log causes an unlinearizable internal dynamics. If we inspect the cylindrical log alone by shutting off the 2-joint arm motion, it should be an unstable system due to the overall center of gravity (CG) at somewhere above the centroid of the log. Therefore, the internal dynamics is unstable so that the entire log-arm robot cannot be a minimum-phase control system. Figure 6.6 shows a snapshot of falling down in a realistic animation study.

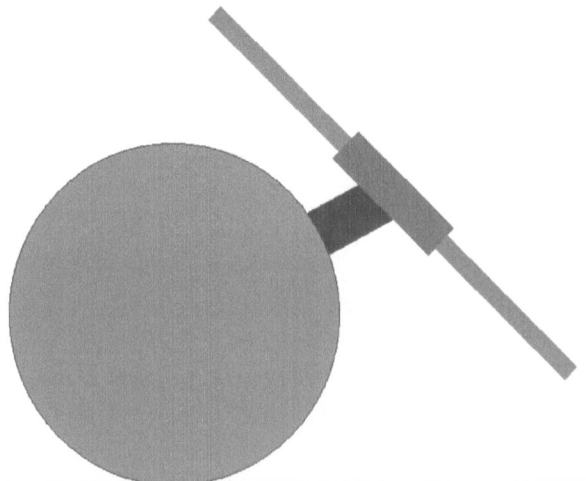

Figure 6.6   The log-arm robot is rolling down due to unstable internal dynamics

However, we can remedy the interference of the unstable internal dynamics by the following two possible methods:

1. Add a rolling friction between the log and the floor; or
2. Lower the mass center of the log by a bias b, as shown in Figure 6.7.

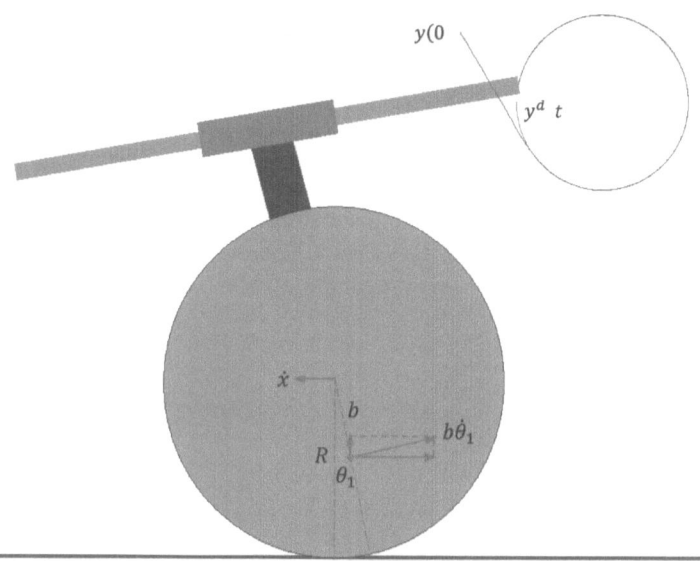

Figure 6.7   The log-arm robot is well-controlled to track a desired trajectory

As we have developed a complete dynamic model for this log-arm robotic system in Chapter 3, the general dynamic equation can be formulated as follows:

$$W\ddot{q} + c(q,\dot{q}) - \tau_g = \begin{pmatrix} 0 \\ \tau_2 \\ f_3 \end{pmatrix},$$

where τ_2 is the joint torque for the second revolute joint θ_2, and f_3 is the joint force the prismatic joint d_3 sliding along the second sleeve-type link. Obviously, the first joint is drive-less as the log is rolling freely on the floor. We may add a viscous-type friction that is proportional to the speed as a natural torque imposed between the log and floor. Namely, replace the zero element of the input on the right-hand side of the above dynamic equation by $-b_v\dot{\theta}_1$ with a certain viscous coefficient $b_v > 0$.

The second way is to lower the mass center of the log down by a bias b. In this case, the kinetic energy K_1 for the log needs to be modified from the original derivation in Chapter 3. Recall that the original kinetic energy K_1 was given by

$$K_1 = \frac{1}{2}m_1\dot{x}^2 + \frac{1}{2}I_1\dot{\theta}_1^2 = \frac{1}{2}(m_1R^2 + I_1)\dot{\theta}_1^2,$$

due to the original mass center matched with the centroid of the log. Now, with a bias $b > 0$, as depicted in Figure 6.7, the new kinetic energy is modified as

$$K_1 = \frac{1}{2}m_1[(b\dot{\theta}_1\cos\theta_1 - R\dot{\theta}_1)^2 + b^2\dot{\theta}_1^2\sin^2\theta_1] + \frac{1}{2}I_1\dot{\theta}_1^2$$

$$= \frac{1}{2}m_1(b^2 + R^2 - 2bR\cos\theta_1)\dot{\theta}_1^2 + \frac{1}{2}I_1\dot{\theta}_1^2.$$

With such a modification, the first term due to the log with the mass m_1 in the first element w_{11} of the 3 by 3 inertial matrix in equation (3.19) is now updated from m_1R^2 to

$$m_1(R^2 + b^2 - 2bR\cos\theta_1),$$

as well as the original moment of inertia I_1 should be updated to $I_1 = I_1 + m_1b^2$ according to the parallel-axis theorem. Any other terms remain unchanged. We adopted this method by adding a bias b to stabilize the internal dynamics in our animation study. Figure 6.7 also demonstrates a successful convergence during the trajectory-tracking control process for this log-arm robotic system.

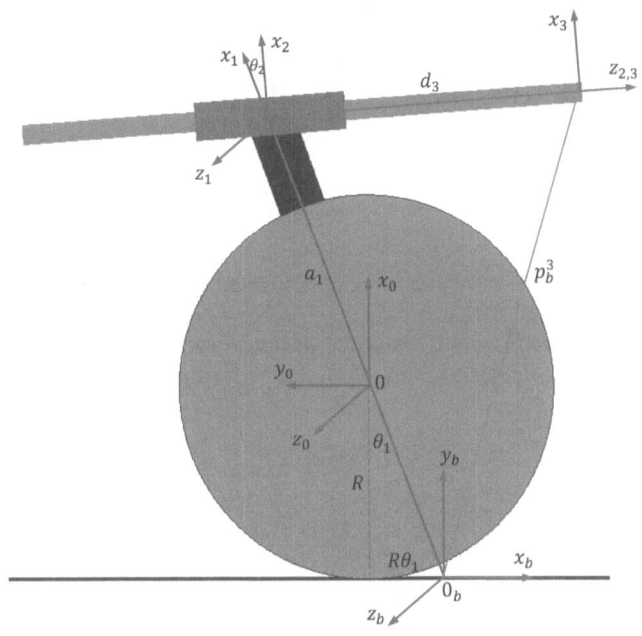

Figure 6.8 The log-arm robot frame assignment based on the D-H convention

Figure 6.8 illustrates a frame assignment based on the D-H convention. Then, its D-H table can be readily determined in Table 6.2.

Under the D-H table, all the one-step homogeneous transformation matrices can be found as follows:

$$A_0^1 = \begin{pmatrix} c_1 & -s_1 & 0 & a_1c_1 \\ s_1 & c_1 & 0 & a_1s_1 \\ 0 & 0 & 1 & 0 \\ 0 & 0 & 0 & 1 \end{pmatrix}, \quad A_1^2 = \begin{pmatrix} c_2 & 0 & s_2 & 0 \\ s_2 & 0 & -c_2 & 0 \\ 0 & 1 & 0 & 0 \\ 0 & 0 & 0 & 1 \end{pmatrix},$$

Joint Angle θ_i	Joint Offset d_i	Twist Angle α_i	Link Length a_i
θ_1	0	0	a_1
θ_2	0	90°	0
0	d_3	0	0

Table 6.2
The D-H parameter table for the log-arm robotic system

$$A_2^3 = \begin{pmatrix} 1 & 0 & 0 & 0 \\ 0 & 1 & 0 & 0 \\ 0 & 0 & 1 & d_3 \\ 0 & 0 & 0 & 1 \end{pmatrix}, \quad \text{and} \quad A_b^0 = \begin{pmatrix} 0 & -1 & 0 & -R\theta_1 \\ 1 & 0 & 0 & R \\ 0 & 0 & 1 & 0 \\ 0 & 0 & 0 & 1 \end{pmatrix},$$

where the last one A_b^0 is the homogeneous transformation between frame 0 and the base frame.

Multiplying all the homogeneous transformations together yields $A_b^3 = A_b^0 A_0^1 A_1^2 A_2^3$. The first two elements of the last column of A_b^3 is just the 2D position vector from its tail point at the base origin to the robot tip point at the origin of frame 3:

$$p_b^3 = \begin{pmatrix} -a_1 s_1 + d_3 c_{12} - R\theta_1 \\ a_1 c_1 + d_3 s_{12} + R \end{pmatrix}.$$

Then, its 2 by 3 Jacobian matrix can be derived as

$$J_b = \frac{\partial p_b^3}{\partial q} = \begin{pmatrix} -a_1 c_1 - d_3 s_{12} - R & -d_3 s_{12} & c_{12} \\ -a_1 s_1 + d_3 c_{12} & d_3 c_{12} & s_{12} \end{pmatrix}.$$

After the above kinematics modeling and preparation, based on the equations for the under-actuated robot dynamics and control formulation given in (6.72) and (6.73), the 2 by 2 decoupling matrix should be $D = J_b B$, where the 3 by 2 matrix B is formed by augmenting the last two columns of the inertial matrix inverse W^{-1} due to the first element of the input torque vector being zero:

$$\tau = \begin{pmatrix} 0 \\ \tau_2 \\ f_3 \end{pmatrix}$$

in this log-arm robot case. Based on equation (6.73), the control law to drive the top revolute-prismatic-joint arm can be determined as

$$u = \begin{pmatrix} \tau_2 \\ f_3 \end{pmatrix} = D^{-1}\ddot{y} - D^{-1}\dot{J}_b\dot{q} + D^{-1}J_b W^{-1}[c(q,\dot{q}) - \tau_g] = \alpha(x) + \beta(x)v.$$

Let the new input $v = \ddot{y}$ be specified to track a desired trajectory $y^d(t)$, i.e.,

$$v = \ddot{y}^d + K_v(\dot{y}^d - \dot{y}) + K_p(y^d - y),$$

where the output $y = p_b^3$. Figure 6.7 exhibits this log-arm system tracking a desired circular trajectory with a large initial deviation between the actual $y(0)$ and desired $y^d(0)$. This evidently shows that *for an under-actuated robotic system, a successful trajectory-tracking control requires a stable internal dynamics, or the entire system be at least minimum-phase.*

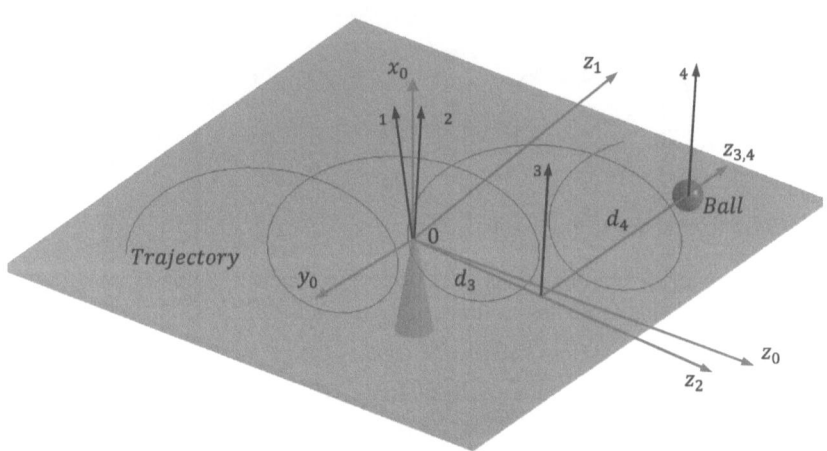

Figure 6.9 A ball-board control system with D-H convention

Example 6.3. In this example, we intend to demonstrate how to model and control a ball-board trajectory-tracking gaming system, which is virtually not a robotic system because it is not manipulable. However, we can utilize the method and procedure of robot kinematics and dynamics modeling and analysis to facilitate the control design for this non-robotic system. Suppose that a board is pivoted at its center point to have a 2-dimensional rotation: pitch and yaw without spin, which can be driven by two joint torques τ_1 and τ_2. With the board 2D orientation change, a marble ball on the board can overcome the gravity to balance and track a desired trajectory on the board surface, as shown in Figure 6.9.

In order to automatically control the orientation of the board, let us first establish a kinematic model for the system by borrowing the modeling procedure from robotics. By directly applying the D-H convention, the ball-board system can be modeled as a 4-joint open serial-chain "robot": the first two revolute joints are rotated for the board orientation changes and the last two prismatic joint to mimic the ball rolling on the 2D surface of the board. According to the frame assignment based on the D-H convention, as illustrated in Figure 6.9, the D-H parameter table can be determined in Table 6.3.

Joint Angle θ_i	Joint Offset d_i	Twist Angle α_i	Link Length a_i
θ_1	0	90°	0
θ_2	0	−90°	0
0	d_3	90°	0
0	d_4	0	0

Table 6.3
The D-H parameter table for the ball-board system

Then, the four one-step homogeneous transformation matrices are formed by

$$A_0^1 = \begin{pmatrix} c_1 & 0 & s_1 & 0 \\ s_1 & 0 & -c_1 & 0 \\ 0 & 1 & 0 & 0 \\ 0 & 0 & 0 & 1 \end{pmatrix}, \quad A_1^2 = \begin{pmatrix} c_2 & 0 & -s_2 & 0 \\ s_2 & 0 & c_2 & 0 \\ 0 & -1 & 0 & 0 \\ 0 & 0 & 0 & 1 \end{pmatrix},$$

$$A_2^3 = \begin{pmatrix} 1 & 0 & 0 & 0 \\ 0 & 0 & -1 & 0 \\ 0 & 1 & 0 & d_3 \\ 0 & 0 & 0 & 1 \end{pmatrix}, \quad A_3^4 = \begin{pmatrix} 1 & 0 & 0 & 0 \\ 0 & 1 & 0 & 0 \\ 0 & 0 & 1 & d_4 \\ 0 & 0 & 0 & 1 \end{pmatrix}.$$

With a number of symbolical multiplications, the homogeneous transformation matrix of frame 4 with respect to the base can be derived as

$$A_0^4 = A_0^1 A_1^2 A_2^3 A_3^4 = \begin{pmatrix} c_1c_2 & -c_1s_2 & s_1 & -d_3c_1s_2 + d_4s_1 \\ s_1c_2 & -s_1s_2 & -c_1 & -d_3s_1s_2 - d_4c_1 \\ s_2 & c_2 & 0 & d_3c_2 \\ 0 & 0 & 0 & 1 \end{pmatrix}.$$

Thus, the position vector of the ball on the 2D board surface with respect to the base is the last column of the above matrix:

$$p_0^4 = \begin{pmatrix} -d_3c_1s_2 + d_4s_1 \\ -d_3s_1s_2 - d_4c_1 \\ d_3c_2 \end{pmatrix}.$$

Since the kinetic energy K_1 for the board is

$$K_1 = \frac{1}{2}I_z\dot{\theta}_1^2 + \frac{1}{2}I_y\dot{\theta}_2^2,$$

and the kinetic energy K_2 for the rolling ball is

$$K_2 = \frac{1}{2}m_2\dot{p}_0^{4T}\dot{p}_0^4 + \frac{1}{2}I_2\left(\frac{\dot{d}_3^2 + \dot{d}_4^2}{r^2}\right),$$

where m_2 and I_2 are the mass and the moment of inertia for the ball, respectively, and r is the ball radius. The time-derivative of the position p_0^4 can be further derived as

$$\dot{p}_0^4 = \begin{pmatrix} -\dot{d}_3 c_1 s_2 + d_3 s_1 s_2 \dot{\theta}_1 - d_3 c_1 c_2 \dot{\theta}_2 + \dot{d}_4 s_1 + d_4 c_1 \dot{\theta}_1 \\ -\dot{d}_3 s_1 s_2 - d_3 c_1 s_2 \dot{\theta}_1 - d_3 s_1 c_2 \dot{\theta}_2 - \dot{d}_4 c_1 + d_4 s_1 \dot{\theta}_1 \\ \dot{d}_3 c_2 - d_3 s_2 \dot{\theta}_2 \end{pmatrix}.$$

With the above preparation, we can now determine the following 4 by 4 inertial matrix W by extracting every coefficient in certain order from the total kinetic energy quadratic form $K = K_1 + K_2 = \frac{1}{2}\dot{q}^T W \dot{q}$, i.e.,

$$W = \begin{pmatrix} m_2 d_4^2 + m_2 d_3^2 s_2^2 + I_z & -m_2 d_3 d_4 c_2 & -m_2 d_4 s_2 & m_2 d_3 s_2 \\ -m_2 d_3 d_4 c_2 & m_2 d_3^2 + I_y & 0 & 0 \\ -m_2 d_4 s_2 & 0 & m_2 + \frac{I_2}{r^2} & 0 \\ m_2 d_3 s_2 & 0 & 0 & m_2 + \frac{I_2}{r^2} \end{pmatrix}.$$

Its derivative matrix W_d that represents the centrifugal and Coriolis terms can also be found by

$$W_d = \begin{pmatrix} \dot{q}^T \frac{\partial W}{\partial \theta_1} \\ \dot{q}^T \frac{\partial W}{\partial \theta_2} \\ \dot{q}^T \frac{\partial W}{\partial d_3} \\ \dot{q}^T \frac{\partial W}{\partial d_4} \end{pmatrix}.$$

Because the first element of p_0^4 is a height $h(q)$ of the ball with respect to the base, according to the frame assignment, the gravitational potential energy P_g becomes

$$P_g = m_2 g h(q) = m_2 g(-d_3 c_1 s_2 + d_4 s_1),$$

with the gravitational acceleration $g = 9.806$ m/sec^2. Thus, the joint torque due to gravity should be

$$\tau_g = -\frac{\partial P_g}{\partial q} = m_2 g \begin{pmatrix} -d_3 s_1 s_2 - d_4 c_1 \\ d_3 c_1 c_2 \\ c_1 s_2 \\ -s_1 \end{pmatrix}.$$

Finally, the dynamics of the ball-board gaming system obeys

$$W \ddot{q} + \left(W_d^T - \frac{1}{2} W_d \right) \dot{q} - \tau_g = \tau, \tag{6.79}$$

where the 4 by 1 joint driving torque vector is $\tau = (\tau_1 \ \tau_2 \ 0 \ 0)^T$ with the last two zero elements due to the fact that the ball is free rolling on the board surface.

Let the 2D output vector y and its Jacobian matrix J be specified by

$$y = h(q) = \begin{pmatrix} d_3 + \theta_2 \\ d_4 - \theta_1 \end{pmatrix}, \quad \text{so that} \quad J = \frac{\partial y}{\partial q} = \begin{pmatrix} 0 & 1 & 1 & 0 \\ -1 & 0 & 0 & 1 \end{pmatrix}. \tag{6.80}$$

Then, according to equation (6.73), we have

$$J\ddot{q} + JW^{-1}\left(W_d^T - \frac{1}{2}W_d\right)\dot{q} - JW^{-1}\tau_g = JBu,$$

where the 4 by 2 matrix $B = W^{-1}(:,1:2)$ is augmented by the first two columns of the inertial matrix inverse W^{-1}. If the 2 by 2 decoupling matrix $D = JB$ is non-singular, by noticing that J is a constant matrix, we obtain the following control law:

$$u = \begin{pmatrix} \tau_1 \\ \tau_2 \end{pmatrix} = D^{-1}v + D^{-1}JW^{-1}\left[\left(W_d^T - \frac{1}{2}W_d\right)\dot{q} - \tau_g\right] = \alpha(x) + \beta(x)v.$$

Likewise, if the new input $v = \ddot{y}$ for tracking a desired trajectory, then,

$$v = \ddot{y}^d + K_v(\dot{y}^d - \dot{y}) + K_p(y^d - y),$$

with a desired trajectory specified by $y^d(t)$ and its time-derivatives.

The output function $y = h(q)$ often plays two major roles in control of the input-output linearized systems with or without an internal dynamics:

1. To be a major part of the decoupling matrix D required for inverse in order to resolve a state-feedback control law;
2. To be an actual output trajectory $y(t)$ in trajectory-tracking control cases.

While an output function could be defined arbitrarily, it must meet the physical requirements as well as to make sure that the decoupling matrix D should be invertible within a large range. In the current example, the definition $y = \begin{pmatrix} d_3 \\ d_4 \end{pmatrix}$ may well-meet the physical requirement in order to uniquely represent an output motion to track a desired trajectory on the 2D board surface. However, if $y = h(q)$ is defined in this way, the first role would be failed. In other words, the Jacobian matrix becomes

$$J = \begin{pmatrix} 0 & 0 & 1 & 0 \\ 0 & 0 & 0 & 1 \end{pmatrix},$$

which could easily lead to a singular decoupling matrix $D = JB$. Then, we may have to add the first two joint variables θ_1 and θ_2 into the output definition. Since the rotation of θ_1 about the z_0-axis makes an opposite direction of the translation d_4, while the rotation of θ_2 about the z_1-axis makes the same direction as the translation d_3, we thus create the output as given in equation (6.80). The tracking performance can be further improved by adjusting a constant k if the output is defined by

$$y = h(q) = \begin{pmatrix} d_3 + k\theta_2 \\ d_4 - k\theta_1 \end{pmatrix}, \quad \text{so that} \quad J = \frac{\partial y}{\partial q} = \begin{pmatrix} 0 & k & 1 & 0 \\ -k & 0 & 0 & 1 \end{pmatrix}.$$

Similar to the last example, the trajectory-tracking control can be assured successfully if the internal dynamics is stable. However, for such a real physical system, intuitively, the free-rolling ball will easily diverge under gravity. Therefore, it cannot be a minimum-phase system in nature. If we add a viscous-type rolling friction between the ball and the board surface, such as let the original 4 by 1 joint torque vector on the right-hand side of equation (6.79) be

$$\tau = \begin{pmatrix} u \\ -b_1 d_3 \\ -b_2 d_4 \end{pmatrix},$$

with certain viscous friction coefficients $b_1 > 0$ and $b_2 > 0$, then, the internal dynamics will be stable and the entire system can be minimum-phase.

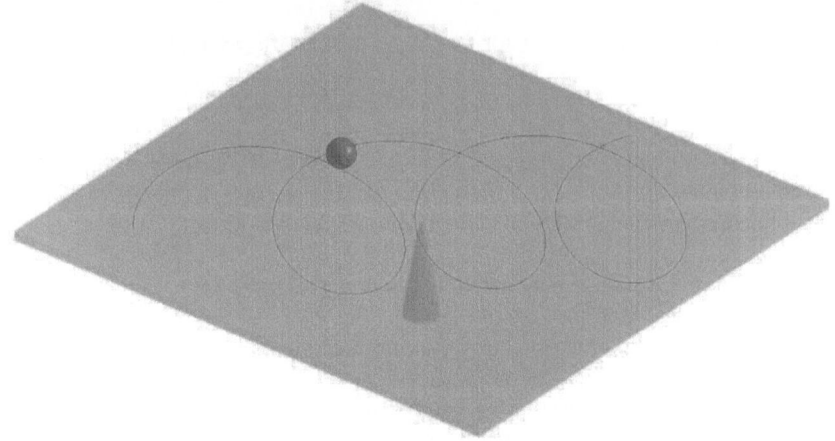

Figure 6.10   A ball-board control system in the animation study

Figure 6.10 exhibits that the ball is successfully tracking a desired screw-type trajectory during an animation study under the condition that the ball has a rolling friction on the board surface. In this simulation study, we set $k = 1.25$ in the latest output definition.

Example 6.4. The last example is a so-called "Ballbot" system to self-control the top box with three wheels riding on a big ball for balance, as shown in Figure 6.11 [12, 13]. It is, though, called a ball-robot, it is not a robotic system due to no manipulation capable. However, we can model and control it through the robot dynamics modeling and control procedures.

Under the coordinate frame assignment based on the D-H convention, we have a D-H table in Table 6.4.

With the D-H table, five one-step homogeneous transformations plus a homogeneous transformation to convert frame 0 to the base can be found below:

$$A_b^0 = \begin{pmatrix} 0 & 0 & 1 & 0 \\ 0 & -1 & 0 & 0 \\ 1 & 0 & 0 & R \\ 0 & 0 & 0 & 1 \end{pmatrix}, \quad A_0^1 = \begin{pmatrix} 1 & 0 & 0 & 0 \\ 0 & 0 & -1 & 0 \\ 0 & 1 & 0 & d_1 \\ 0 & 0 & 0 & 1 \end{pmatrix},$$

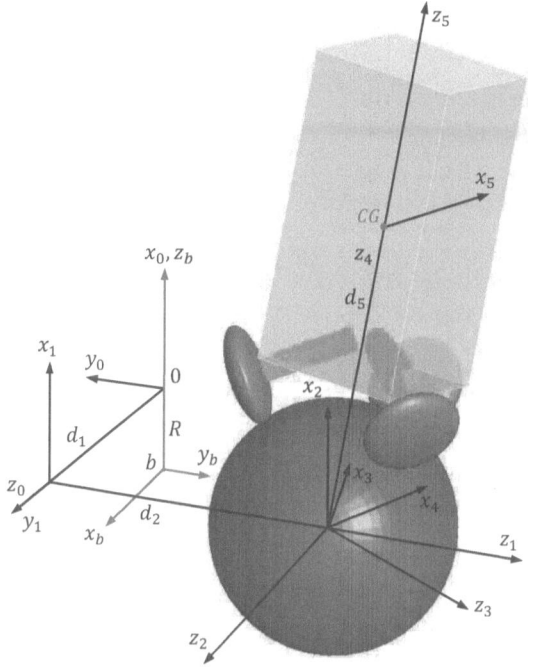

Figure 6.11 A ballbot system with D-H convention

Joint Angle θ_i	Joint Offset d_i	Twist Angle α_i	Link Length a_i	Home Position
0	d_1	90°	0	$d_1 = d_1(0)$
0	d_2	−90°	0	$d_2 = d_2(0)$
θ_3	0	90°	0	$\theta_3 = 0$
θ_4	0	−90°	0	$\theta_4 = -90°$
θ_5	d_5	0	0	$\theta_5 = 0$

Table 6.4
The D-H parameter table for the ballbot system

$$A_1^2 = \begin{pmatrix} 1 & 0 & 0 & 0 \\ 0 & 0 & 1 & 0 \\ 0 & -1 & 0 & d_2 \\ 0 & 0 & 0 & 1 \end{pmatrix}, \quad A_2^3 = \begin{pmatrix} c_3 & 0 & s_3 & 0 \\ s_3 & 0 & -c_3 & 0 \\ 0 & 1 & 0 & 0 \\ 0 & 0 & 0 & 1 \end{pmatrix},$$

$$A_3^4 = \begin{pmatrix} c_4 & 0 & -s_4 & 0 \\ s_4 & 0 & c_4 & 0 \\ 0 & -1 & 0 & 0 \\ 0 & 0 & 0 & 1 \end{pmatrix}, \quad A_4^5 = \begin{pmatrix} c_5 & -s_5 & 0 & 0 \\ s_5 & c_5 & 0 & 0 \\ 0 & 0 & 1 & d_5 \\ 0 & 0 & 0 & 1 \end{pmatrix},$$

where R is the big ball radius.

We can thus derive the homogeneous transformation of frame 5 with respect to the base $A_b^5 = A_b^0 A_0^1 A_1^2 A_2^3 A_3^4 A_4^5$. Namely,

$$A_b^5 = \begin{pmatrix} s_4 c_5 & -s_4 s_5 & c_4 & d_1 + d_5 c_4 \\ -s_3 c_4 c_5 - c_3 s_5 & s_3 c_4 s_5 - c_3 c_5 & s_3 s_4 & d_2 + d_5 s_3 s_4 \\ c_3 c_4 c_5 - s_3 s_5 & -c_3 c_4 s_5 - s_3 c_5 & -c_3 s_4 & R - d_5 c_3 s_4 \\ 0 & 0 & 0 & 1 \end{pmatrix}. \tag{6.81}$$

The last column of A_b^5 should be the position vector p_b^5, i.e.,

$$p_b^5 = \begin{pmatrix} d_1 + d_5 c_4 \\ d_2 + d_5 s_3 s_4 \\ R - d_5 c_3 s_4 \end{pmatrix},$$

and its Jacobian matrix turns out to be

$$J_b = \frac{\partial p_b^5}{\partial q} = \begin{pmatrix} 1 & 0 & 0 & -d_5 s_4 & 0 \\ 0 & 1 & d_5 c_3 s_4 & d_5 s_3 c_4 & 0 \\ 0 & 0 & d_5 s_3 s_4 & -d_5 c_3 c_4 & 0 \end{pmatrix}.$$

In order to go for a dynamic model, let us first seek the kinetic energy of the ballbot system. Since this system consists of only two bodies with nonzero masses: one is the big ball and the other one is the top box. If the origin of frame 5 is just at the mass center of the box, then the kinetic energy for the box is

$$K_2 = \frac{1}{2} m_2 (\dot{p}_b^5)^T \dot{p}_b^5 + \frac{1}{2}(I_{x2}\dot{\theta}_3^2 + I_{y2}\dot{\theta}_4^2 + I_{z2}\dot{\theta}_5^2).$$

The kinetic energy for the big ball is

$$K_1 = \frac{1}{2} m_1 (\dot{d}_1^2 + \dot{d}_2^2) + \frac{1}{2}\left(I_{y1}\frac{\dot{d}_1^2}{R^2} + I_{x1}\frac{\dot{d}_2^2}{R^2}\right).$$

By noticing that $\dot{p}_b^5 = J_b \dot{q}$ with the joint position vector $q = (d_1\ d_2\ \theta_3\ \theta_4\ \theta_5)^T$, the first term of K_2 should be

$$\frac{1}{2} m_2 (\dot{p}_b^5)^T \dot{p}_b^5 = \frac{1}{2} m_2 \dot{q}^T J_b^T J_b \dot{q}.$$

Finally, the 5 by 5 inertial matrix W for this ballbot system can be symbolically derived as follows:

$$W = \begin{pmatrix} m_1 + m_2 + \frac{I_{y1}}{R^2} & 0 & 0 & -m_2 d_5 s_4 & 0 \\ 0 & m_1 + m_2 + \frac{I_{x1}}{R^2} & m_2 d_5 c_3 s_4 & m_2 d_5 s_3 c_4 & 0 \\ 0 & m_2 d_5 c_3 s_4 & m_2 d_5^2 s_4^2 + I_{x2} & 0 & 0 \\ -m_2 d_5 s_4 & m_2 d_5 s_3 c_4 & 0 & m_2 d_5^2 + I_{y2} & 0 \\ 0 & 0 & 0 & 0 & I_{z2} \end{pmatrix}. \tag{6.82}$$

The derivative matrix W_d for the inertial matrix W can be found as well, while the gravitational potential energy is based on the height of the mass center of the box, which should be the last element of p_b^5, i.e., $h = R - d_5 c_3 s_4$. Therefore, the joint torque due to gravity can be determined by

$$\tau_g = -\frac{\partial P_g}{\partial q} = m_2 g \begin{pmatrix} 0 \\ 0 \\ -d_3 s_3 s_4 \\ d_5 c_3 c_4 \\ 0 \end{pmatrix}.$$

Because θ_5 is actually a spin angle for the box, which does not affect the position p_b^5, the last column of the Jacobian matrix J_b is zero. Thus, let us define an output function more appropriately by replacing the z-component of p_b^5 by the spin angle θ_5. Namely, the new output function and its Jacobian matrix for dynamics modeling become

$$y = h(q) = \begin{pmatrix} d_1 + d_5 c_4 \\ d_2 + d_5 s_3 s_4 \\ \theta_5 \end{pmatrix}, \quad \text{and} \quad J = \frac{\partial h(q)}{\partial q} = \begin{pmatrix} 1 & 0 & 0 & -d_5 s_4 & 0 \\ 0 & 1 & d_5 c_3 s_4 & d_5 s_3 c_4 & 0 \\ 0 & 0 & 0 & 0 & 1 \end{pmatrix}.$$

Also, we can readily calculate the time-derivative of J required by the input-output linearization control law below:

$$\dot{J} = \begin{pmatrix} 0 & 0 & 0 & -d_5 c_4 \dot{\theta}_4 & 0 \\ 0 & 0 & -d_5 s_3 s_4 \dot{\theta}_3 + d_5 c_3 c_4 \dot{\theta}_4 & d_5 c_3 c_4 \dot{\theta}_3 - d_5 s_3 s_4 \dot{\theta}_4 & 0 \\ 0 & 0 & 0 & 0 & 0 \end{pmatrix}.$$

Once again, based on equation (6.73), we have

$$\ddot{y} - \dot{J}\dot{q} + JW^{-1}\left(W_d^T - \frac{1}{2}W_d\right)\dot{q} - JW^{-1}\tau_g = JBu,$$

where the 5 by 3 matrix $B = W^{-1}(:, 3:5)$ is augmented by the last three columns of the inertial matrix inverse W^{-1}. If the 3 by 3 decoupling matrix $D = JB$ is non-singular, we obtain the following control law:

$$u = \begin{pmatrix} \tau_3 \\ \tau_4 \\ \tau_5 \end{pmatrix} = D^{-1}v - D^{-1}\dot{J}\dot{q} + D^{-1}JW^{-1}\left[\left(W_d^T - \frac{1}{2}W_d\right)\dot{q} - \tau_g\right] = \alpha(x) + \beta(x)v.$$

Likewise, if the new input $v = \ddot{y}$ for tracking a desired trajectory, then,

$$v = \ddot{y}^d + K_v(\dot{y}^d - \dot{y}) + K_p(y^d - y),$$

with a desired trajectory specified by $y^d(t)$ and its time-derivatives. Since the control objective for this particular ballbot system is to stabilize the top box avoiding fall down, we can set a simple desired output y^d. For instance, let $y^d = (d_1 \ d_2 \ \theta_5)^T$.

This system also has a non-trivial internal dynamics that is the big ball rolling on the ground. Intuitively, such kind of internal dynamics should be stable on a flat ground surface, and thus, we don't need to add any friction between the ball and ground surface.

Since the objective of the ballbot system is to control the box for stabilizing itself, every movement is relatively in a small neighborhood of the initial "home" position. We can also use the Lyapunov indirect stabilization method, as stated in Theorem 6.5, i.e., the local linearization to control the system for balance without considering the issue of internal dynamics stability.

To do so, let us first perform a coordinate shift. Since only $\theta_4 = -90°$ at the balanced home position, we have to define $\theta_4' = \theta_4 + 90°$ such that $\theta_4 = \theta_4' - 90°$. Then, each $s_4 = -c_4' \approx -1$ and $c_4 = s_4' \approx \theta_4'$. After the coordinate shift, as we have applied for the rotating inverted pendulum system in Figure 6.2, let every second-order term be zero and let each $\sin\theta \approx \theta$ and $\cos\theta \approx 1$ for both the inertial matrix W and τ_g. The centrifugal and Coriolis terms in $\left(W_d^T - \frac{1}{2}W_d\right)\dot{q}$ will all be zero due to each term in the second-order. Thus, based on equation (6.82), the locally linearized inertial matrix becomes

$$
\hat{W} = \begin{pmatrix}
m_1 + m_2 + \frac{I_{y1}}{R^2} & 0 & 0 & m_2 d_5 & 0 \\
0 & m_1 + m_2 + \frac{I_{x1}}{R^2} & -m_2 d_5 & 0 & 0 \\
0 & -m_2 d_5 & m_2 d_5^2 + I_{x2} & 0 & 0 \\
m_2 d_5 & 0 & 0 & m_2 d_5^2 + I_{y2} & 0 \\
0 & 0 & 0 & 0 & I_{z2}
\end{pmatrix},
$$

and the locally linearized joint torque due to gravity is

$$
\hat{\tau}_g = m_2 g \begin{pmatrix}
0 \\
0 \\
d_5 \theta_3 \\
d_5 \theta_4' \\
0
\end{pmatrix}.
$$

Once again, all the joint angles $\theta_3 = 0$, $\theta_4' = 0$ and $\theta_5 = 0$ at the balanced home position after the coordinate shift.

After the local linearization around the "home" equilibrium point, the linearized dynamic equation is reduced to

$$
\hat{W}\ddot{q} - \hat{\tau}_g = \tau.
$$

Let $x_1 = q = (d_1 \; d_2 \; \theta_3 \; \theta_4' \; \theta_5)^T$, and $u = (\tau_3 \; \tau_4 \; \tau_5)^T$. Then,

$$
\begin{cases}
\dot{x}_1 = x_2 \\
\dot{x}_2 = \hat{W}^{-1}\hat{\tau}_g + \hat{W}^{-1}\tau,
\end{cases}
$$

where

$$\tau = \begin{pmatrix} 0 \\ 0 \\ \tau_3 \\ \tau_4 \\ \tau_5 \end{pmatrix} = \begin{pmatrix} 0 \\ 0 \\ u \end{pmatrix}.$$

From the above state-space equation, it can be easily seen that the 10 by 10 constant matrix A and 10 by 3 constant matrix B are formed by

$$A = \begin{pmatrix} O_{5\times2} & O_{5\times3} & I_{5\times5} \\ O_{5\times2} & m_2 g d_5 \xi_{3-4} & O_{5\times6} \end{pmatrix}, \quad \text{and} \quad B = \begin{pmatrix} O_{5\times3} \\ \xi_{3-5} \end{pmatrix},$$

where the 5 by 2 matrix ξ_{3-4} is augmented by the 3rd and 4th columns of the linearized inertial matrix inverse $\hat{W}^{-1}$, and the 5 by 3 matrix ξ_{3-5} is augmented by the 3rd, 4th and 5th columns of $\hat{W}^{-1}$. Once the locally linearized system equation $\dot{x} = Ax + Bu$ is completed, the Lyapunov indirect stabilization procedure, similar to the rotating inverted pendulum control system in Figure 6.2, can readily be followed up. However, one more step to make up towards an implementation of the control procedure is to transform the three joint torques τ_3, τ_4 and τ_5 into the three wheels dragging torques on the top surface of the big ball [12, 13].

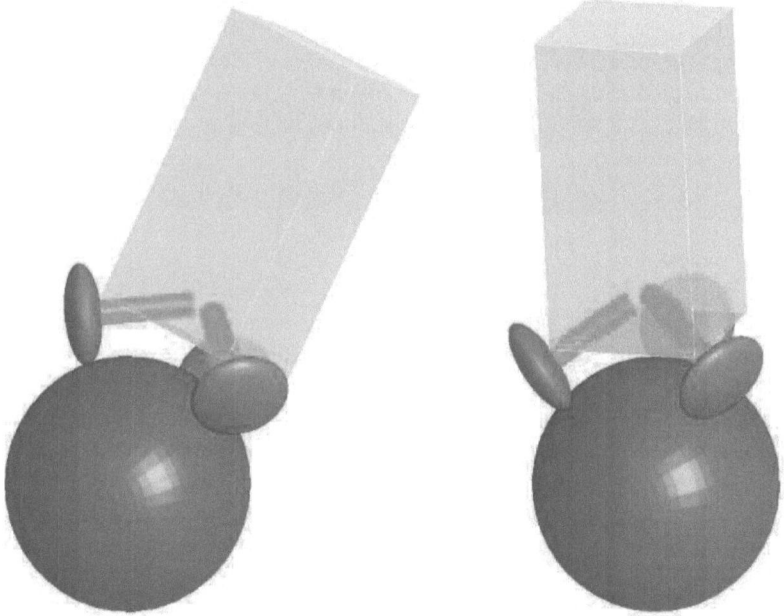

Figure 6.12 The ballbot system in animation study: the left is an initial posture, and the right is controlled to struggle for balance

Actually, the best and most efficient way to control the top box for fall avoidance over the above input-output feedback control and local stabilization is the *isometric*

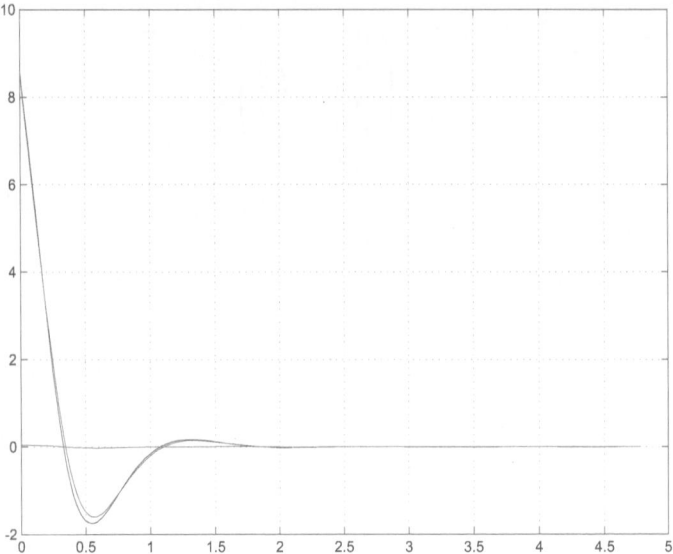

Figure 6.13   The plot of three joint torques vs. time. The vertical axis is in Newton-meter, and the horizontal axis is in seconds

embedding control scheme, which is, indeed, a unique linearization-free method. As we have studied and discussed in fair details in Chapter 4, an isometric embedding for the top box, as a dominant body of the entire ballbot system, can be defined in the following form:

$$z = \zeta(q) = \begin{pmatrix} \sqrt{m_2}\, p_b^5 \\ b_1 r_x \\ b_2 r_y \\ b_3 r_z \end{pmatrix},$$

which is 12-dimensional, where b_1, b_2 and b_3 are the dynamic parameters that can be found by equation (4.7) in terms of the three moments of inertia I_x, I_y and I_z of the box, and r_x, r_y and r_z are the three columns of the rotation matrix R_b^5 that is the upper-left 3 by 3 corner of the homogeneous transformation A_b^5 in (6.81) and uniquely represents the orientation of the box with respect to the base. Namely,

$$r_x = \begin{pmatrix} s_4 c_5 \\ -s_3 c_4 c_5 - c_3 s_5 \\ c_3 c_4 c_5 - s_3 s_5 \end{pmatrix}, \quad r_y = \begin{pmatrix} -s_4 s_5 \\ s_3 c_4 s_5 - c_3 c_5 \\ -c_3 c_4 s_5 - s_3 c_5 \end{pmatrix}, \quad r_z = \begin{pmatrix} c_4 \\ s_3 s_4 \\ -c_3 s_4 \end{pmatrix}.$$

Because the isometric embedding vector $z = \zeta(q)$ is an explicit function of q, we can symbolically derive its 12 by 5 Jacobian matrix $J_z = \frac{\partial \zeta}{\partial q}$, as an accurate genuine mathematical Jacobian matrix. Then, a compact dynamic equation for the ballbot system can be quickly achieved:

$$\tau = J_z^T \ddot{z} - \tau_g.$$

Now, let the acceleration vector be defined as

$$\ddot{z} = \ddot{z}^d + K_2(\dot{z}^d - \dot{z}) + K_1(z^d - z).$$

Then, the above compact dynamic equation can control the system to follow a desired trajectory specified by z^d and its time-derivatives. In our simulation study, since the home balanced position and orientation of the box, as the dominant body, should be

$$p_b^5(home) = \begin{pmatrix} d_1 \\ d_2 \\ R+d_5 \end{pmatrix}, \quad \text{and} \quad R_b^5(home) = \begin{pmatrix} -1 & 0 & 0 \\ 0 & -1 & 0 \\ 0 & 0 & 1 \end{pmatrix},$$

let them be adopted to define the 12-dimensional z^d with the same dynamic parameters as z does. With the two positive control gain constants K_1 and K_2, the entire ballbot system is successfully controlled by the three joint torques τ_3, τ_4 and τ_5 for a quick balance. Figure 6.12 shows two snapshots during the animation study: the left one is the initial posture with a large deviation from the home balanced position and orientation, and the right one is about two seconds later, the box is being controlled to struggle for balance. The three joint torques, as control inputs, versus time are plotted in Figure 6.13. As can be clearly seen, all the three joint torques are getting small as the box approaches to its home balanced posture.

Through this typical under-actuated ballbot dynamic system, we have practiced and tested three different modeling and control schemes to stabilize the top box for fall avoidance. In summary, a comparative study among the three nonlinear control schemes can be outlined in Table 6.5.

Modeling and Control Schemes	Input-Output Linearization	Local Linearization for Stabilization	Isometric Embedding Control Scheme
Required Conditions	Dynamic Equation Output Function	Dynamic Equation Needed Only	Need Only Kinematic Model
Efforts and Major Drawbacks	Feedback Control and Stability of Internal Dynamics	Locally Linearize To Find A and B Limited Range	Derive a Large-size Jacobian Matrix For Dominant Body
Advantages	Well-Procedurized	Relatively Simple	Compact, Linearization-Free

Table 6.5

A comparative study among the three major modeling and control schemes

REFERENCES

1. Vidyasagar, M., (1993) Nonlinear Systems Analysis, 2nd Edition. Prentice-Hall, Englewood Cliffs, N.J.

2. Slotine, J. and Li, W., (1991) Applied Nonlinear Control. Prentice Hall, New Jersey.

3. Khalil, H., (1996) Nonlinear Systems, 2nd Edition. Prentice Hall, New Jersey.

4. Banks, S., (1988) Mathematical Theories of Nonlinear Systems. Prentice Hall, New York.

5. Isidori, A., (1995) Nonlinear Control Systems: An Introduction, 3rd Edition. Springer-Verlag, New York.

6. Nijmeijer, H. and Van der Schaft, A., (1990) Nonlinear Dynamical Control Systems. Springer-Verlag, New York.

7. Ogata, K., (2010) Modern Control Engineering, 5th Edition. Prentice Hall, New York.

8. Dorf, R. and Bishop, R., (2017) Modern Control Systems, 13th Edition. Prentice Hall, Upper Saddle River, New Jersey.

9. Gu, Edward Y.L., (1993) A Direct Adaptive Control Scheme for Under-Actuated Dynamic Systems. Proc. 32nd IEEE Conference on Decision and Control. San Antonio, TX, Dec., pp. 1625-1627.

10. Gu, E. and Xu, Y., (1994) A Normal Form Augmentation Approach to Adaptive Control of Space Robot Systems. Journal of Dynamics and Control, Kluwer Academic Publishers, June.

11. Gu, Edward Y.L., (2013) A Journey from Robot to Digital Human. Springer, Heidelberg, New York.

12. Lauwers, T., Kantor, G. and Hollis, R., (2005) "One is Enough!", *The 12th International Symposium on Robotics Research*, San Francisco, CA, Oct. 12-15, pp. 327-336.

13. Lauwers, T., Kantor, G. and Hollis, R., (2006) "A Dynamically Stable Single-Wheeled Mobile Robot with Inverse Mouse-Ball Drive", *Proc. IEEE International Conference on Robotics and Automation*, Orlando, FL., May 15-19, pp. 2884-2889.

7 Adaptive Control of Robotic Systems

7.1 THE CONTROL LAW AND ADAPTATION LAW

Adaptive control plays a pivotal role in stabilizing a system with parameter uncertainty [1–3, 7, 8]. Although we can design a very robust controller to overwhelm the error due to the parameter deviation, such as using a high control gain to accelerate the attraction rate, etc., we cannot achieve a satisfactory convergence in many cases, especially in a nonlinear system. Therefore, in addition to designing a control law, it often requires a parameter **adaptation law** acting simultaneously on the system in a real-time fashion. A general parameter-emphasized system model is given as follows:

$$\dot{x} = f(x) + F(x)\xi + \sum_{i=1}^{r} g_i(x)u_i, \tag{7.1}$$

where $\xi \in \mathbb{R}^k$ is a parameter column vector to be adapted. We assume that all the coefficient $g_i(x)$'s of the input $u \in \mathbb{R}^r$ temporarily do not contain any uncertain parameters.

In addition, it is important that all the available nonlinear adaptive control methods thus far are under a *linear parametric condition*, i.e., the system equation is linear in parameters, as seen in (7.1) [9, 10]. A typical example is a nonlinear-spring mechanical system, called the Duffing's system, with a control input u added into its second equation, i.e.,

$$\begin{cases} \dot{x}_1 = x_2 \\ \dot{x}_2 = -\mu x_2 + \alpha x_1 - \beta x_1^3 + u. \end{cases} \tag{7.2}$$

In comparison with (7.1), it can be seen that

$$f(x) = \begin{pmatrix} x_2 \\ 0 \end{pmatrix}, \quad F(x) = \begin{pmatrix} 0 & 0 & 0 \\ -x_2 & x_1 & -x_1^3 \end{pmatrix}, \quad \xi = \begin{pmatrix} \mu \\ \alpha \\ \beta \end{pmatrix}, \quad g(x) = \begin{pmatrix} 0 \\ 1 \end{pmatrix}.$$

In this particular case, the parameter vector ξ is 3 by 1. Since its equilibrium point is $x = \begin{pmatrix} x_1 \\ x_2 \end{pmatrix} = 0$ under $u = 0$, the simplest Lyapunov candidate can be defined as $V(x) = \frac{1}{2}x^T x$. However, we have to add another positive-definite term to represent the parameter ξ for a possible convergence between the modeled parameter vector ξ_m and the actual but unknown parameter vector ξ. Let a parameter error vector be $\Delta\xi = \xi - \xi_m$. Therefore, a completed Lyapunov candidate for adaptive control should be

$$V(x) = \frac{1}{2}x^T x + \frac{1}{2}\Delta\xi^T \Delta\xi \tag{7.3}$$

that is clearly strongly positive-definite if the state vector x is augmented by the parameter error vector $\Delta\xi$. Taking a time-derivative on it yields

$$\dot{V} = x^T \dot{x} + \Delta\xi^T \Delta\dot{\xi} = x^T (f(x) + F(x)\xi + g(x)u) + \Delta\xi^T \Delta\dot{\xi}.$$

By noticing that the actual parameter $\xi = \Delta\xi + \xi_m$, we obtain

$$\dot{V} = x^T f(x) + x^T F(x)\Delta\xi + x^T F(x)\xi_m + x^T g(x)u + \Delta\xi^T \Delta\dot{\xi}.$$

If we define an adaptation law in the following form

$$\Delta\dot{\xi} = -F^T(x)x,$$

then,

$$\dot{V} = x^T f(x) + x^T F(x)\xi_m + x^T g(x)u \leq -\varepsilon(x, \Delta\xi) \leq 0,$$

where $\varepsilon(x, \Delta\xi)$ is a scalar positive or positive-definite function of the state x and the parameter-error $\Delta\xi$. Since $x^T f(x) = x_1 x_2$, $x^T F(x) = x_2(-x_2 \ x_1 \ -x_1^3)$ and $x^T g(x) = x_2$, by setting $\varepsilon(x, \Delta\xi) = cx_2^2$ for $c > 0$, we obtain a control law

$$u = \alpha(x, \xi_m) = -x_1 - cx_2 - (-x_2 \ x_1 \ -x_1^3)\xi_m,$$

which requires the modeled parameter vector ξ_m and is quite reasonable. Also, for this particular system, the adaptation law becomes

$$\Delta\dot{\xi} = -F^T(x)x = x_2 \begin{pmatrix} x_2 \\ -x_1 \\ x_1^3 \end{pmatrix}.$$

Even if $\varepsilon(x, \Delta\xi)$ is just positive but not positive-definite, it can be shown that the above control law plus the adaptation law $\Delta\dot{\xi} = -F^T(x)x$ can still asymptotically stabilize the system based on the La Salle's Theorem. However, the choice of $\varepsilon(x, \Delta\xi)$ does not take care of the parameter adaptation, and it is virtually impossible to include any $\Delta\xi^T \Delta\xi$ term into the definition of $\varepsilon(x, \Delta\xi)$. Therefore, the modeled parameter ξ_m is not guaranteed to converge to the actual real parameter ξ, but at least it does not diverge according to the La Salle's Theorem. Nevertheless, in most application cases, we only care about the convergence of the state x and do not really need each parameter convergence. In fact, since the parameters are uncertain and also immeasurable, the desired target for each parameter is virtually unavailable.

Moreover, when one implements both the control and adaptation laws into an autonomous (time-invariant) system, it should be understood that in $\Delta\xi = \xi - \xi_m$, the actual real parameter ξ is fixed, i.e., $\dot{\xi} = 0$ for every autonomous system, while the modeled parameter ξ_m is being adjusted during the process of adaptation so that $\Delta\dot\xi = -\dot\xi_m$. In other words, the adaptation law can be rewritten as

$$\dot\xi_m = -\Delta\dot\xi = F^T(x)x.$$

The above adaptive control strategy on a single system can also be extended to a cascaded control system by using an *adaptive backstepping procedure*, and the reader may refer to [3] for a detailed discussion.

Another example is the following 3rd-order nonlinear system:

$$\begin{cases} \dot x_1 = \xi_1 x_1 + \xi_2 x_1 x_2 + u \\ \dot x_2 = -x_2 + \xi_3 x_3 \\ \dot x_3 = \xi_1 x_1^2 - \xi_4 x_3 + u. \end{cases}$$

Comparing with (7.1), we obtain

$$f(x) = \begin{pmatrix} 0 \\ -x_2 \\ 0 \end{pmatrix}, \quad F(x) = \begin{pmatrix} x_1 & x_1 x_2 & 0 & 0 \\ 0 & 0 & x_3 & 0 \\ x_1^2 & 0 & 0 & -x_3 \end{pmatrix}, \quad \xi = \begin{pmatrix} \xi_1 \\ \xi_2 \\ \xi_3 \\ \xi_4 \end{pmatrix}, \quad g(x) = \begin{pmatrix} 1 \\ 0 \\ 1 \end{pmatrix}.$$

Since the equilibrium point of the system is also at the origin $x = 0$, we can define the same Lyapunov candidate as (7.3). Because in this case,

$$x^T f(x) = -x_2^2, \quad x^T F(x) = (x_1^2 + x_1^2 x_3 \ \ x_1^2 x_2 \ \ x_2 x_3 \ \ -x_3^2), \quad \text{and } x^T g(x) = x_1 + x_3,$$

we define an adaptation law

$$\Delta\dot\xi = -\dot\xi_m = -F^T(x)x = - \begin{pmatrix} x_1^2 + x_1^2 x_3 \\ x_1^2 x_2 \\ x_2 x_3 \\ -x_3^2 \end{pmatrix},$$

such that

$$\dot V = x^T f(x) + x^T F(x)\xi_m + x^T g(x)u$$
$$= -x_2^2 + (x_1^2 + x_1^2 x_3 \ \ x_1^2 x_2 \ \ x_2 x_3 \ \ -x_3^2)\xi_m + (x_1 + x_3)u \le -\varepsilon(x, \Delta\xi).$$

Thus, let the control law be

$$u = \alpha(x, \xi_m) = -\frac{(x_1^2 + x_1^2 x_3 \ \ x_1^2 x_2 \ \ x_2 x_3 \ \ -x_3^2)\xi_m}{x_1 + x_3},$$

then, $\varepsilon(x, \Delta\xi) = x_2^2$. Based on La Salle's Theorem 6.2, the set S with $x_2 = 0$ can be shown to only contain the trivial equilibrium point $x = 0$ so that the system can be asymptotically stable at $x = 0$, but not at $\Delta\xi = 0$.

A simulation study has been performed to verify the above justification for both the control law and adaptation law of the 3rd-order system. We defined its actual

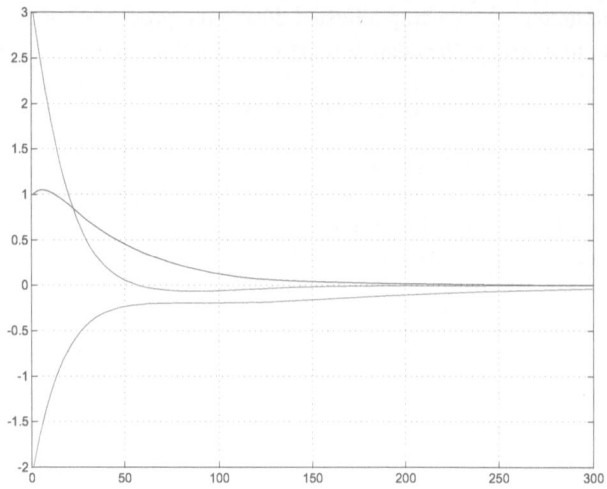

Figure 7.1 A simulation study on adaptive control with the plot for three states converging to the equilibrium point

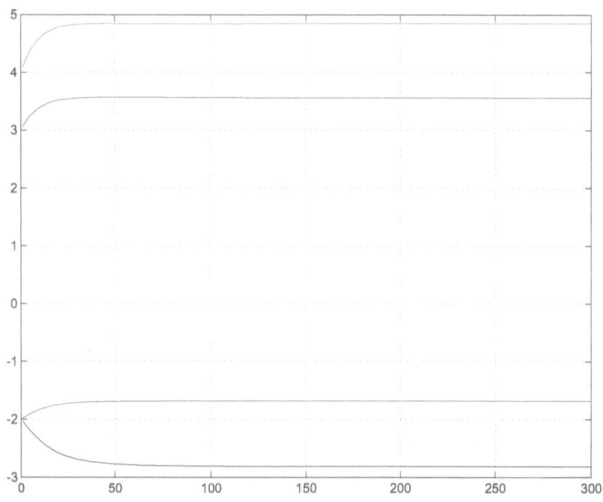

Figure 7.2 A simulation study on adaptive control with the plot of parameter adaptation, where the vertical axis is $\Delta\xi = \xi - \xi_m$, and the horizontal axis is the sampling points with $\Delta t = 0.01$ sec.

parameter to be $\xi = (-2 \ -2 \ 3 \ 4)^T$, while the initial modeled parameter was set as a zero vector $\xi_m = 0$. With the initial state specified as $x(0) = (1 \ -2 \ 3)^T$, all the three state components have converged to the equilibrium point $x = 0$, as shown in Figure 7.1. However, all the four modeled parameters ξ_{m1} through ξ_{m4} were not converging to the actual parameter values in ξ, but their errors in $\Delta\xi = \xi - \xi_m$ are all approaching to nonzero constants, as shown in Figure 7.2.

We now turn to investigate how to adapt a state-feedback control system after linearization. Recalling the state-feedback control law in (6.55), each of $\alpha(x)$ and $\beta(x)$ contains a number of parameters, which may need adaptation if they are uncertain.

Let $\alpha_m(x)$ and $\beta_m(x)$ be the two functions of x but using all the modeled parameters in ξ_m, and $\alpha(x)$ and $\beta(x)$ without the subscript m denote the functions with the actual parameters in ξ. Thus, a control law must be determined and updated by $u = \alpha_m(x) + \beta_m(x)v$ with a new input v. For a trajectory-tracking control task, if each output channel of the system has a relative degree $r_i = 2$, then v can be defined by equation (6.61), i.e.,

$$v = \ddot{y}^d + K_v\Delta\dot{y} + K_p\Delta y. \tag{7.4}$$

On the other hand, the control law will be implemented to excite the real system, which is described by $u = \alpha(x) + \beta(x)\ddot{y}$ with the actual parameters ξ and the output acceleration vector $\ddot{y}$. Hence,

$$\alpha_m(x) + \beta_m(x)v = u = \alpha(x) + \beta(x)\ddot{y}.$$

Rearranging this equation and noticing that $v - \ddot{y} = \ddot{e} + K_v\dot{e} + K_pe$ due to (7.4), where $e = \Delta y = y^d - y$ is the output tracking error vector, we have

$$\ddot{e} + K_v\dot{e} + K_pe = \beta_m^{-1}(x)[\alpha(x) - \alpha_m(x) + (\beta(x) - \beta_m(x))\ddot{y}].$$

This clearly shows that if the parameters are all precisely known, then the right-hand side of the above equation will vanish, and the left-hand side guarantees that the output tracking error e asymptotically converges to zero if both $K_p > 0$ and $K_v > 0$. However, due to the parameter uncertainty, the right-hand side cannot be zero. Instead, it will be a non-zero function of x, ξ and even $\ddot{y}$, which may jeopardize the tracking convergence.

If the system is nonlinear in x, but is linear in the parameter ξ, i.e., the linear parametric condition holds, the right-hand side of the above error equation can be factorized to

$$\ddot{e} + K_v\dot{e} + K_pe = F(x, \xi_m, \ddot{y})\Delta\xi, \tag{7.5}$$

for an m by k matrix F that is a function of the state x, modeled parameter ξ_m and the output acceleration $\ddot{y}$. Since we wish both $e \to 0$ and $\dot{e} \to 0$ as $t \to \infty$, and also wish $\Delta\xi \to 0$, it is reasonable to construct an s.p.d.f. Lyapunov candidate as follows:

$$V(x) = \frac{1}{2}(\dot{e} + K_ve)^T(\dot{e} + K_ve) + \frac{1}{2}e^TK_pe + \frac{1}{2}\Delta\xi^TK_a^{-1}\Delta\xi, \tag{7.6}$$

where $K_a > 0$ is a k by k adaptation gain constant symmetric matrix. Thus,

$$\dot{V} = (\dot{e} + K_ve)^T(\ddot{e} + K_v\dot{e}) + e^TK_p\dot{e} + \Delta\xi^TK_a^{-1}\Delta\dot{\xi}.$$

According to (7.5), $\ddot{e} + K_v\dot{e} = F(x, \xi_m, \ddot{y})\Delta\xi - K_pe$. If we define an adaptation law:

$$\Delta\dot{\xi} = -\dot{\xi}_m = -K_aF^T(x, \xi_m, \ddot{y})(\dot{e} + K_ve), \tag{7.7}$$

then,

$$\dot{V} = -e^T K_v K_p e \le 0,$$

provided that $K_v K_p > 0$. Equation (7.7) offers a general form of adaptation law that requires factorizing all the parameters from the state-feedback equation to find the matrix function F, and then to adapt (adjust and update) the modeled parameter ξ_m by (7.7) at each sampling point. Namely, under the first-order approximation,

$$\xi_m(t + \Delta t) = \xi_m(t) + \Delta \xi_m \approx \xi_m(t) + K_a F(t)^T (\dot{e}(t) + K_v e(t)) \Delta t.$$

The above discussion sounds successful in achieving an adaptation law as a value-added result to the state-feedback control for an input-output linearizable system. However, such an adaptation law given by (7.7) is actually unfeasible, because it requires us to know the output acceleration $\ddot{y}$ that is unavailable. In fact, the above Lyapunov function is constructed without invoking any dynamics of the system if it is a dynamic system. Therefore, we have to seek a more feasible adaptation law for a class of high-dimensional and high-nonlinear complex dynamic control systems, such as a robotic system.

Recall the isometric embedding model that directly comes from the kinematic structure for a robotic system and transforms itself to a dynamic model with dynamic parameters. Namely, let an isometric embedding be formed by both the position and orientation of a given robotic system and coated by its dynamic parameters:

$$z = \zeta(q) = \begin{pmatrix} a p_0^c \\ b_1 r_x \\ b_2 r_y \\ b_3 r_z \end{pmatrix},$$

where a, b_1, b_2 and b_3 are all the dynamic parameters, p_0^c is the position vector of the mass center of a dominant body for the robot, while r_x, r_y and r_z are the three columns of the rotation matrix R_0^c for the dominant body. Since this function $z = \zeta(q)$ satisfies that $\frac{1}{2}\dot{z}^T \dot{z} = K$ the kinetic energy, z has already been part of the robot dynamics. For an open serial-chain robotic system, its dynamics obeys $\tau + \tau_g = J^T \ddot{z}$, where $J = \frac{\partial \zeta(q)}{\partial q}$ is the genuine mathematical Jacobian matrix of the isometric embedding $z = \zeta(q)$. Due to the special structure, this compact dynamic equation can be easily factorized to left-extract all the dynamic parameters from the variables, i.e.,

$$z = \zeta(q) = \begin{pmatrix} a I_3 & 0 & 0 & 0 \\ 0 & b_1 I_3 & 0 & 0 \\ 0 & 0 & b_2 I_3 & 0 \\ 0 & 0 & 0 & b_3 I_3 \end{pmatrix} \begin{pmatrix} p_0^c \\ r_x \\ r_y \\ r_z \end{pmatrix} = \Gamma z_0,$$

where each I_3 is the 3 by 3 identity, Γ is a 12 by 12 diagonal parameter matrix and z_0 is the embedding vector without parameters. Obviously,

$$\ddot{z} = \Gamma \ddot{z}_0, \quad \text{and} \quad J = \Gamma \frac{\partial z_0}{\partial q} = \Gamma J_0, \tag{7.8}$$

where $\ddot{z}_0$ and J_0 are the acceleration vector and Jacobian matrix of the isometric embedding without dynamic parameters, respectively.

In fact, the vector z can also be factorized to right-extract the parameters from the variables, i.e.,

$$z = \zeta(q) = \begin{pmatrix} p_0^c & 0 & 0 & 0 \\ 0 & r_x & 0 & 0 \\ 0 & 0 & r_y & 0 \\ 0 & 0 & 0 & r_z \end{pmatrix} \begin{pmatrix} a \\ b_1 \\ b_2 \\ b_3 \end{pmatrix} = Q_0 \xi, \tag{7.9}$$

where Q_0 is a 12 by 4 kinematic structure matrix of the embedding and ξ is the dynamic parameter vector. With such an easy parameter extraction feature, the trajectory-tracking control scheme that is often used to replace $\ddot{z}$ in the compact dynamic equation $u = \tau + \tau_g = J^T \ddot{z}$ can be written as follows:

$$\ddot{z}^d + K_v(\dot{z}^d - \dot{z}) + K_p(z^d - z) = Q_{v0}\xi.$$

Now, let us reconsider the Lyapunov candidate $V(x)$ in equation (7.6) again, but defining each error $e = z^d - z$ to be a deviation between the desired and actual isometric embedding vectors. Then,

$$\dot{V} = (\dot{e} + K_v e)^T (\ddot{e} + K_v \dot{e}) + e^T K_p \dot{e} + \Delta\xi^T K_a^{-1} \Delta\dot{\xi}$$

$$= (\dot{e} + K_v e)^T (\ddot{e} + K_v \dot{e} + K_p e - K_p e) + e^T K_p \dot{e} + \Delta\xi^T K_a^{-1} \Delta\dot{\xi}$$

$$= (\dot{e} + K_v e)^T (\ddot{z}^d + K_v \dot{e} + K_p e - \ddot{z}) - e^T K_v K_p e + \Delta\xi^T K_a^{-1} \Delta\dot{\xi}.$$

Since $\ddot{z}^d + K_v \dot{e} + K_p e = Q_{v0}\xi$, and $\ddot{z}$ in the compact dynamic equation $u = J^T \ddot{z}$ is replaced by the modeled trajectory-tracking control scheme

$$\ddot{z} = (\ddot{z}^d + K_v \dot{e} + K_p e)_m = Q_{v0}\xi_m, \tag{7.10}$$

we have

$$\dot{V} = (\dot{e} + K_v e)^T Q_{v0}(\xi - \xi_m) - e^T K_v K_p e + \Delta\xi^T K_a^{-1} \Delta\dot{\xi}$$

$$= (\dot{e} + K_v e)^T Q_{v0}\Delta\xi - e^T K_v K_p e + \Delta\xi^T K_a^{-1} \Delta\dot{\xi}.$$

Let now an adaptation law be defined as

$$\Delta\dot{\xi} = -\dot{\xi}_m = -K_a Q_{v0}^T(\dot{e} + K_v e)_m = -K_a Q_{v0}^T Q_{e0}\xi_m, \tag{7.11}$$

where $(\dot{e} + K_v e)_m = Q_{e0}\xi_m$ as a parameter right-extraction in the modeled case and so is $\dot{e} + K_v e = Q_{e0}\xi$ in the actual case. Then,

$$\dot{V} = \Delta\xi^T Q_{e0}^T Q_{v0}\Delta\xi - e^T K_v K_p e. \tag{7.12}$$

It is clearly observable that if we would know $\dot{e} + K_v e$ in the actual parameter case, the adaptation law could completely cancel the first term so that the above $\dot{V} = -e^T K_v K_p e$ would be strictly negative. However, the reality is not always so ideal, and the positiveness or negativeness of the first term $\Delta\xi^T Q_{e0}^T Q_{v0}\Delta\xi$ is uncertain. Therefore, in order to guarantee $\dot{V} \leq -\varepsilon(e,\Delta\xi) \leq 0$, either the first term has a smaller absolute value, or the second negative term $-e^T K_v K_p e$ has to be large enough, which may require higher control gains for both K_v and K_p.

In fact, if we add one more term $K_c(\dot{e}+K_v e)_m$ into equation (7.10), i.e.,

$$\ddot{z} = (\ddot{z}^d + K_v\dot{e} + K_p e)_m + K_c(\dot{e}+K_v e)_m = Q_{v0}\xi_m + K_c Q_{e0}\xi_m,$$

then, $\dot{V}$ in (7.12) will have an additional negative term, i.e.,

$$\dot{V} = \Delta\xi^T Q_{e0}^T Q_{v0}\Delta\xi - e^T K_v K_p e - (\dot{e}+K_v e)^T K_c(\dot{e}+K_v e)_m. \tag{7.13}$$

Although $(\dot{e}+K_v e) \neq (\dot{e}+K_v e)_m$, they should keep the same sign if the adaptation range is not too large so that this additional term can always be negative. Thus, the new trajectory-tracking control scheme becomes

$$\ddot{z} = [\ddot{z}^d + (K_v + K_c)\dot{e} + (K_p + K_c K_v)e]_m = (Q_{v0} + K_c Q_{e0})\xi_m, \tag{7.14}$$

where $K_c > 0$ is a control gain constant.

Once the tracking factor $\dot{e}+K_v e$ has a dynamic model involved in the above Lyapunov stability justification, it will also bring the dynamic parameters ξ in, which may result in an uncertain term $\Delta\xi^T Q_{e0}^T Q_{v0}\Delta\xi$ imposed on $\dot{V}$. Therefore, let us now return back to the joint space for the tracking error definition $\dot{e}+K_v e$, which is known as the adaptive control algorithm by Slotine and Li [4–6]. Namely, let $\dot{e}+K_v e = \dot{q}^d - \dot{q} + K_v(q^d - q) = s$, and $\eta = \dot{q}^d + K_v(q^d - q)$ so that $s = \eta - \dot{q}$, which are now independent of the dynamic parameters.

For the dynamic model of an n-joint robotic system, we have developed a compact formulation in Chapter 4 through the isometric embedding approach. Namely, if $z = \zeta(q)$ is an isometric embedding that sends the system configuration manifold (C-manifold) M^n into an ambient Euclidean m-space $\mathbb{R}^m$, and its Jacobian matrix is $J = \frac{\partial\zeta}{\partial q}$, then, the inertial matrix of the robot will be $W = J^T J$, while the centrifugal and Coriolis terms will be

$$C(q,\dot{q})\dot{q} = \left(W_d^T - \frac{1}{2}W_d\right)\dot{q} = J^T \dot{J}\dot{q}.$$

Now, let a new Lyapunov candidate be defined by involving the robot dynamics:

$$V(x) = \frac{1}{2}s^T W s + \frac{1}{2}\Delta\phi^T K_a^{-1}\Delta\phi = \frac{1}{2}s^T J^T J s + \frac{1}{2}\Delta\phi^T K_a^{-1}\Delta\phi,$$

where $\Delta\phi = \phi - \phi_m$ is the new dynamic parameter error vector between the actual and modeled cases. Thus,

$$\dot{V} = s^T J^T J\dot{s} + s^T J^T \dot{J}s + \Delta\phi^T K_a^{-1}\Delta\dot{\phi}$$

$$= s^T J^T J(\dot{\eta} - \ddot{q}) + s^T J^T \dot{J}(\eta - \dot{q}) + \Delta\phi^T K_a^{-1}\Delta\dot{\phi}.$$

Notice that the compact dynamic equation is $u = \tau + \tau_g = J^T \ddot{z} = J^T(J\ddot{q} + \dot{J}\dot{q}) = J^T J\ddot{q} + J^T \dot{J}\dot{q}$. Substituting it into $\dot{V}$ yields

$$\dot{V} = s^T J^T J\dot{\eta} + s^T J^T \dot{J}\eta - s^T u + \Delta\phi^T K_a^{-1}\Delta\dot{\phi}.$$

Let a control law be defined by

$$u = J_m^T J_m \dot{\eta} + J_m^T \dot{J}_m \eta + Ks \tag{7.15}$$

with a constant gain matrix $K > 0$, and let an adaptation law be defined by

$$\Delta\dot{\phi} = -\dot{\phi}_m = -K_a Q^T s, \tag{7.16}$$

where Q is an n by k matrix that is determined by the dynamic parameters right-extraction procedure. Namely, according to the left-extraction in (7.8), $J = \Gamma J_0$, and $J^T J = J_0^T \Gamma^2 J_0$. Then, based on the right-extraction in (7.9), $J^T J\dot{\eta} = Q_1\phi$, where each element of ϕ is the square of the corresponding parameter in ξ, i.e., each $\phi_i = \xi_i^2$ for $i = 1, \cdots, k$. Similarly, $J^T \dot{J}\eta = Q_2\phi$. Therefore,

$$(J^T J - J_m^T J_m)\dot{\eta} + (J^T \dot{J} - J_m^T \dot{J}_m)\eta = (Q_1 + Q_2)\Delta\phi = Q\Delta\phi.$$

Obviously, the matrix $Q = Q_1 + Q_2$ is now a function of q, $\dot{q}$, q_d, $\dot{q}_d$ and $\ddot{q}_d$, and is independent of the actual acceleration $\ddot{q}$ and all the dynamic parameters. Finally, with the control law in (7.15) and adaptation law in (7.16), we achieve

$$\dot{V} = -s^T Ks \leq 0,$$

and $s^T Ks$ is a p.d.f. in the $\Delta x = \begin{pmatrix} e \\ \dot{e} \end{pmatrix}$ state error space, but is not included in the $\Delta\phi$ parameter error space. By further testing both the control law in (7.15) and adaptation law in (7.16), it can be concluded that they are feasible.

The above adaptive control justification under the compact dynamic model $u = J^T \ddot{z}$ offers two different suggestions on both the control law and adaptation law. Regarding the control law, the first suggestion is to replace $\ddot{z}$ in the compact dynamic model by equation (7.14). According to equation (7.8), the complete modeled dynamic control input in the first suggestion should be

$$u = J_m^T \ddot{z}_m = J_0^T \Gamma_m^2 [\ddot{z}_0^d + (K_v + K_c)\dot{e}_0 + (K_p + K_c K_v)e_0] = J_0^T (Q_{v0} + K_c Q_{e0})\phi_m, \tag{7.17}$$

where $e_0 = z_0^d - z_0$, and each element of the dynamic parameter vector ϕ_m is the square of the corresponding element of ξ_m.

The second suggestion is given by equation (7.15), which requires more computations than the first one in (7.17), including the time-derivative of the Jacobian matrix J_m, i.e., $\dot{J}_m = \Gamma_m \dot{J}_0$. In fact, the major difference between (7.15) and (7.17) is $J_m\dot{\eta} + \dot{J}_m\eta$ with $\eta = \dot{q}^d + K_v(q^d - q)$ versus $(\ddot{z}^d + K_v\dot{e} + K_p e)_m$ with $e = z^d - z$. If there is no parameter deviation, η in the former formula should be $\dot{q}$ so that $J\ddot{q} + \dot{J}\dot{q} = \frac{d}{dt}(J\dot{q}) = \ddot{z}$, which is back to the same as the latter formula $\ddot{z}^d + K_v\dot{e} + K_p e = \ddot{z}$.

Comparison Aspects	Adaptive Control Algorithm By Slotine and Li	Adaptive Control Algorithm with Isometric Embedding
The Control Law	Equation (7.15)	Equation (7.17)
The Adaptation Law	Equation (7.16)	Equation (7.11)
Initial $\Delta\xi$ Allowed	Can be very large	Has to be smaller
Computation Complexity	Very high with IK required	Reasonably low, no IK needed
Applications to Both Position/Orient. Control	Position control is OK, but orientation only in joint space	Can control both position and orientation without IK

Table 7.1
A comparison between two adaptive control algorithms

Regarding the adaptation law, both equations (7.11) and (7.16) have the similar formats, but the former one requires $(\dot{e} + K_v e)_m$ that depends on the dynamic parameters, while the latter one needs $s = \dot{e} + K_v e$ in joint space, which is independent of the dynamic parameter. The connection between them lies on the inverse kinematics (IK) of the robot between Cartesian and joint spaces. Therefore, in order for the former version of adaptation to be successful, it may require a better pre-knowledge of the dynamic parameters so that the initial $\Delta\xi(0)$ could be as small as possible.

7.2 APPLICATIONS AND SIMULATION/ANIMATION STUDIES

After we have discussed in details the adaptive control algorithms with robotic applications, let us now make a comparative study between the well-known algorithm by Slotine and Li and the algorithm that we have just developed based on the isometric embedding (IE) theory.

As we can be clearly seen, Table 7.1 outlines the advantages and disadvantages for both two adaptive control algorithms. The Slotine and Li algorithm is very effective and allows a larger initial dynamic parameter deviation, especially in the application cases of lacking knowledge of dynamic parameters. However, its computational complexity is very high, which requires all the dynamics formula and inverse kinematics (IK) if an orientation control is required. In contrast, the isometric embedding based adaptive control algorithm requires much less computations and easy dynamic parameter extraction from the variables. The orientation control can also be directly implemented in Cartesian space without IK needed. The only weakness of the isometric embedding based algorithm is that the initial parameter deviation $\Delta\xi(0)$ cannot be too large. In other words, it is required to have a certain knowledge of dynamic parameters before implementing the adaptive controller into a robotic system.

Example 7.1. Let us revisit the Ballbot system, as we have introduced and discussed in Chapter 6, see Figure 6.11, for our first case study of the adaptive control algorithms.

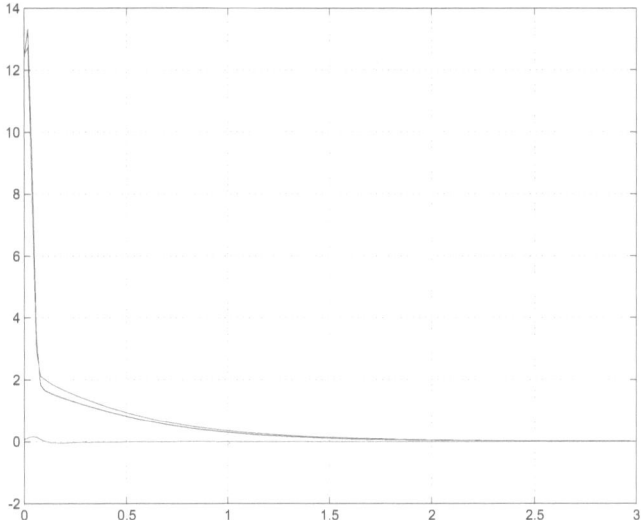

Figure 7.3 A simulation study on adaptive control of the Ballbot system with the plot for three joint torques. The vertical axis is τ in N-m, and the horizontal axis is time in sec.

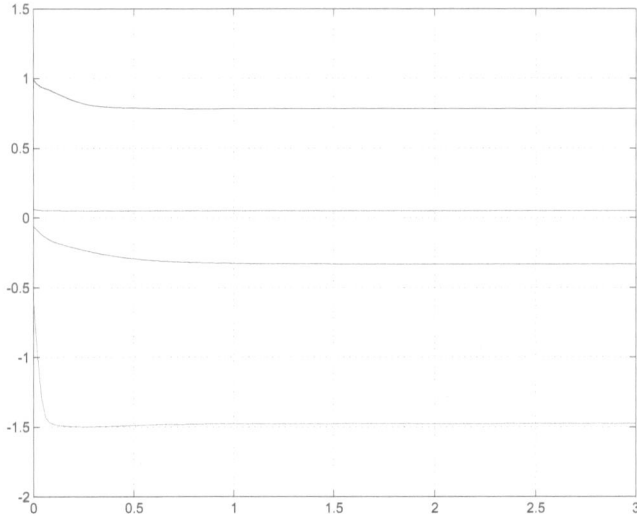

Figure 7.4 A simulation study on adaptive control of the Ballbot system with the plot of four dynamic parameter adaptation, where the vertical axis is $\Delta\xi = \xi - \xi_m$, and the horizontal axis is time in sec.

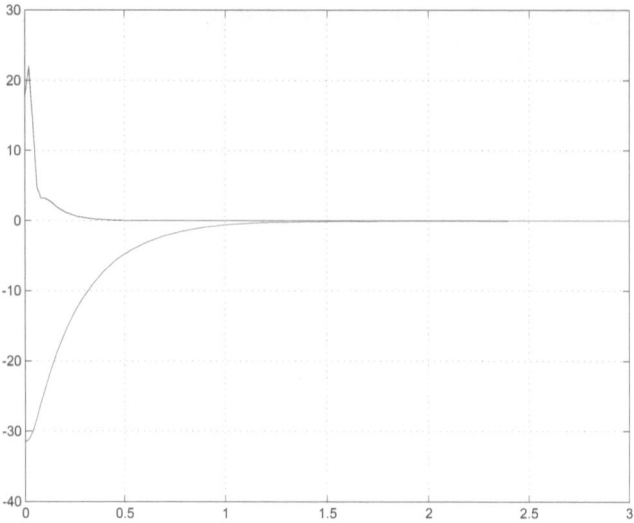

Figure 7.5 A simulation study on adaptive control of the Ballbot system with the plot for the first and second terms of $\dot{V}(x)$ in equation (7.12)

In the ballbot control system, we actually keep the same control law as Example 6.4 in Chapter 6, i.e.,

$$u = \tau = J_m^T(\ddot{z}^d + K_v(\dot{z}^d - \dot{z}) + K_p(z^d - z))_m - \tau_g,$$

and just allow all the modeled dynamic parameters $\sqrt{m}$, b_1, b_2 and b_3 to form a 4 by 1 parameter vector ξ_m and then to be adapted by the adaptation law in (7.11), i.e.,

$$\xi_m(t + \Delta t) \approx \xi_m(t) + \dot{\xi}_m(t)\Delta t = \xi_m(t) + K_a Q_{v0}^T(t)Q_{e0}(t)\xi_m(t)\Delta t,$$

where the adaptation gain constant was selected as $K_a = 0.5$ in the simulation study.

The modeled initial parameter vector was randomly defined as $\xi_m(0) = (0.3\sqrt{m} \quad 1.6b_1 \quad 0.4b_2 \quad 1.8b_3)^T$, in comparison to the actual parameter vector $\xi = (\sqrt{m} \quad b_1 \quad b_2 \quad b_3)^T$. With a large initial position/orientation deviation for the dominant body of the ballbot system, the control system can be stabilized to converge back to the desired posture. Figure 7.3 plots the three joint torques versus time, which shows that all the joint torques are converging to zero due to the balance posture getting reached. Figure 7.4 depicts the errors of four dynamic parameters $\Delta\xi = \xi - \xi_m$ are approaching to their respective constants, each of which is, however, not equal to zero. This means that each parameter is stabilized, but keeping a constant distance away from each corresponding actual parameter.

Figure 7.5 reveals the values of the two terms of the Lyapunov function time-derivative $\dot{V}(x)$ in equation (7.12) versus time. The positive curve represents the first term $\Delta\xi^T Q_{e0}^T Q_{v0}\Delta\xi$ while the negative curve is the second term $-e^T K_v K_p e$. Obviously, with the adaptation gain constant $K_a = 0.5$ and some pre-knowledge of the dynamic parameters, the negative curve is dominant over the positive one in $\dot{V}(x)$ so that the entire adaptive control process works successfully. However, if $K_a = 1$, the adaptive control will diverge due to $\dot{V}(x)$ turning to be positive.

In this case study, both position and orientation of the top box are controlled adaptively. This is one of the major advantages by adopting the isometric embedding based control strategy. If we switch to adopting the adaptive control algorithm by Slotine and Li, since the 3D orientation cannot uniquely be represented by a 3D output vector y, the orientation control has to be performed in the joint space, then an IK algorithm is necessary in order to describe each orientation in the joint space. In addition to the orientation control issue, the parameter extraction is another big challenging by the Slotine and Li algorithm. Therefore, the isometric embedding based adaptive control algorithm has an exclusive algorithmic advantage.

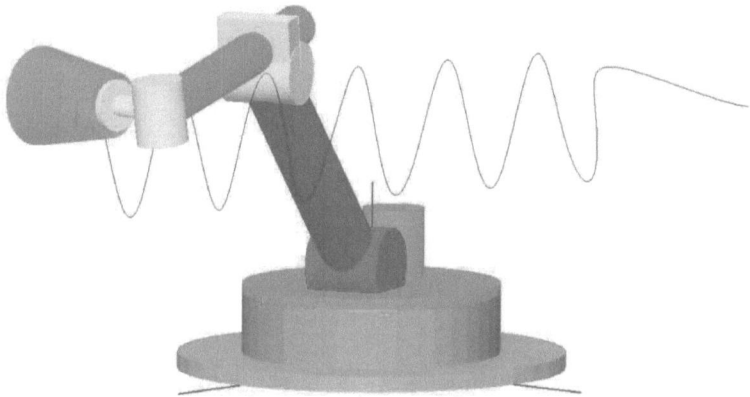

Figure 7.6 A simulation study on adaptive control of the industrial robot manipulator carrying a heavy tool

Example 7.2. The second case study is the industrial robot manipulator with 6 revolute joints, as we have studied in Chapter 4, see Figures 4.3 and 4.5. Its end-effector carries a heavy tool or load as a dominant object, as shown in Figure 7.6. The dynamic model and control strategy based on the isometric embedding (IE) theory have been well developed as well in Chapter 4, see the block-diagram in Figure 4.4 with the animation study. This time, instead of using the regular trajectory-tracking control scheme based on equation (4.9), we apply the adaptive control law given in equation (7.17) along with the adaptation law in equation (7.11) on this 6-joint industrial robot for tracking both position and orientation trajectories.

With a relatively large initial tracking deviation between the desired and actual starting points, the adaptation gain constant in the adaptation law (7.11) was set to be $K_a = 0.1$. The initial modeled dynamic parameters were set to be about $\pm 70 - 80\%$ over the actual parameters. Within the first second of control process, the entire adaptive control system may have to be struggled to overcome both the large initial

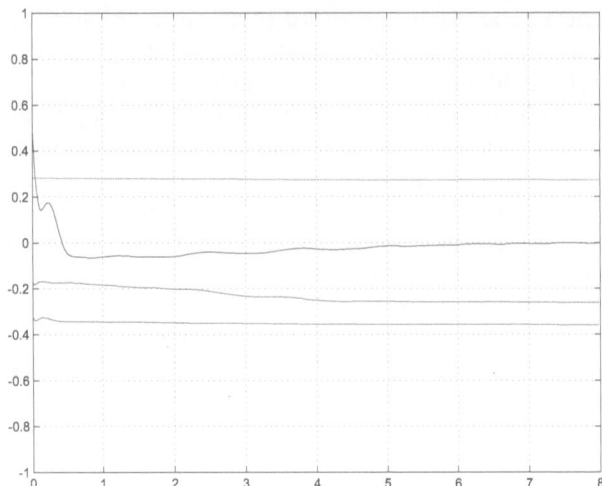

Figure 7.7 A simulation study on adaptive control of the industrial robot with the plot of four dynamic parameter adaptation, where the vertical axis is $\Delta\xi = \xi - \xi_m$, and the horizontal axis is time in sec.

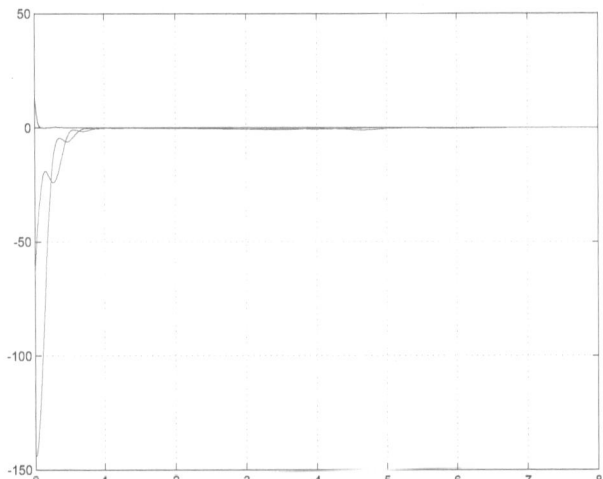

Figure 7.8 A simulation study on adaptive control of the industrial robot with the plot for the first, second and third terms of $\dot{V}(x)$ in equation (7.13) vs. time in sec.

tracking error and dynamic parameter deviation. However, a short moment later, the trajectory-tracking task is getting stabilized, as can be observed in Figure 7.7, where every component of the parameter deviation $\Delta\xi = \xi - \xi_m$ is getting flat, though they do not approach to zero.

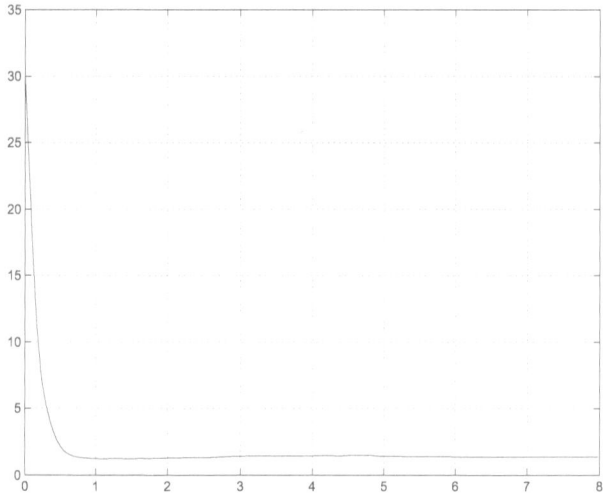

Figure 7.9 A simulation study on adaptive control of the industrial robot with the plot for the Lyapunov function value $V(x)$ in equation (7.6) vs. time in sec.

We have also closely monitored the time-derivative of the Lyapunov function $\dot{V}$ based on equation (7.13), which comprises three terms versus time: the first one is $\Delta\xi^T Q_{e0}^T Q_{v0} \Delta\xi$, the second one is $-e^T K_v K_p e$ and the third one is $-(\dot{e} + K_v e)^T K_c (\dot{e} + K_v e)_m$, as plotted in Figure 7.8. It can be realistically seen that within the first second, both the second and third terms of $\dot{V}$ are keeping negative against the first positive term. After less than a second of adjustment, they are all quickly approaching to zero. We set $K_c = 4$, $K_v = 4$ and $K_p = 49$ in the control law (7.17). The overall negative value of $\dot{V}$ is the key to keep the "energy" $V(x)$ monotonically going down in order to stabilize the entire adaptive control system with its tracking error quickly converging to zero. Figure 7.9 depicts the Lyapunov function value $V(x)$ versus time. It can be evidently seen that the value of $V(x)$ is dropped down sharply, and then it is keeping flat at a nonzero constant due to the fact that the adaptation law (7.11) cannot guarantee the dynamic parameter deviation $\Delta\xi$ approaching to zero.

The entire adaptive control process for this 6-joint industrial robot, as demonstrated in Figure 7.6, is tracking both a desired sine wave by the mass center of the tool and a desired orientation trajectory as well. The animation study evidently shows a success of the IE-based (isometric embedding) adaptive trajectory-tracking control scheme with complete 6 d.o.f. dynamic motion.

Example 7.3. We now turn to revisit the dynamic control of Stewart platform for such a typical closed parallel-chain system by adding a dynamic parameter adaptation. According to **the Principle of Duality**, as completely listed in Table 5.2 from Chapter 5, particularly under the IE (isometric embedding) control strategy, the dynamic formulations are different between open serial-chain and closed parallel-chain robotic systems, i.e.,

$$u = J^T \ddot{z} - \tau_g \quad \text{vs.} \quad \ddot{z} = J(u + \tau_g). \tag{7.18}$$

Figure 7.10 A simulation study on adaptive control of the closed parallel-chain Stewart platform system

The previous two examples: the ballbot balancing system and the 6-revolute-joint industrial robot, both are of open serial-chain mechanism. They deserve to have the first equation of (7.18) as their dynamic model. Whereas the dynamics of a Stewart platform system has to be governed by the second equation of (7.18). However, the second dynamic equation has an intrinsic issue that must be addressed before implementing the adaptive control scheme. Because the control input u is behind the Jacobian matrix in such a special formula of the second equation in (7.18), when resolving the control law, we have to multiply the left-hand side $\ddot{z}$ by the pseudo-inverse of the Jacobian matrix so that

$$u + \tau_g = J^{+}\ddot{z} = (J^T J)^{-1} J^T \ddot{z}.$$

Because the Jacobian matrix J for the isometric embedding z is a 12 by 6 tall matrix, the above resolved control formula is a least-square solution for the second dynamic equation in (7.18) based on linear algebra.

However, when the least-square solution is substituted back into the dynamic equation, $J(u + \tau_g)$ will no longer be equal exactly to $\ddot{z}$ due to $JJ^{+} = J(J^T J)^{-1}J^T \neq I$. This fact will cause a noticeable trajectory-tracking error. The main reason is obviously due to the commonly recognizable fact that a 3D orientation or rotation cannot be uniquely represented by a 3-dimensional so-called "orientation vector". This mathematical reality compels us to seek a higher dimensional isometric embedding to replace the impractical 3D "orientation vector", such as to split the three columns

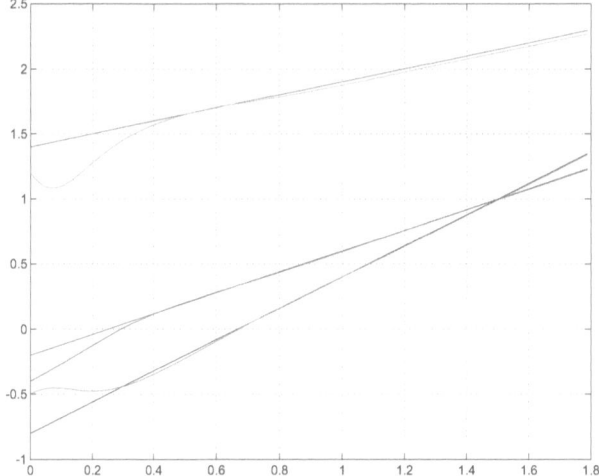

Figure 7.11 A simulation study on adaptive control of the Stewart platform with the plot for the desired and actual position trajectories. The vertical axis is position in meters and the horizontal axis is time in sec.

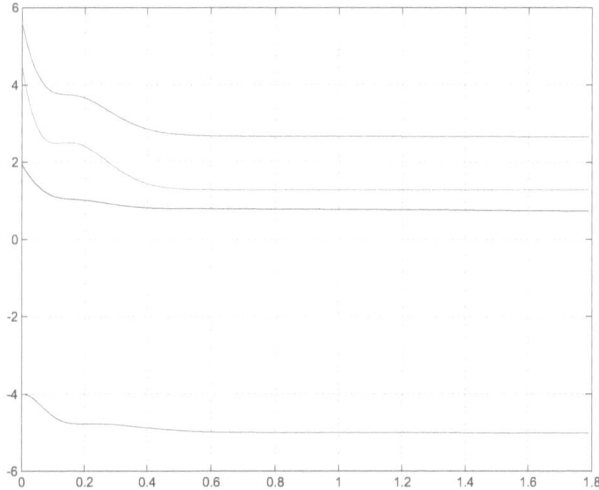

Figure 7.12 A simulation study on adaptive control of the Stewart platform with the plot for the dynamic parameter adaptation errors vs. time in sec.

of a rotation matrix to augment them and form a 9-dimensional vector with dynamic parameters coated. The detailed theoretical development of the isometric embedding can be found in Chapter 4 with the applications in Chapter 5.

Due to the above intrinsic issue in the dynamic formulation and solution for the closed parallel-chain robotic systems, such as a Stewart platform, it would be pre-

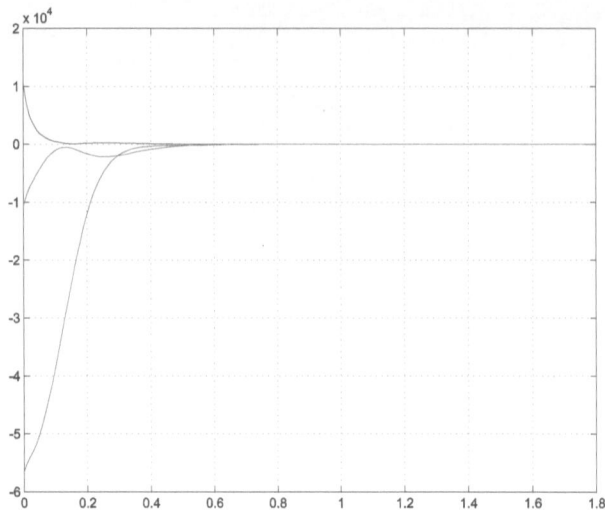

Figure 7.13 A simulation study on adaptive control of the Stewart platform with the plot for the three components of the time-derivative of Lyapunov function $\dot{V}(x)$ given in equation (7.13) vs. time in sec.

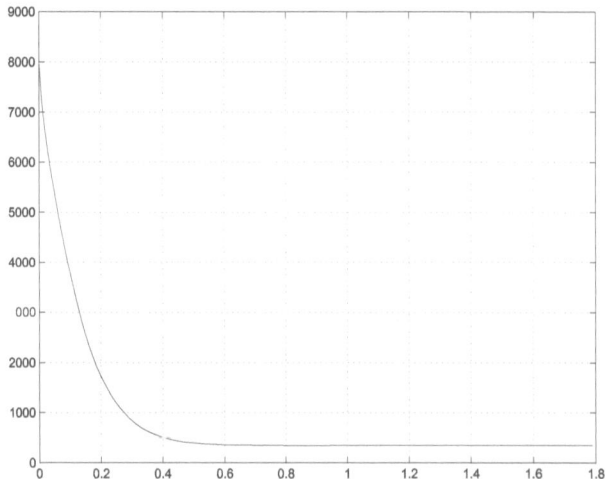

Figure 7.14 A simulation study on adaptive control of the Stewart platform with the plot for the Lyapunov function value $V(x)$ in equation (7.6) vs. time in sec.

dictably failed if we would directly impose an adaptation law on the Stewart platform system for its adaptive control trial. Let us verify our postulation by an animation study for the 6-leg Stewart platform, as shown in Figure 7.10.

In the animation study, if we adopt the following control law as a solution of the second dynamic equation in (7.18):

$$u = J^+\ddot{z} - \tau_g = J^+[\ddot{z}^d + K_v(\dot{z}^d - \dot{z}) + K_p(z^d - z)] - \tau_g,$$

then, it is difficult to add an adaptation law. As a result, the adaptive control trial is unsuccessful.

However, if we directly apply

$$Ju = [\ddot{z}^d + K_v(\dot{z}^d - \dot{z}) + K_p(z^d - z)] - J\tau_g$$

on the Stewart platform system, no matter it is feasible or not, the adaptive control animation result turns to be quite successful. Figure 7.11 plots both the desired and actual trajectories for the three components of the top panel mass center position versus time to see their quick convergence from their initial deviations. Figure 7.12 exhibits the dynamic parameter error $\Delta\xi$ versus time, where they are all approaching to nonzero constants. Figure 7.13 plots the three components of the time-derivative of the Lyapunov function $\dot{V}(x)$ versus time based on equation (7.13), while Figure 7.14 shows the trace of the Lyapunov function value $V(x)$ versus time. As can be clearly seen, the $V(x)$ value is quickly diving down to approach to a small positive constant, which is, however, not going to zero due to the nonzero constant parameter deviation, while the tracking error for both position and orientation can converge to zero. In the animation study, we set a small adaptation gain constant $K_a = 0.05$ for gradually adjusting the parameters, while the control gain constants are set as $K_c = 4$, $K_v = 6$ and $K_p = 49$.

REFERENCES

1. Sastry, S. and Bodson, M., (1989) Adaptive Control: Stability, Convergence, and Robustness. Prentice Hall, Englewood Cliffs, NJ.
2. Narendra, K.S. and A.M. Annaswamy, (1989) Stable Adaptive Systems, Prentice Hall, Englewood Cliffs, New Jersey.
3. Kristic, M., Kanellakopoulos, I. and Kokotovic, P., (1995) Nonlinear and Adaptive Control Design. John Wiley & Sons, New York.
4. Slotine, J. and Li, W., (1988) Adaptive Manipulator Control: A Case Study. IEEE Transactions on Automatic Control, Vol. 33-11.
5. Slotine, J. and Li, W., (1989) Composite Adaptive Control of Robot Manipulators. Automatica, Vol. 25-4.
6. Slotine, J. and Li, W., (1991) Applied Nonlinear Control. Prentice Hall, New Jersey.
7. Landau, I.D., Lozano, R. and M'Saad, M., (1998) Adaptive Control. Springer-Verlag, New York, NY.
8. Tao, G., (2003) Adaptive Control Design and Analysis. Wiley-Interscience, Hoboken, NJ.
9. Gu, E.Y.L., (1993) A Direct Adaptive Control Scheme for Under-Actuated Dynamic Systems. *Proc. 32nd IEEE Conference on Decision and Control.* San Antonio, TX, Dec., pp. 1625-1627.
10. Gu, E.Y.L. and Xu, Y., (1994) A Normal Form Augmentation Approach to Adaptive Control of Space Robot Systems. *Journal of Dynamics and Control*, Kluwer Academic Publishers, June.

8 Dynamics Modeling and Control of Cascaded Systems

8.1 DYNAMIC INTERACTIONS BETWEEN ROBOT AND ENVIRONMENT

A robotic system in many application environments is not standing alone. It often interacts with other work pieces, such as a tool or a fixture, etc. For many mobile robots, humanoid robots or walking robots, there are more chances of interactions between the robots and environments. How to model an interaction both kinematically and dynamically in order to more effectively control the robot for better work performance becomes a critical challenging topic in robotics research.

Figure 8.1 A human-machine interaction (HMI) model

Let us first look at a typical example where a humanoid robot is sitting on the driver seat to make itself ready for driving the vehicle, as shown in Figure 8.1. The humanoid robot is interacting with the car seat by its hip and two thighs, and also with the seat back by its torso back. The car seat is then interacting with the chassis as well as through the car suspension system and wheels to the road surface. This is a typical cascaded interaction between the humanoid robot and the environments. Let us start modeling it before developing its effective control strategies.

In general, a k-cascaded MIMO (Multi-Input Multi-Output) dynamic system can be modeled as a serial linkage and formulated in the following form:

$$\dot{x} = f(x) + G(x)y_1$$

$$y_1 = h_1(x, \xi_1)$$

$$\dot{\xi}_1 = \eta_1(x, \xi_1) + \Gamma_1(x, \xi_1)y_2$$

$$y_2 = h_2(\xi_1, \xi_2)$$

$$\vdots$$

$$\dot{\xi}_{k-1} = \eta_{k-1}(\xi_{k-2}, \xi_{k-1}) + \Gamma_{k-1}(\xi_{k-2}, \xi_{k-1})y_k$$

$$y_k = h_k(\xi_{k-1}, \xi_k)$$

$$\dot{\xi}_k = \eta_k(\xi_{k-1}, \xi_k) + \Gamma_k(\xi_{k-1}, \xi_k)u, \tag{8.1}$$

where $x \in \mathbb{R}^n$, $y_i \in \mathbb{R}^{l_i}$, $\xi_i \in \mathbb{R}^{m_i}$ for $i = 1, \cdots, k$, and $u \in \mathbb{R}^r$ is the overall control input. Since for the entire cascaded system, each stage is, in general, an MIMO nonlinear subsystem, the coefficient $G(x)$ for the input y_1 in (8.1) is an n by l_1 matrix, i.e., $G(x) = (g_1(x) \cdots g_{l_1}(x))$, where each column is an n-dimensional vector field of x. Also, each $\Gamma_i(\xi_{i-1}, \xi_i)$ is an m_i by l_{i+1} matrix. We may intentionally define each y_i and the overall input u in (8.1) to have the same dimension, i.e., $y_i \in \mathbb{R}^l$ and $u \in \mathbb{R}^l$ to make each stage be a square MIMO subsystem [5,6].

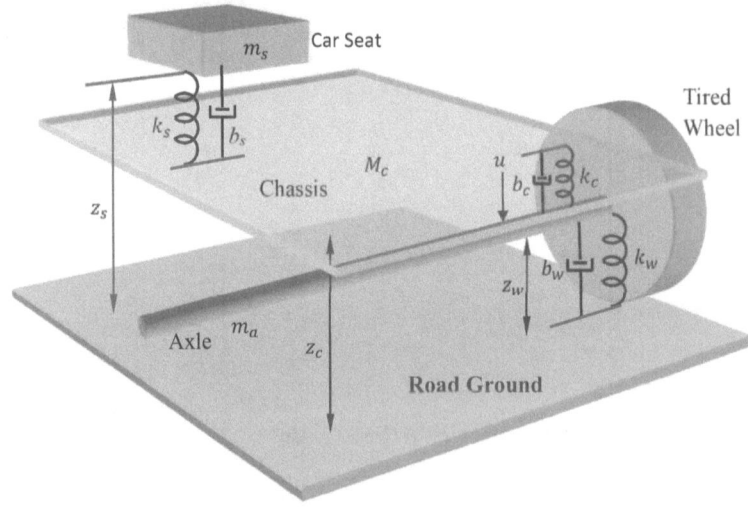

Figure 8.2 A car seat, chassis and wheel suspension systems model

As a detailed analysis, from Figure 8.2, we can readily find the following linear dynamic equations: for the car chassis,

$$M_c\ddot{z}_c + b_c(\dot{z}_c - \dot{z}_w) + k_c(z_c - z_w) + b_s(\dot{z}_c - \dot{z}_s) + k_s(z_c - z_s) = 0,$$

and also, for the car seat sat by a humanoid robot driver,

$$m_s \ddot{z}_s + b_s(\dot{z}_s - \dot{z}_c) + k_s(z_s - z_c) = f_h(z_s, \dot{z}_s),$$

and for the axle/wheel,

$$m_a \ddot{z}_w + b_w \dot{z}_w + k_w z_w + b_c(\dot{z}_w - \dot{z}_c) + k_c(z_w - z_c) = u,$$

where, b_w and k_w, b_c and k_c, and b_s and k_s are the positive damping coefficients and spring constants for the tired wheel, active chassis suspension system, and the car seat, respectively, M_c, m_a and m_s are the effective masses for the car chassis, axle/wheel and seat, respectively, and $f_h(z_s, \dot{z}_s)$ is the reacting force from the seat to the humanoid robot driver. This is a typical quarter-car active suspension control system model. To further study it and design an active suspension control strategy, we can now adopt the k-cascaded model that is formulated in equation (8.1).

Let $\xi_{11} = z_s$, $\xi_{12} = \dot{z}_s$, $\xi_{21} = z_c$, $\xi_{22} = \dot{z}_c$, $\xi_{31} = z_w$ and $\xi_{32} = \dot{z}_w$. Then, the above linear dynamic models can directly be converted into the following set of state-space equations:

$$\dot{x} = f(x) + G(x)y_1$$
$$y_1 = J^T(x)f_h(\xi_{11}, \xi_{12})$$

$$\begin{cases} \dot{\xi}_{11} = \xi_{12} \\ \dot{\xi}_{12} = -\frac{b_s}{m_s}\xi_{12} - \frac{k_s}{m_s}\xi_{11} + \frac{1}{m_s}f_h(\xi_{11}, \xi_{12}) + y_2 \end{cases}$$

$$y_2 = \frac{b_s}{m_s}\xi_{22} + \frac{k_s}{m_s}\xi_{21}$$

$$\begin{cases} \dot{\xi}_{21} = \xi_{22} \\ \dot{\xi}_{22} = -\frac{b_w}{M_c}\xi_{22} - \frac{k_w}{M_c}\xi_{21} - \frac{b_s}{M_c}(\xi_{22} - \xi_{12}) - \frac{k_s}{M_c}(\xi_{21} - \xi_{11}) + y_3 \end{cases}$$

$$y_3 = \frac{b_w}{M_c}\xi_{32} + \frac{k_w}{M_c}\xi_{31}$$

$$\begin{cases} \dot{\xi}_{31} = \xi_{32} \\ \dot{\xi}_{32} = -\frac{b_w}{m_a}\xi_{32} - \frac{k_w}{m_a}\xi_{31} - \frac{b_c}{m_a}(\xi_{32} - \xi_{22}) - \frac{k_c}{m_a}(\xi_{31} - \xi_{21}) + \frac{1}{m_a}u. \end{cases} \qquad (8.2)$$

The first x-equation represents the humanoid robot dynamics. The joint torque input $y_1 = \tau$ is due to a reacting force f_h from the car seat through its kinematic Jacobian matrix J based on the robot statics equation. This reacting force $f_h(\xi_{11}, \xi_{12})$ is obviously a function of the car seat displacement ξ_{11} and its velocity ξ_{12} with respect to the ground. The second ξ_1-equation of (8.2) is the car seat dynamics with an input y_2 that is a reacting force from the car chassis. The third ξ_2-equation describes the car chassis dynamics with an input y_3 that is a reacting force from the axle/tired wheel. The fourth ξ_3-equation represents the active suspension dynamics as well as the axle/tired wheel dynamics with an input u produced by the active suspension system to control the chassis for better absorbing any bump from the road ground. This input u now becomes an overall control for this $k = 4$-cascaded vehicle active suspension system.

In the above driver/active suspension example modeled as a $k = 4$-cascaded control system, only the humanoid robot driver x-equation is nonlinear, and the ξ_1, ξ_2 and ξ_3-equations are all linear. Actually, one can add any nonlinearity into those subsystems. Therefore, the k-cascaded system model given in (8.1) is a very general modeling formulation, as long as each subsystem can be connected to its pre-stage and post-stage subsystems together by their inputs and outputs. This is also an ideal model to represent a robot interaction with any complex environments. The control objective for such a k-cascaded system is to design the overall control law u to meet a set of desired specifications for the top x-system $\dot{x} = f(x) + G(x)y_1$.

8.2 CASCADED DYNAMICS MODELS WITH BACKSTEPPING CONTROL RECURSION

8.2.1 CONTROL DESIGN WITH THE LYAPUNOV DIRECT METHOD

We have studied the exact linearization procedures and static state-feedback control schemes in Chapter 6. The procedures are quite effective in many applications. However, they have to walk a common way through the *nonlinearity cancellation* between the system itself and the state-feedback control. In other words, the state-feedback control law given in either equation (6.39) or equation (6.55) is virtually an inverse of the system itself. However, sometimes we wish to design a control law directly without cancellation of all the nonlinearities. Motivated by this notion, we now develop a direct approach to the control design without going through the state-feedback linearization procedure.

For numerous control applications, we may be able to directly design a control law during the course of testing the stability for a given control system via the well-known Lyapunov direct method. For a better illustration, let us first look at a following so-called schistosomiasis infection and disease control model [8]:

$$\begin{cases} \dot{x}_1 = ax_2 - bx_1 + u \\ \dot{x}_2 = cx_1(1 - x_2) - \mu x_2, \end{cases} \tag{8.3}$$

where x_1 is the average number of schistosome worms per person and x_2 is the prevalence of patent infections in snails, and all the coefficients a, b, c and μ are positive constant parameters.

As can be observed, the origin $x = 0$ of the 2D state phase plane is an equilibrium point. Let us define a Lyapunov candidate $V(x) = \frac{1}{2}(x_1^2 + x_2^2)$ that is obviously an s.p.d.f. Its time-derivative becomes

$$\dot{V}(x) = x_1\dot{x}_1 + x_2\dot{x}_2 = ax_1x_2 - bx_1^2 + cx_1x_2 - cx_1x_2^2 - \mu x_2^2 + x_1u.$$

If the disease control input

$$u = -(a+c)x_2 + cx_2^2,$$

then, $\dot{V}(x) = -bx_1^2 - \mu x_2^2$. Clearly, $-\dot{V}(x)$ is an s.p.d.f. This means that the above control input can asymptotically stabilize the disease transmission system.

Another disease spreading and control model was studied in Chapter 6. Let us revisit it here again. Suppose that there is a fatal disease spreading across a village. If any one is infected, then this person will either die, or be cured with permanent immunity from the disease.

Since more infected people may cause either more deaths that reduce x_1, or more cures that increase x_2, the decreasing rate $-\dot{x}_1$ of the infected people should have a relation: $-\dot{x}_1 = ax_1 + \dot{x}_2$, where $a > 0$ is a proportional constant. Moreover, since both the larger number of the originally non-infected people x_2 and the larger number of the infected people x_1 will lead to more people being infected, the decreasing rate $-\dot{x}_2$ should be proportional to both x_1 and x_2 so that $\dot{x}_2 = -bx_1x_2$, where $b > 0$ is another proportional constant. Therefore, the entire disease spreading system reaches a final model in the following state-equation form:

$$\begin{cases} \dot{x}_1 = -ax_1 + bx_1x_2 \\ \dot{x}_2 = -bx_1x_2. \end{cases} \tag{8.4}$$

Once again, the state vector x of this disease-spreading system is

$$x = \begin{pmatrix} x_1 \\ x_2 \end{pmatrix}, \quad \text{and} \quad f(x) = \begin{pmatrix} -ax_1 + bx_1x_2 \\ -bx_1x_2 \end{pmatrix}.$$

Clearly, the above system model is nonlinear without any input. We may thus add an input excitation u on the first equation of (8.4) to achieve a so-called *disease control system model*:

$$\begin{cases} \dot{x}_1 = -ax_1 + bx_1x_2 + u \\ \dot{x}_2 = -bx_1x_2. \end{cases} \tag{8.5}$$

It can be easily found that the equilibrium point of the disease-control system under $u = 0$ is $x_1 = 0$. In other words, x_2 can be any arbitrary positive number, and thus the system has infinite number of equilibrium points that are continuously distributed on the positive half of the x_2-axis in the 2D phase plane.

Because $x_1(0) + x_2(0)$ represents the total number of initial lives at $t = 0$ in the village, during the disease spreading and developing period, the amount of people alive, including both infected and non-infected people, could be decreasing due to the deaths. Therefore, we may set $x_1^d = x_1(0) + x_2(0)$ as a desired target to control the system towards the healthy people $x_2 \to x_2^d$ as $t \to \infty$. Under such a control objective, the Lyapunov candidate can be defined as follows:

$$V(x) = \frac{1}{2}(x_1^2 + (x_2 - x_2^d)^2),$$

which is obviously an s.p.d.f. at the equilibrium point $x_e = \begin{pmatrix} 0 \\ x_2^d \end{pmatrix}$. Hence,

$$\dot{V} = x_1 \dot{x}_1 + (x_2 - x_2^d)\dot{x}_2 = -ax_1^2 + bx_1^2 x_2 + x_1 u - b(x_2 - x_2^d)x_1 x_2.$$

If we define a control law

$$u = -bx_1 x_2 + b(x_2 - x_2^d)x_2 - kx_1 \tag{8.6}$$

with a gain constant $k > 0$, then $\dot{V} = -(a+k)x_1^2 \le 0$. Furthermore, since the first equation of (8.5) after invoking the above control law (8.6) becomes $\dot{x}_1 = -(a+k)x_1 + b(x_2 - x_2^d)x_2$, by setting $x_1 \equiv 0$, we can see that only $x_2 = x_2^d$ or $x_2 = 0$ is in the set $S = \{x \in \mathbb{R}^2 \mid \dot{V}(x) \equiv 0\}$. Therefore, based on the La Salle's Theorem 6.2, the disease control system (8.5) under the control law (8.6) has an asymptotically stable equilibrium point x_e.

After the above two disease-control systems in equations (8.3) and (8.5) are re-solved by the Lyapunov direct control design procedure, a more general question arises: can the above control design procedure based on the Lyapunov direct method always work for every nonlinear control system? To answer this fundamental question, we need a further study of the theories behind the procedure.

First, for a general nonlinear system

$$\dot{x} = f(x, t), \quad x \in \mathbb{R}^n, \tag{8.7}$$

let us introduce the following two well-known theorems regarding the stability:

Theorem 8.1. – (La Salle-Yoshizawa) *Suppose that $x = 0$ is an equilibrium point of (8.7) and f is locally Lipschitz in x uniformly over t. Let $V(x)$ be an s.p.d.f. such that*

$$\dot{V} = \frac{\partial V}{\partial x} f(x, t) \le -\varepsilon(x) \le 0, \quad \forall t \ge 0, \ \forall x \in \mathbb{R}^n,$$

where $\varepsilon(\cdot)$ is a continuous function. Then, all solutions of (8.7) are globally uniformly bounded and satisfy

$$\lim_{t \to \infty} \varepsilon(x(t)) = 0.$$

In addition, if $\varepsilon(x)$ is an s.p.d.f., then the equilibrium point $x = 0$ is globally uniformly asymptotically stable.

Before we state the second theorem, let us give a definition of **invariant set** for an autonomous system:

$$\dot{x} = f(x), \quad x \in \mathbb{R}^n. \tag{8.8}$$

Definition 8.1. A set M is called an invariant set of (8.8) if any solution $x(t)$ that belongs to M at some time instant t_1 must also belong to M for all the future and past time, i.e.,

$$x(t_1) \in M \Rightarrow x(t) \in M, \quad \forall t \in \mathbb{R}.$$

It is called a positively invariant set if only for the future time:

$$x(t_1) \in M \Rightarrow x(t) \in M, \quad \forall t \ge t_1.$$

Theorem 8.2. – (La Salle) *Let Ω be a positively invariant set of (8.8). Let $V : \Omega \rightarrow \mathbb{R}_+$ be a differentiable function such that $\dot{V}(x) \leq 0$, $\forall x \in \Omega$. Let $S = \{x \in \Omega \mid V(x) = 0\}$, and M be the largest invariant set in S. Then, every bounded solution $x(t)$ starting in Ω converges to M as $t \rightarrow \infty$.*

Actually, this was the original theorem by La Salle, and Theorem 6.2 was its corollary. After introducing the above two fundamental theorems to reinforce the Lyapunov stability theory, we now consider a nonlinear system with a control input to see how we can take advantage of the control input to stabilize or improve the stability of the system. Given a single-input nonlinear control system

$$\dot{x} = f(x, u), \quad x \in \mathbb{R}^n, \quad u \in \mathbb{R}, \quad f(0, 0) = 0. \tag{8.9}$$

Definition 8.2. A smooth s.p.d.f. $V(x)$ is called a control Lyapunov function (c.l.f.) for (8.9) if

$$\inf_{u \in \mathbb{R}} \left\{ \frac{\partial V}{\partial x} f(x, u) \right\} < 0, \quad \forall x \neq 0.$$

This mathematical definition promotes the general s.p.d.f. requirement to a c.l.f. by adding a new "hope" condition to find an input u such that $\dot{V}$ is at least non-positive [1], even if the "hope" is little.

If a general single-input system is also affine of the input, then equation (8.9) is narrowed down to

$$\dot{x} = f(x) + g(x)u, \quad f(0) = 0. \tag{8.10}$$

We now start with (8.10) to interpret the above two fundamental stability theorems along with the c.l.f. definition.

Consider a state-feedback control law $u = \alpha(x)$ available for the system (8.10). Then, a scalar function $V(x)$ to be c.l.f. implies that

$$\frac{\partial V}{\partial x} f(x) + \frac{\partial V}{\partial x} g(x)\alpha(x) \leq -\varepsilon(x) \tag{8.11}$$

for a non-negative function or a p.d.f $\varepsilon(x)$. The equilibrium point $x = 0$ is asymptotically stable under the control law $u = \alpha(x)$ if $\varepsilon(x)$ can be at least a p.d.f. Therefore, the c.l.f. definition imposes a new requirement on the s.p.d.f. condition for a Lyapunov candidate $V(x)$. In other words, $V(x)$ is not only an s.p.d.f., but it is also able to offer the "hope" of finding a state-feedback control $u = \alpha(x)$ such that the scalar function $\varepsilon(x)$ in (8.11) will be at lease non-negative.

For example, let us test a scalar system given by

$$\dot{x} = xp(x) + u$$

for some continuous function $p(x)$. Since $x = 0$ is the equilibrium point for the system at the origin, let us define $V(x) = \frac{1}{2}x^2$ that is an s.p.d.f. for this scalar system. Then, $\dot{V} = x\dot{x} = x^2 p(x) + xu \leq -\varepsilon(x)$. Suppose that we choose $\varepsilon(x) = ax^2$ with a constant $a > 0$. This function $\varepsilon(x) = ax^2$ is obviously an s.p.d.f., and $\dot{V} = x^2 p(x) + xu \leq$

$-\varepsilon(x) = -ax^2$. Let us solve this inequality for the input u by treating it as an equation to find a control law:

$$u = -ax - xp(x) = \alpha(x).$$

Thus, the Lyapunov function $V(x) = \frac{1}{2}x^2$ for this particular scalar system is also a c.l.f.

Another example is a 2D system with a single input:

$$\begin{cases} \dot{x}_1 = x_1^2 x_2^3 + x_2^2 \\ \dot{x}_2 = -x_1^2 + x_2^2 + u. \end{cases} \tag{8.12}$$

Because the equilibrium point is at $x = \begin{pmatrix} x_1 \\ x_2 \end{pmatrix} = 0$, let $V(x) = \frac{1}{2}(x_1^2 + x_2^2)$ that is an s.p.d.f. Then,

$$\dot{V} = x_1 \dot{x}_1 + x_2 \dot{x}_2 = x_1^3 x_2^3 + x_1 x_2^2 - x_1^2 x_2 + x_2^3 + x_2 u \leq -\varepsilon(x).$$

Let $\varepsilon(x) = ax_2^2$ with a constant $a > 0$. By equating $\dot{V} = x_1^3 x_2^3 + x_1 x_2^2 - x_1^2 x_2 + x_2^3 + x_2 u = -\varepsilon(x) = -ax_2^2$, a control law can be immediately resolved by

$$u = \alpha(x) = -x_1^3 x_2^2 - x_1 x_2 + x_1^2 - x_2^2 - ax_2.$$

However, in this case, $\varepsilon(x) = ax_2^2$ is just a positive function, instead of a p.d.f. Therefore, the control law can at least stabilize the 2D system, but whether it is asymptotic has to be answered by the La Salle's Theorem.

If the number of inputs in a system is more than one, say $u \in \mathbb{R}^r$ for $r > 1$, then we have the following general equation that is also affine of u:

$$\dot{x} = f(x) + \sum_{i=1}^{r} g_i(x)u_i, \quad f(0) = 0. \tag{8.13}$$

The implication of the c.l.f. condition now becomes

$$\frac{\partial V}{\partial x} f(x) + \sum_{i=1}^{r} \frac{\partial V}{\partial x} g_i(x)\alpha_i(x) \leq -\varepsilon(x),$$

for each component of a control law $u_i = \alpha_i(x)$, $i = 1, \cdots, r$. Once a scalar function $\varepsilon(x) \geq 0$ is chosen, the solutions of the inputs u_i's may not be unique in many cases, and one may select them to avoid any state variable x_i appearing in the denominator, because the equilibrium point is at $x_i = 0$ for each $i = 1, \cdots, n$. If finding a state-feedback control $u = \alpha(x)$ is hopeless, either the Lyapunov candidate $V(x)$ has not been a c.l.f. yet, or the system is uncontrollable. In the former case, one may either try a different non-negative function $\varepsilon(x)$ to reduce its positiveness, or redefine a new Lyapunov candidate to try it again, or both. However, in the latter case, we have to remodel the system to be controllable. Regarding the controllability test procedure for nonlinear systems, one may refer to Chapter 6, and also the literature [2–4].

For example, a 2-input system is given as follows:

$$\begin{cases} \dot{x}_1 = -x_3 + e^{x_2}u_1 \\ \dot{x}_2 = x_1 + x_2^2 + e^{x_2}u_2 \\ \dot{x}_3 = x_1 - x_2 - x_3. \end{cases} \qquad (8.14)$$

Clearly, the equilibrium point of this system is $x = 0$. Let

$$V(x) = \frac{1}{2}(x_1^2 + x_2^2 + x_3^2),$$

which is obviously an s.p.d.f. Then,

$$\dot{V} = x_1 x_2 - x_2 x_3 + x_3^3 - x_3^2 + x_1 e^{x_2}u_1 + x_2 e^{x_2}u_2 \le -\varepsilon(x)$$

for some function $\varepsilon(x) \ge 0$. Let us more ambitiously choose an s.p.d.f. $\varepsilon(x) = x_1^2 + x_2^2 + x_3^2$. We can thus solve the two inputs by

$$u_1 = \alpha_1(x) = -e^{-x_2}(x_1 + x_2) \text{ and } u_2 = \alpha_2(x) = e^{-x_2}(x_3 - x_2^2 - x_2),$$

and none of them has any x_i appearing in denominator. Since this $\varepsilon(x)$ is an s.p.d.f., the control law $u = (\alpha_1(x) \ \alpha_2(x))^T$ can globally asymptotically stabilize the system.

8.2.2 BACKSTEPPING RECURSIONS IN CONTROL DESIGN

A backstepping approach for control design is to recursively run "step back" until a control law is determined. This control design procedure is often used to deal with a cascaded control system [1]. It will be found very useful in modeling and control of robot-environment dynamic interactive systems. To introduce the concepts and formulations of this interesting control design approach in more progressive manner, let us start studying a simple backstepping case.

First, let a single-input system in the form of equation (8.10) be augmented by an integrator:

$$\begin{cases} \dot{x} = f(x) + g(x)\xi \\ \dot{\xi} = u, \end{cases} \qquad (8.15)$$

and let the first equation of (8.15) have a differentiable state-feedback virtual control law $\xi = \sigma(x)$. Namely, there are two p.d.f.'s $V(x)$ and $\varepsilon(x)$ such that

$$\frac{\partial V}{\partial x}f(x) + \frac{\partial V}{\partial x}g(x)\sigma(x) \le -\varepsilon(x).$$

Then, by defining an error function $e(x, \xi) = \xi - \sigma(x)$, a new Lyapunov candidate is proposed by

$$V_a(x, \xi) = V(x) + \frac{1}{2}e^2 = V(x) + \frac{1}{2}(\xi - \alpha(x))^2,$$

which should be a c.l.f. for the augmented system (8.15). In other words, there exists a feedback control law $u = \alpha(x, \xi)$ such that $x = 0$ along with $e = \xi - \sigma(x) = 0$ is an asymptotically stable equilibrium point of (8.15).

In fact, $\xi = e(x, \xi) + \sigma(x)$, and

$$\dot{V}_a(x, \xi) = \dot{V}(x) + e(\dot{\xi} - \dot{\sigma}(x)) = \dot{V}(x) + e(u - \dot{\sigma}(x)). \tag{8.16}$$

Notice that

$$\dot{V}(x) = \frac{\partial V}{\partial x} f(x) + \frac{\partial V}{\partial x} g(x)(e + \sigma) \le -\varepsilon(x) + \frac{\partial V}{\partial x} g(x)e,$$

and

$$\dot{\sigma}(x) = \frac{\partial \sigma}{\partial x} f(x) + \frac{\partial \sigma}{\partial x} g(x)\xi.$$

Substituting them into (8.16), it can be clearly seen that if

$$u = \alpha(x, \xi) = -c(\xi - \sigma(x)) + \frac{\partial \sigma(x)}{\partial x}(f(x) + g(x)\xi) - \frac{\partial V(x)}{\partial x} g(x), \tag{8.17}$$

with a constant $c > 0$, then the time-derivative of the new Lyapunov candidate becomes

$$\dot{V}_a(x, \xi) \le -\varepsilon(x) - ce^2,$$

and obviously, $\varepsilon(x) + ce^2$ is also a p.d.f.

Let us now look at an example to illustrate the procedure of finding the control law $u = \alpha(x, \xi)$:

$$\begin{cases} \dot{x} = x^2 + x\xi \\ \dot{\xi} = u. \end{cases} \tag{8.18}$$

In the first equation, we have an s.p.d.f. $V(x) = \frac{1}{2}x^2$ defined and $\dot{V} = x\dot{x} = x^3 + x^2\xi \le -\varepsilon(x)$. Let $\varepsilon(x) = x^2$ such that the *virtual control law* is $\xi = \sigma(x) = -x - 1$.

Now, defining an error state $e = \xi - \sigma(x) = \xi + x + 1$, we have

$$\begin{cases} \dot{x} = x^2 + x\xi = x^2 + x(e - x - 1) \\ \dot{e} = u + x^2 + x(e - x - 1). \end{cases} \tag{8.19}$$

Since the new state equation for both x and e has $x = 0$ and $e = 0$ as its equilibrium point, we choose $V_a = \frac{1}{2}(x^2 + e^2)$ that is an s.p.d.f. Thus,

$$\dot{V}_a = x\dot{x} + e\dot{e} = -x^2 + eu + x^2e + xe^2 - xe \le -\varepsilon_a(x, e).$$

Let $\varepsilon_a(x, e) = x^2 + e^2$. We finally obtain a control law for (8.18):

$$u = \alpha(x, \xi) = -x^2 - xe + x - e = -2x^2 - x\xi - \xi - x - 1.$$

If we use the general form of feedback control law in (8.17) and pick $c = 1$, since $\frac{\partial \sigma}{\partial x} = -1$ and $\frac{\partial V(x)}{\partial x} = x$, we reach

$$u = -(\xi + x + 1) - (x^2 + x\xi) - x^2 = -2x^2 - x\xi - \xi - x - 1,$$

which is the same as the above $\alpha(x, \xi)$.

Secondly, we extend the backstepping design procedure for more complex cascaded control systems. Let the single-input system (8.10) be augmented by a linear system through its output:

$$\begin{cases} \dot{x} = f(x) + g(x)y, \quad f(0) = 0, \ x \in \mathbb{R}^n, \ y \in \mathbb{R} \\ \dot{\xi} = A\xi + Bu, \quad y = C\xi, \ \xi \in \mathbb{R}^k, \ u \in \mathbb{R}. \end{cases} \tag{8.20}$$

If the (second) linear system is minimum-phase with relative degree $r_d = 1$, i.e., $CB \neq 0$, and the first system satisfies the condition (8.11) for both s.p.d.f.'s $V(x)$ and $\varepsilon(x)$ with a smooth virtual control law $y = \sigma(x)$, then there exists a feedback control $u = \alpha(x, \xi)$ to globally asymptotically stabilize the equilibrium point at $x = 0$ and $\xi = 0$ for the entire system (8.20).

Similarly, by defining an error state

$$e(x, \xi) = y - \sigma(x) = C\xi - \sigma(x),$$

a new Lyapunov candidate can be proposed by

$$V_a(x, \xi) = V(x) + \frac{1}{2}e^2 = V(x) + \frac{1}{2}(C\xi - \sigma(x))^2.$$

Since the time-derivative of the output is $\dot{y} = C\dot{\xi} = CA\xi + CBu$, and the decoupling matrix $D = CB$ in this SISO case is a non-zero scalar due to $r_d = 1$ so that

$$u = \frac{1}{CB}(\dot{y} - CA\xi) = \frac{1}{CB}(v - CA\xi), \tag{8.21}$$

where a new input $v = \dot{y}$ is defined. After taking the time-derivative on the new Lyapunov candidate, we obtain almost the same result as (8.16), except that u in the last term of (8.16) is replaced by $\dot{y} = v$. Therefore, the new input v should have exactly the same formula as (8.17). Finally, substituting it into (8.21) yields

$$\begin{aligned} u &= \alpha(x, \xi) = \frac{1}{CB}[-c(C\xi - \sigma(x)) - CA\xi + \\ &+ \frac{\partial \sigma(x)}{\partial x}(f(x) + g(x)C\xi) - \frac{\partial V(x)}{\partial x}g(x)], \quad c > 0. \end{aligned} \tag{8.22}$$

As a more general case, the system (8.10) is now augmented by a nonlinear system, instead of a linear system, through its output, i.e.,

$$\begin{cases} \dot{x} = f(x) + g(x)y, \quad f(0) = 0, \ x \in \mathbb{R}^n, \ y \in \mathbb{R} \\ \dot{\xi} = \beta(x, \xi) + \gamma(x, \xi)u, \quad y = h(\xi), \ h(0) = 0, \ \xi \in \mathbb{R}^k, \ u \in \mathbb{R}. \end{cases} \tag{8.23}$$

Under the similar conditions, the (second) nonlinear system has a constant relative degree $r_d = 1$ and its zero dynamics is asymptotically stable with respect to ξ and y as its output, while the first system satisfies the condition (8.11) for both s.p.d.f.'s $V(x)$ and $\varepsilon(x)$ with a smooth virtual control law $y = \sigma(x)$. Then, there exists a feedback control $u = \alpha(x, \xi)$ to globally asymptotically stabilize the equilibrium point at $x = 0$ and $\xi = 0$ for the cascaded system (8.23).

By following the same procedure as the above linear augmentation case in equation (8.20),

$$\dot{y} = L_\beta h(\xi) + (L_\gamma h(\xi))u = \frac{\partial h(\xi)}{\partial \xi}\beta(x,\xi) + \frac{\partial h(\xi)}{\partial \xi}\gamma(x,\xi)u.$$

Since $L_\gamma h(\xi) \neq 0$ due to the condition $r_d = 1$, we can resolve the input u by

$$u = \left(\frac{\partial h(\xi)}{\partial \xi}\gamma(x,\xi)\right)^{-1}\left(v - \frac{\partial h(\xi)}{\partial \xi}\beta(x,\xi)\right),$$

where a new input $v = \dot{y}$ is defined here, too.

Because the remaining steps, including to define a new Lyapunov candidate and take its time-derivative, are exactly the same as the second case, we finally achieve a control law for the nonlinear cascaded system (8.23):

$$u = \left(\frac{\partial h(\xi)}{\partial \xi}\gamma(x,\xi)\right)^{-1}\left[-c(h(\xi) - \sigma(x)) - \frac{\partial h(\xi)}{\partial \xi}\beta(x,\xi) + \right.$$
$$\left. + \frac{\partial \sigma(x)}{\partial x}(f(x) + g(x)h(\xi)) - \frac{\partial V(x)}{\partial x}g(x)\right], \quad c > 0. \qquad (8.24)$$

As a real example, let the 2D system in (8.12) be the first nonlinear equation of (8.23). Then, it is cascaded by another 2D nonlinear simple-pendulum mechanical system through its output, i.e.,

$$\begin{cases} \dot{x}_1 = x_1^2 x_2^3 + x_2^2 \\ \dot{x}_2 = -x_1^2 + x_2^2 + y, \end{cases} \quad y = a_1\xi_1 + a_2\xi_2, \quad \begin{cases} \dot{\xi}_1 = \xi_2 \\ \dot{\xi}_2 = -b\xi_2 - k\sin\xi_1 + u, \end{cases} \qquad (8.25)$$

where a_1, a_2, b and k are all positive real constants. First, the x-system has been analyzed in the previous example to have a c.l.f.

$$V(x) = \frac{1}{2}(x_1^2 + x_2^2)$$

and a virtual control law

$$y = \sigma(x) = -x_1^3 x_2^2 - x_1 x_2 + x_1^2 - x_2^2 - cx_2$$

for a constant $c > 0$. However, $\varepsilon(x) = cx_2^2$ is just non-negative but is not a p.d.f., lets alone an s.p.d.f. However, we can apply the La Salle's Theorem to find that it is also asymptotically stabilized by the virtual control law. Now, we turn to investigate the ξ-system with its output $y = h(\xi)$. Under this output, the relative degree $r_d = 1$, because

$$L_\gamma h(\xi) = (a_1 \ a_2) \begin{pmatrix} 0 \\ 1 \end{pmatrix} = a_2 \neq 0.$$

Thus, the exactly linearized system is only 1-dimensional, i.e., $\dot{y} = v$ for a new input v based on the input-output linearization procedure in Chapter 6.

As a counterpart, the 1-dimensional internal dynamics that is unlinearizable can be defined by

$$\dot{\xi}_1 = \xi_2 = \frac{1}{a_2}(y - a_1\xi_1).$$

By turning off the output, i.e., $y = 0$, we obtain a zero dynamics

$$\dot{\xi}_1 = -\frac{a_1}{a_2}\xi_1$$

that is obviously asymptotically stable due to $a_1/a_2 > 0$.

After both the two cascaded systems are tested and all the conditions for (8.23) hold, we can utilize the suggested control law (8.24) to determine the input $u = \alpha(x, \xi)$ for the entire cascaded system given in (8.25), where

$$\beta(x, \xi) = \begin{pmatrix} \xi_2 \\ -b\xi_2 - k\sin\xi_1 \end{pmatrix}, \quad \gamma(x, \xi) = \begin{pmatrix} 0 \\ 1 \end{pmatrix}, \quad \frac{\partial h(\xi)}{\partial \xi} = (a_1 \ a_2),$$

and

$$\frac{\partial \sigma(x)}{\partial x} = (-3x_1^2 x_2^2 - x_2 + 2x_1 \quad -2x_1^3 x_2 - x_1 - 2x_2 - c), \quad \text{and} \quad \frac{\partial V(x)}{\partial x} = (x_1 \ x_2).$$

As a summary, a complete flowchart of the backstepping control design approach is depicted in Figure 8.3. The nonlinear augmented systems control design through the backstepping procedure can be extended to a nonlinear k-cascaded MIMO (Multi-Input Multi-Output) dynamic system given in equation (8.1).

It can be interpreted that for the i-th ξ-subsystem, $y_i = h_i(\xi_{i-1}, \xi_i)$ is an output function for the equation $\dot{\xi}_i = \eta_i(\xi_{i-1}, \xi_i) + \Gamma_i(\xi_{i-1}, \xi_i)y_{i+1}$ with an input y_{i+1}. We assume that:

1. For the top x-subsystem $\dot{x} = f(x) + G(x)y_1$ in (8.1), there exists a virtual control law $y_1 = \sigma_1(x)$ to asymptotically stabilize the x-subsystem at its equilibrium point $x = 0$; and
2. For the i-th input-output ξ-subsystem, the relative degree of each output channel is $r_{dj}^i = 1$ for $j = 1, \cdots, l_i$ so that the total relative degree is $r_d^i = l_i$, the dimension of the output y_i, and it is also an asymptotically minimum-phase subsystem for $i = 1, \cdots, k$.

The first assumption implies that there is a Lyapunov candidate $V_1(x)$ that is an s.p.d.f. such that

$$\dot{V}_1(x) = \frac{\partial V_1}{\partial x}f(x) + \frac{\partial V_1}{\partial x}G(x)\sigma_1(x) \leq -\varepsilon_1(x) \tag{8.26}$$

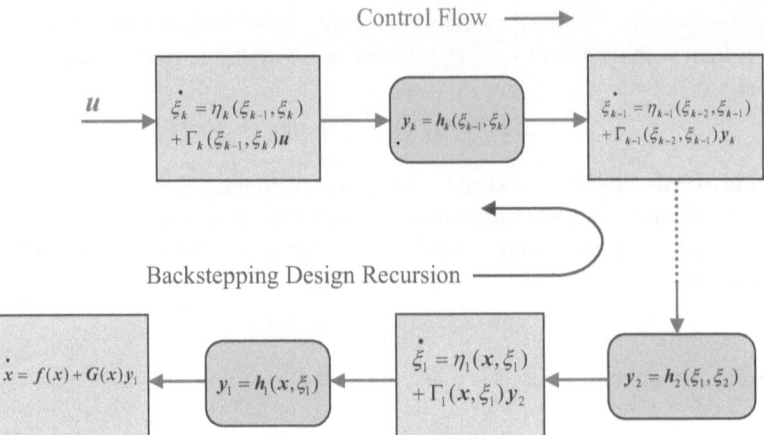

Figure 8.3 A flowchart of backstepping control design recursion for a k-cascaded dynamic system

for an s.p.d.f. $\varepsilon_1(x)$, or just a semi-positive-definite function $\varepsilon_1(x) \geq 0$ if every condition of the La Salle's Theorem holds. The second assumption imposed on each input-output ξ-subsystem ensures that its feedback linearization requires no higher than the first-order time-derivative of the output plus its internal dynamics is asymptotically stable.

After the above interpretation, based on the theory of the state-feedback linearization, the first time-derivative of the output of the first ξ-subsystem ($i = 1$) is given by

$$\dot{y}_1 = \frac{\partial h_1}{\partial x} f(x) + \frac{\partial h_1}{\partial x} G(x) y_1 + \frac{\partial h_1}{\partial \xi_1} \eta_1(x, \xi_1) + \left(\frac{\partial h_1}{\partial \xi_1} \Gamma_1(x, \xi_1) \right) y_2, \qquad (8.27)$$

and the coefficient of the input y_2 here must be an l by l non-singular matrix due to the second assumption, which is known as a decoupling matrix $D(x, \xi_1)$ in a certain neighborhood of the equilibrium point at $x = 0$ and $\xi_1 = 0$ for the first ξ-subsystem in (8.1). Therefore, the input of the first ξ-subsystem can be resolved by

$$y_2 = \left(\frac{\partial h_1}{\partial \xi_1} \Gamma_1(x, \xi_1) \right)^{-1} \left[v_1 - \frac{\partial h_1}{\partial x} f(x) - \frac{\partial h_1}{\partial x} G(x) y_1 - \frac{\partial h_1}{\partial \xi_1} \eta_1(x, \xi_1) \right] \qquad (8.28)$$

with a new input $v_1 = \dot{y}_1$ to be further determined.

Since the output y_1 of the first ξ-subsystem is also the input of the top x-subsystem, we wish this output $y_1 = h_1(x, \xi_1)$ could be as close to the virtual control $\sigma_1(x)$ as possible. Namely, by defining an error function

$$e_1(x, \xi_1) = y_1 - \sigma_1(x) = h_1(x, \xi_1) - \sigma_1(x),$$

we wish e_1 would go to zero asymptotically. Thus, an *extended Lyapunov candidate* for the first set ($i = 1$) of cascaded subsystems can be defined as follows:

$$V_2(x, \xi_1) = V_1(x) + \frac{1}{2} e_1^T e_1. \qquad (8.29)$$

Its time-derivative becomes

$$
\begin{aligned}
\dot{V}_2 &= \dot{V}_1 + e_1^T \dot{e}_1 = \frac{\partial V_1}{\partial x} f(x) + \frac{\partial V_1}{\partial x} G(x) y_1 + e_1^T (\dot{y}_1 - \dot{\sigma}_1) \\
&= \frac{\partial V_1}{\partial x} f(x) + \frac{\partial V_1}{\partial x} G(x)(\sigma_1 + e_1) + e_1^T (\dot{y}_1 - \dot{\sigma}_1) \\
&\leq -\varepsilon_1(x) + \frac{\partial V_1}{\partial x} G(x) e_1 + e_1^T (\dot{y}_1 - \dot{\sigma}_1).
\end{aligned}
\tag{8.30}
$$

Now, let

$$
v_1 = \dot{y}_1 = -a_1 e_1 + \dot{\sigma}_1 - G^T \left(\frac{\partial V_1}{\partial x} \right)^T.
\tag{8.31}
$$

Substituting it into (8.30), the time-derivative of the extended Lyapunov function becomes

$$
\dot{V}_2 \leq -\varepsilon_1(x) - a_1 e_1^T e_1 \leq -\varepsilon_2(x, \xi_1)
$$

with an s.p.d.f. $\varepsilon_2(x, \xi_1)$ and a positive gain constant $a_1 > 0$. Substituting equation (8.31) into (8.28), we obtain a new virtual control $y_2 = \sigma_2(x, \xi_1)$ for the first set of cascaded subsystem.

By repeating the same procedure for the second set ($i = 2$) of ξ-subsystem again, first, let the error

$$
e_2(x, \xi_1, \xi_2) = y_2 - \sigma_2(x, \xi_1) = h_2(\xi_1, \xi_2) - \sigma(x, \xi_1).
$$

Then, similar to (8.29), we can define a new extended Lyapunov candidate as follows:

$$
V_3(x, \xi_1, \xi_2) = V_2(x, \xi_1) + \frac{1}{2} e_2^T e_2 = V_1(x) + \frac{1}{2} e_1^T e_1 + \frac{1}{2} e_2^T e_2.
\tag{8.32}
$$

Likewise, without need of further explanation, the new input y_3 to the second set of ξ-subsystem turns out to be

$$
y_3 = \left(\frac{\partial h_2}{\partial \xi_2} \Gamma_2(\xi_1, \xi_2) \right)^{-1} \left[v_2 - \frac{\partial h_2}{\partial \xi_1} \eta_1(x, \xi_1) - \frac{\partial h_2}{\partial \xi_1} \Gamma_1(x, \xi_1) y_2 - \frac{\partial h_2}{\partial \xi_2} \eta_2(\xi_1, \xi_2) \right]
\tag{8.33}
$$

with a new input $v_2 = \dot{y}_2$, once again, to be further determined. To do so, similar to (8.30),

$$
\dot{V}_3 \leq -\varepsilon_2(x, \xi_1) + \frac{\partial V_2}{\partial \xi_1} \Gamma_1(x, \xi_1) e_2 + e_2^T (\dot{y}_2 - \dot{\sigma}_2).
\tag{8.34}
$$

Let the new input

$$
v_2 = \dot{y}_2 = -a_2 e_2 + \dot{\sigma}_2 - \Gamma_1^T \left(\frac{\partial V_2}{\partial \xi_1} \right)^T.
\tag{8.35}
$$

Substituting it into (8.34) yields

$$
\dot{V}_3 \leq -\varepsilon_2(x, \xi_1) - a_2 e_2^T e_2 \leq -\varepsilon_3(x, \xi_1, \xi_2)
$$

with another p.d.f. $\varepsilon_3(x,\xi_1,\xi_2)$ and another positive gain constant $a_2 > 0$. Substituting (8.35) into (8.33), we achieve a new virtual control $y_3 = \sigma_3(x,\xi_1,\xi_2)$ for the second set ($i = 2$) of cascaded subsystems.

Continuing the above recursion up to k steps, we will finally be able to reach and resolve the overall input u for the k-cascaded system control design. The final Lyapunov function is constructed by

$$V_{k+1}(x,\xi_1,\cdots,\xi_k) = V_1(x) + \frac{1}{2}\sum_{i=1}^{k} e_i^T e_i \tag{8.36}$$

with each error function $e_i = y_i - \sigma_i(x,\xi_1,\cdots,\xi_{i-1})$. It can also be observed from (8.30) and (8.34) that at each recursive step, the time-derivative of each extended Lyapunov function is always in the following form:

$$\dot{V}_{i+1} \leq -\varepsilon_i(x,\xi_1,\cdots,\xi_{i-1}) + \frac{\partial V_i}{\partial \xi_{i-1}}\Gamma_{i-1}(\xi_{i-2},\xi_{i-1})e_i + e_i^T(\dot{y}_i - \dot{\sigma}_i)$$

so that each new input is defined by

$$v_i = \dot{y}_i = -a_i e_i + \dot{\sigma}_i - \Gamma_{i-1}^T\left(\frac{\partial V_i}{\partial \xi_{i-1}}\right)^T.$$

This is due to the common fact that the "distance" between each output y_i and the corresponding virtual control σ_i demands converging to zero. In summary, Figure 8.3 shows the overall backstepping control design procedure for a k-cascaded dynamic system.

The k-cascaded systems modeling and backstepping control design procedure will be very useful in analysis and control of robot-environment dynamic interaction systems [5–7]. In next section, we will extensively discuss how to model a humanoid robot either statically or dynamically interacting with the environments, as well as how to design an overall control to meet certain desired objectives or criteria. The example in the early section, as shown in Figures 8.1 and 8.2, is one of the most representative applications to modeling a humanoid-environment interaction system and design an active vehicle suspension control system to enhance the vehicle ride quality for more effective performance improvement. More research and investigation on robot-environment interaction can be further referred to the literature [9–13].

8.3 MODELING AND INTERACTIVE CONTROL OF ROBOT-ENVIRONMENT SYSTEMS

Example 8.1. One of the most typical examples for robots interacting with the environments is operating a picking/placing task, as shown in Figure 8.4, where a dual-arm industrial robot with 17 revolute-joints is to carry a heavy block and place it on a table. When the two grippers press the two opposite side surfaces of the load and lift it up, the two end-effectors of the dual-arm robot act a vertical lift-up force to

Figure 8.4 A dual-arm industrial robot is picking up and moving a load block

overcome the load gravity. If the movement also has additional accelerations, more forces in different directions will be imposed on the two end-effectors of the robot. Suppose that the 17-joint dual-arm robot has the following dynamic equation:

$$\dot{x} = f(x) + G(x)y, \tag{8.37}$$

where based on equation (6.52),

$$f(x) = \begin{pmatrix} \dot{q} \\ W^{-1}((\tfrac{1}{2}W_d - W_d^T)\dot{q} + \tau_g) \end{pmatrix} \quad \text{and} \quad G(x) = \begin{pmatrix} O \\ W^{-1} \end{pmatrix},$$

along with y as the joint torque vector for both static and dynamic balances. Then, this additional joint torque vector y can be given by

$$y = J^T F, \quad \text{or} \quad y = J_z^T \ddot{z}. \tag{8.38}$$

The first equation is a model of statics, while the second equation is a model of dynamics in the isometric embedding form. Namely, if the load block is lifted up by the two grippers of the dual-arm robot without movement, then the joint torque vector y follows the first equation of statics. If the load block is also moving to have both position and orientation changes in the 3D task space in addition to being lifted, then, it should follow the second equation of dynamics. Between the two equations in (8.38) for the joint torque vector y of the picking/placing task-operating robot, the Jacobian matrix J in the statics equation is just a regular 12 by 17 augmented Jacobian matrix as a tangent-space transformation from the 17 joint velocities $\dot{q}_i$'s to the $6 \times 2 = 12$-dimensional wrench F for the two end-effectors of the dual-arm manipulator. In contrast, the matrix J_z in the dynamics equation is a genuine mathematical Jacobian matrix $J_z = \frac{\partial \zeta(q)}{\partial q}$ for the isometric embedding $z = \zeta(q)$ of the load block associated with the dual-arm robot. The dimension of the isometric embedding $z = \zeta(q)$ can be either 9 by 1 if equation (4.4) is adopted or 12 by 1 if equation (4.6) is employed, as developed in Chapter 4.

If the statics equation from (8.38) is adopted in the motionless case, then no more state-space equation associated with the static wrench F as a ξ-subsystem is cascaded with the dual-arm robot x-equation (8.37). However, if the load block has both position and orientation changes, the dynamics equation from (8.38) has to be employed. Let the acceleration $\ddot{z} = F_z(z, \dot{z})$ as a function of the isometric embedding z and its velocity $\dot{z}$ to represent a generalized force/torque required for the dual-arm robot to drive the load block for such position and orientation changes. Based on the Newton's Third Law, the cascaded load block should also be acted by the reacting force/torque, and the entire dual-arm robot-load block system obeys the following cascaded equation:

$$\dot{x} = f(x) + G(x)y, \quad y = J_z^T(q)F_z(z_1, z_2), \quad \begin{cases} \dot{z}_1 = z_2 \\ \dot{z}_2 = F_z(z_1, z_2) + u, \end{cases}$$

where u is an additional control input if necessary. The above z-equation actually follows the Newton's Second Law $\ddot{z} = F_z(z, \dot{z}) + u$ in Euclidean space.

Moreover, if the load block is just a single rigid body and the two end-effectors of the dual-arm robot is gripping the load block on two opposite faces, as shown in Figure 8.4, then, we can readily find the positive vector p_0^c of the load mass center with respect to the base. Namely, if the left and right two end-effectors of the dual-arm robot have position vectors p_0^a and p_0^b, then $p_0^c = \lambda p_0^a + (1 - \lambda)p_0^b$ as a convex combination. For instance, if $\lambda = 0.5$, p_0^c is just arrow-pointing at the middle center between two gripping points. In contrast, finding the instantaneous orientation for the load block is more straightforward than finding the position vector, because it is just a rotation matrix R_0^c and is independent of the gripping point location. Since both p_0^c and R_0^c are functions of all the 17 joint angles θ_i's for the dual-arm manipulator, while the isometric embedding vector $z = \zeta(q)$ is 12-dimensional if equation (4.6) is adopted, then, the Jacobian matrix $J_z = \frac{\partial \zeta(q)}{\partial q}$ will be 12 by 17, becoming a short matrix, instead of a tall matrix. The main reason is due to the fact that the two grippers are now linearly combined together plus the kinematic redundancy for such a 17-joint dual-arm industrial robot. Because of the kinematic redundancy, it has a potential to adjust and optimize the instantaneous posture by adding a null-solution control of this dual-arm robot during the picking/placing task operation.

Example 8.2. A robot-environment interaction system, in addition to being described by the k-cascaded serial-linkage dynamic formulation, as given in (8.1), can also be modeled in the following k-parallel connection:

$$\dot{x} = f(x) + G(x)y$$

$$y = a_1y_1 + \cdots + a_ky_k$$

$$y_1 = h_1(x,\xi_1), \qquad \cdots, \qquad y_k = h_k(x,\xi_k) \qquad\qquad (8.39)$$

$$\dot{\xi}_1 = \eta_1(x,\xi_1) + \Gamma_1(x,\xi_1)u_1, \quad \cdots, \quad \dot{\xi}_k = \eta_k(x,\xi_k) + \Gamma_k(x,\xi_k)u_k$$

In such a k-parallel interaction model, any one of the k inputs u_i can be designed to meet the top x-system control objectives via the corresponding input-output y_i channel. Each coefficient a_i of the input-output vector y for $i = 1, \cdots, k$ in equation (8.39) is virtually a weight for y_i over the total y.

Figure 8.5   A humanoid robot is pushing a big heavy log to roll it forward

A typical example of the k-parallel interaction model is shown in Figure 8.5, where the 40-joint humanoid robot, as studied and kinematically modeled in Chapter 2, see Figure 2.7 in Example 2.3, is pushing a big cylindrical log to roll it forward. The two hands of the humanoid robot provide the big log with a push force f_h, while its two feet are receiving two reacting forces f_l and f_r from the ground. Since the humanoid robot is an open parallel-serial hybrid-chain system with four "end-effectors": two hands and two feet. Now, each "end-effector" is exerted by a force, and clearly in this particular case, each of f_l on the left foot and f_r on the right foot can be modeled as *a hard spring-damper force* in the following form:

$$\begin{cases} \dot{\xi}_{i1} = \xi_{i2} \\ \dot{\xi}_{i2} = -k_i\xi_{i1} - b_i\xi_{i2} + f_i(\xi_{i1},\xi_{i2}) + u_i, \end{cases}$$

for $i = 1, 2$ to represent each of the left foot and right foot interaction models. Whereas for the two hands to provide the big log with a pushing force f_h, the subsystem that links to the humanoid robot x-system can be modeled as

$$\begin{cases} \dot{\xi}_{31} = \xi_{32} \\ \dot{\xi}_{32} = f_h(\xi_{31}, \xi_{32}) + u_3. \end{cases}$$

Then, the input-output equation can now be parallel-connected with all the three interacting forces:

$$y = a_1 y_1 + a_2 y_2 + a_3 y_3$$

$$= a_1 J_{lf}^T(q) f_1(\xi_{11}, \xi_{12}) + a_2 J_{rf}^T(q) f_2(\xi_{21}, \xi_{22}) + a_3 J_z^T(q) f_h(\xi_{31}, \xi_{32}),$$

where $J_{lf}(q)$ and $J_{rf}(q)$ are the kinematic Jacobian matrices for the left and right feet, respectively, as discussed in section 2.3.2 of Chapter 2, while $J_z(q)$ is the mathematical Jacobian matrix for the isometric embedding $z = \zeta(q)$, and $f_h(z, \dot{z}) = \ddot{z}$ reflects the dynamic movement of rolling the cylindrical log forward.

In the above hard spring-damper force model, each of f_1 and f_2 should be in the normal direction to the ground. However, due to the friction that is proportional to the normal pressure force, both f_l and f_r are the resultant forces between the normal pressure and friction so that each magnitude can be a multiple of the normal pressure, but the direction is leaning towards pushing each leg of the humanoid robot forward, as depicted in Figure 8.5.

Once the overall joint torque distribution $y = a_1 y_1 + a_2 y_2 + a_3 y_3$ for the humanoid robot is determined at each time instant, it may follow a best-posture control algorithm to optimize the instantaneous posture of the robot, such as using the potential function $p(q) = \text{tr}(JJ^T)$ to minimize the norm $\|y\|$ of the joint torque vector y during the course of null-solution control, as discussed in section 2.3.3 of Chapter 2.

REFERENCES

1. Kristic, M., Kanellakopoulos, I. and Kokotovic, P., (1995) Nonlinear and Adaptive Control Design. John Wiley & Sons, New York.
2. Isidori, A., (1995) Nonlinear Control Systems: An Introduction, 3rd Edition. Springer-Verlag, New York.
3. Khalil, H., (1996) Nonlinear Systems, 2nd Edition. Prentice Hall, New Jersey.
4. Nijmeijer, H. and Van der Schaft, A., (1990) Nonlinear Dynamical Control Systems. Springer-Verlag, New York.
5. Gu, Edward Y.L., (2012) Modeling of Human-Vehicle Dynamic Interactions and Control of Vehicle Active Systems. *International Journal on Vehicle Autonomous Systems*, Vol.10, No. 4, pp. 297-314.
6. Gu, E. and Das, M., (2013) Backstepping Control Design for Vehicle Active Restraint Systems. Transactions: *ASME Journal of Dynamic Systems, Measurement and Control*. Vol. 135(1), No. 1, 011012, January.
7. Gu, Edward Y.L., (2013) A Journey from Robot to Digital Human. Springer, Heidelberg, New York.

8. Seto, Edmund Y.W. and Carlton, E.J., (2009) "Disease Transmission Models for Public Health Decision-Making: Designing Intervention Strategies for Schistosoma Japonicum", Modelling Parasite Transmission and Control, Edited by Michael, E. and Spear, R.C. Landes Bioscience, Landes Bioscience, Austin, Texas.

9. Vukobratovic, M., Surdilovic, D., Ekalo, Y. and Katic, D., (2009) Dynamics and Robust Control of Robot-Environment Interaction. World Scientific, New Jersey, London.

10. Prats, M., del Pobil, A. and Sanz, P., (2013) Robot Physical Interaction through the Combination of Vision, Tactile and Force Feedback. Springer, Berlin, Heidelberg.

11. Xiong, G.L., Chen, H.C., Zhang, R.H. and Liang, F.Y., (2012) Robot-Environment Interaction Control of a Flexible Joint Light Weight Robot Manipulator, *International Journal of Advanced Robotic Systems*, Vol.9, Issue 3, September.

12. Sam, S.Z., Li, Y.N. and Wang, C., (2014) Impedance Adaptation for Optimal Robot–Environment Interaction. *International Journal of Control*, Vol.87, No. 2, pp. 249-263.

13. Romanelli, F., (2011) Advanced Methods for Robot-Environment Interaction towards an Industrial Robot Aware of Its Volume. *Journal of Robotics*, Volume 2011, Article ID 389158.

Index